ELECTRON

This book marks the centenary of the discovery of the electron by J. J. Thomson in 1897, an event which occurred at a great turning point in the history of scientific ideas, and the impact of which on the development of science in the 20th century has been profound.

The electron was the first elementary particle to be discovered. It sets the basic scales of energy and length in chemistry and materials science, and its ubiquitous presence to drive electrical and electronic devices in everyday life is familiar to everyone. In this book the discussion moves from the historical context of the discovery of the electron and its basic properties, to the Dirac equation, bonding in condensed matter, Fermi and non-Fermi liquids, quantum order, superconductivity, heavy, coherent and composite electrons, and the role of the electron in the cosmos. Each chapter is written by a leading figure in the field, and the book focuses on conceptual issues, exploring our perception of the nature of the electron and its interactions 100 years after its discovery.

The book will be of interest to advanced students and researchers in physics working in any area where the nature and properties of the electron are manifest. In particular, researchers working in one area of physics will appreciate a clear and succinct summary of achievements in other fields.

ELECTRON

a centenary volume

Edited by

MICHAEL SPRINGFORD

H. H. Wills Physics Laboratory, University of Bristol

CAMBRIDGE
UNIVERSITY PRESS

CAMBRIDGE UNIVERSITY PRESS
Cambridge, New York, Melbourne, Madrid, Cape Town, Singapore, São Paulo

Cambridge University Press
The Edinburgh Building, Cambridge CB2 8RU, UK

Published in the United States of America by Cambridge University Press, New York

www.cambridge.org
Information on this title: www.cambridge.org/9780521561303

First published 1997
This digitally printed version 2008

A catalogue record for this publication is available from the British Library

Library of Congress Cataloguing in Publication data

Electron : a centenary volume / edited by Michael Springford.
       p.   cm.
    Includes index.
    ISBN 0 521 56130 2
    1. Electron. I. Springford Michael, 1936–   .
QC793.5.E62E39   1997
539.7′2112–dc20   96-44744 CIP

ISBN 978-0-521-56130-3 hardback
ISBN 978-0-521-07889-4 paperback

Contents

List of contributors   *page* ix

Preface   xi

1   J. J. Thomson and the discovery of the electron   *A. B. Pippard*   1

2   The isolated electron   *W. N. Cottingham*   24

3   The relativistic electron   *D. I. Olive*   39

4   The electron glue   *B. L. Gyorffy*   60

5   The electron fluid   *P. Coleman*   88

6   The magnetic electron   *G. G. Lonzarich*   109

7   The paired electron   *A. J. Leggett*   148

8   The heavy electron   *M. Springford*   182

9   The coherent electron   *Y. Imry and M. Peshkin*   208

10   The composite electron   *R. J. Nicholas*   237

11   The electron in the cosmos   *M. S. Longair*   257

References   315

Index   329

Contributors

P. Coleman
Serin Laboratory, Rutgers University, Piscataway, New Jersey 08855-0849, USA

W. N. Cottingham
H. H. Wills Physics Laboratory, University of Bristol, Bristol BS8 1TL, UK

B. L. Gyorffy
H. H. Wills Physics Laboratory, University of Bristol, Bristol BS8 1TL, UK

Y. Imry
Condensed Matter Department, The Weizmann Institute of Science, Rehovot 76100, Israel

A. J. Leggett
Physics Department, University of Illinois, Urbana, Illinois 61801, USA

M. S. Longair
The Cavendish Laboratory, University of Cambridge, Cambridge CB3 0HE, UK

G. G. Lonzarich
The Cavendish Laboratory, University of Cambridge, Cambridge CB3 0HE, UK

R. J. Nicholas
Physics Department, Clarendon Laboratory, Oxford University, Parks Road, Oxford OX1 3PU, UK

D. I. Olive
Department of Physics, University of Wales, Swansea SA2 8PP, UK

M. Peshkin
Physics Division, Argonne National Laboratory, Argonne, Illinois, USA

A. B. Pippard
The Cavendish Laboratory, University of Cambridge, Cambridge CB3 0HE, UK

M. Springford
H. H. Wills Physics Laboratory, University of Bristol, Bristol BS8 1TL, UK

Thus on this view we have in the cathode rays matter in a new state, a state in which the subdivision of matter is carried very much further than in the ordinary gaseous state: a state in which all matter – that is, matter derived from different sources such as hydrogen, oxygen &c. – is one and the same kind; this matter being the substance from which all the chemical elements are built up.

J. J. Thomson, *Philosophical Magazine*, S.5, vol. 44, no. 269, October 1897.

Preface

Centenaries provide an opportunity for reflection or celebration altogether different in character from other anniversaries. Comparable with a lifespan, the century is above all the human unit of time, whose history we embrace as our own. It is also a time sufficient for us to judge the impact of events. The discovery of the electron by J. J. Thomson in 1897 was an event whose influence on 20th century science, and indirectly on our own lives, is hard to exaggerate.

Discoveries are rarely solo performances, as the history of science amply testifies, and the discovery of the first elementary particle, the electron, was no exception. Plaudits must in some measure go to several others, including Stoney, Wiechert, Zeeman and Lorentz, but it was undoubtedly Thomson to whom the crown belongs, as it was he who demonstrated for the first time the existence of a free particle which was also a common constituent of matter. Within a few years, the discovery inspired new theories of atoms, molecules and solids, which, together with the advent of the theory of relativity and quantum mechanics, yielded new insights into the nature of matter. These developments, however, not only enriched and extended the fields of physics, astronomy, chemistry, materials science and later biology, they were also the genesis for practical devices such as the thermionic valve, the transistor, the integrated circuit and the computer. Thereby the processes of scientific investigation were themselves revolutionised, as indeed also were the ways that we live.

The ubiquitous electron has been put to use in a multitude of ways, and is undoubtedly the blue-collar worker amongst fundamental particles. Transmission, scanning and tunnelling electron microscopies, spectroscopies such as electron spin resonance, direct and inverse photo-emission and others, as well as high-energy electron beams, are the daily business of science at the end of the 20th century. Faced with such riches as this opportunity presents, we are keenly aware of the inadequacy of a single-volume tribute. But to recognise the behaviour and role of the electron in all its richness would be to encompass much of modern science. Being forced therefore to be highly selective, we hope only to avoid the kind of contempt with which Millikan's *Electron* was received by Rutherford when he remarked that the book was not about the electron at all, but about its charge. Our choice here has been to focus largely on conceptual matters. As the following essays testify, 100 years on we are much preoccupied with the subtleties inherent in elucidating the behaviour of interacting electrons. Their complexity often defies our ability to analyse and understand them with what concepts and many-body theories we have to hand. These are areas which are current and which therefore most clearly reveal both our progress and our ignorance.

I am indebted to my colleague R. G. Chambers for his careful and critical reading of the book in proof.

M. Springford

Bristol
May, 1996

1

J. J. Thomson and the discovery of the electron

A. B. PIPPARD

University of Cambridge

Most physicists, especially English speakers, believe that J. J. Thomson discovered the electron, and many of them know it happened in 1897, although a recent edition of the usually reliable *Encyclopaedia Britannica* offers 1895 as an alternative. It also states, quite erroneously, that he gave it its name. In fact, the word *electron* was coined in 1891 by Johnstone Stoney, and Thomson (who was JJ to all in his lifetime, and will be so called from now on) referred to it as a *corpuscle*. As for the actual discovery, it was made independently by several physicists, of whom Emil Wiechert of Königsberg was the first to publish. Nevertheless, JJ was the most assiduous in following up his discovery, and it is probably right to single him out as the pioneer, provided we remember the contributions of other distinguished physicists – Wiechert, Kaufmann, Zeeman and Lorentz – as well as the strong group that assembled round JJ at precisely the right moment.

When JJ announced, early in 1897, the discovery of a negatively charged corpuscle, much lighter than a hydrogen atom, he was 40 years old and had been Cavendish Professor of Experimental Physics in Cambridge for 12 years. His election to the chair at the age of 28 caused considerable surprise, and not only because of his youth. Like his predecessors, Maxwell and Rayleigh, JJ had been educated as a mathematician, but he differed from them in having little experience of laboratory work and in lacking manual dexterity; it was said that his assistants did their best to keep him from handling the apparatus. Moreover (to complete the score of his failings), some of the most influential of his early theoretical papers were wrong in matters of detail, and he was also liable to make arithmetical mistakes in his undoubtedly brilliant lectures, mistakes that called for juggling of positive and negative signs, to his students' vexation. Nevertheless he was admired for his physical insight and his imaginative powers, and as time went on he was recognised as almost unrivalled in the devising and interpreting of experiments. Whittaker (1951, p. 366), who belongs to that era, refers to

Thomson's reputation as the first of living experimental physicists; and it was in this aspect that he was generally regarded during the latter half of his life.

It is no small achievement to have been ranked above Rutherford.

To appreciate the particular problems that JJ had to face, we must remember that, though one of the founders of Modern Physics, he was already well set in his ways, the ways of a classical physicist educated in the severe and conservative school of Cambridge mathematics. He was never at ease with relativity or quantum theory, but was extremely familiar with Maxwell's (1873) *A Treatise on Electricity and Magnetism*, and even intended his *Notes on Recent Researches in Electricity and Magnetism* (Thomson 1893b) as a third volume to follow Maxwell's two. In its time the *Treatise*, however authoritative, was considered very difficult, and modern readers will agree. Maxwell's emphasis, which he developed from his deep study of Faraday, was on the aether as the carrier of all electric and magnetic processes, the medium which transmitted the influence of one particle on another. Energy (and momentum, according to JJ) were located in the aether, which might also be the seat of mass, though this was disputable. There seemed to be little room in Maxwell's thought for many things we have grown to accept without question, electrons and protons, particles with charge and mass. Indeed, as JJ remarked in the reminiscences he wrote (Thomson 1936, p. 92) in his last years,

Helmholtz, who was a supporter of the theory, said he should be puzzled to explain what an electric charge was on Maxwell's theory beyond being the recipient of a symbol.

As far back as 1885, JJ felt the same.

Helmholtz indeed had deep respect for Maxwell, but he was a German brought up in a very different scientific environment. The continental tradition, as Whittaker (1951, chap. 7) has described in valuable detail, developed the basic conceptions of Coulomb and Ampère, which started from charges and current elements and sought to formulate the forces between them without worrying too much about the mechanism of transmission. One side of Faraday's work which Helmholtz felt had been neglected by Maxwell was electrochemistry, whose laws convinced him that there was a basic unit of charge, the charge carried by a monovalent ion in solution. Where Maxwell seems to have been perplexed by the nature of a current in a metallic wire, Helmholtz and those who thought like him saw in it nothing beyond the motion of free charged particles, possibly ions. They had no difficulty interpreting Rowland's (1878,

1889) demonstration that a spinning charged disc creates a magnetic field – the moving charges behave exactly like currents. JJ, on the other hand, had to associate the charges with tubes of electrical force, whose motion generated the magnetic field; and it was not straightforward to satisfy Maxwell's demand that the displacement and convection currents, taken together, must be non-divergent. This was a thorny problem for JJ, whose earliest substantial paper on electromagnetism (Thomson 1881) had considered the fields around a rather slowly moving charged sphere, and incidentally the force exerted by a steady magnetic field on the charge. As Fitzgerald (1881) pointed out, he had not achieved continuity of current; Heaviside (1889) later gave the right answer, and extended the theory to particles approaching the speed of light. JJ had found that the force was $(1/2)ev \times \mathbf{B}$, but Heaviside's result (now known as the Lorentz force) was $ev \times \mathbf{B}$. When JJ wrote his *Notes* in 1893, he gave yet another derivation, based again on the motion of lines of force, and this time arrived at $(1/3)ev \times \mathbf{B}$. We may sympathise with Arthur Schuster (1897), who had studied under Helmholtz, and who wrote just before JJ's great discovery

I confess I find it difficult to follow the method of 'moving tubes' employed in that investigation. I speak with diffidence on the subject, but the investigation . . . seems to me to be obscure and incomplete.

Very true.

Schuster, who had long accepted Heaviside's analysis, wrote in ignorance of JJ's rethinking of the question, which must have taken place with little time to spare before he used the correct formula to analyse his measurements of the magnetic deflection of cathode rays. I can find no explanation in his published papers, and if he had come to regret giving tubes of force a central importance, his change of heart was incomplete. He was looking back to his earliest aethereal ventures when he wrote (Thomson 1907, p. 2)

The corpuscular theory of matter with its assumptions of electrical charges and the forces between them is not nearly so fundamental as the vortex atom theory of matter.

The analogy between tubes of force and vortices in the aether was deeply embedded in his mind.

One must not depreciate JJ's knowledge of continental ideas. He wrote a long British Association report (Thomson 1885) on electrical theories from Ampère on, pointing out their assumptions and weaknesses, and settling for Maxwell, with Helmholtz's extensions, as the best available. His paper was rather too early to include the important synthesis by H. A. Lorentz, who achieved compatibility between Maxwell's aether and charged particles (Whittaker 1951, chap. 13). A bald statement like this seriously underestimates the depth and range of Lorentz's thought, much of which is nowadays enshrined in equations that students of physics accept without question. For the present discussion it is enough to say that he imagined matter to be made up of electrically charged particles, moving through an aether that was not entrained by the particles, but served to carry their interactions in accordance with Maxwell's equations. It was essential to include distortions of length and mass suffered by fast-moving particles; these, originally conceived by Heaviside, were adopted by Fitzgerald and, with considerable elaboration, by Lorentz in a mathematical structure that became an ingredient of Einstein's special theory of relativity. By 1897, Lorentz was well free of the conflicts between Maxwell and particle theories that still troubled JJ. There was, however, no more thought in Lorentz's mind than in JJ's of discarding the aether. Wholehearted acceptance of Maxwell's theory of light demanded belief in a medium for carrying electromagnetic waves. But once it had ceased to be entrained by moving charges it could be relegated to the sidelines of physics – Maxwell's equations could be used with never a thought of their metaphysical meaning.

Before moving on to the experiments which revealed the electron, it is worth noting once more that the name preceded the discovery. Johnstone Stoney (1891, 1894) (whose earlier papers show his delight in inventing names; *electron* is the only survivor) did not believe that his electron was a particle capable of independent existence, nor was JJ the only one to reject the name when his newly discovered particle had to be christened. Millikan (1935) continued to deplore the usage, and mentioned a number of authoritative writers (all, as it happens, products of JJ's school) who agreed with him. He seemed unaware that JJ had long since begun to forsake his *corpuscle* and conform to popular usage (Thomson 1914), as I shall do for most of what follows.

1.1 The discovery of the electron*

The cathode rays, emitted by the negative electrode of a low-pressure discharge tube, were discovered by Plücker in 1858 as one practical result of Geissler's invention of a mercury-operated air pump. They caused fluorescence of the glass walls and in the gas itself; in the presence of a magnet they followed the

---

* A very thorough study by Isobel Falconer was the basis of her Ph.D. dissertation and of a later paper (Falconer 1987). The Cavendish library copy of the dissertation was in too great demand by other research students for me to read it until this chapter was written, and I was unfortunately ignorant of her paper. I think that in general we agree, but I recommend her account as something much more substantial and well-referenced than any I could have managed.

magnetic lines. They could cast shadows and even exert a force on a vane in their path. The story (Thomson 1893b, p. 132; 1903, chap. 17) of investigations and explanations need not be told here, but it seems something of a paradox that in Germany, the home of particle theories of electrical phenomena, a popular opinion was that cathode rays were an aetherial disturbance, while the British disciples of aetherial electromagnetic waves were convinced cathode rays were charged particles. There is little doubt that Hertz's failure to deflect cathode rays with an electric field aided the aetherial point of view, but JJ (Thomson 1893b, p. 121) recognised that the cathode rays made the surrounding gas a conductor which would become polarised to screen them from the field. This was restated by Fitzgerald (1896), and it is odd that the account given in JJ's reminiscences (Thomson 1936, p. 334) implies that only when he set out to use electrostatic deflection in 1897 did he find how much care he must take to reduce the pressure till the conductivity of the residual gas was not intrusive.

Electric deflection, however, had no part in his Friday evening discourse (Thomson 1897a) at the Royal Institution on April 30, 1897, when he made his first announcement. The centrepiece was an idea he had borrowed from Perrin (1895) and significantly improved. Some of his apparatus survives and is illustrated in Fig. 1.1, together with the diagram he showed in his discourse. The cathode in the bulb B emits rays which pass through the collimator into the main bulb. The collector at the bottom receives nothing except stray radiation until a magnetic field bends the beam so that it hits the collector, when the electrometer records the arrival of negative charge. It became clear, what was previously in dispute, that the charge is associated with whatever in the beam is responsible for fluorescence. In addition to the charge collector, JJ arranged a bolometer which would heat up and so measure the kinetic energy brought by the particles. Thus, for n particles collected, the measured charge is ne and the kinetic energy is $(1/2)nmv^2$; n being unknown, this part of the experiment measures mv^2/e. From the beam curvature in a known magnetic field, JJ found mv/e, and the two together gave v and m/e. The smallness of m/e, about 600 times less than that for a hydrogen atom, might tempt one to think, he said, that e was exceptionally large. But he discussed in some detail the earlier observations of Lenard, who had found cathode rays to travel about 10^5 times as far through a gas or a solid film as one would expect for a molecule. This strongly supported the view that cathode rays consisted of very small particles. Moreover, Lenard had found that the stopping power of gases and solids was determined by their density, as if all atoms were composed of the same small particles; this would explain why the same value of m/e was obtained for every gas (air, H_2 and CO_2) used in the discharge tube. JJ ended by noting that

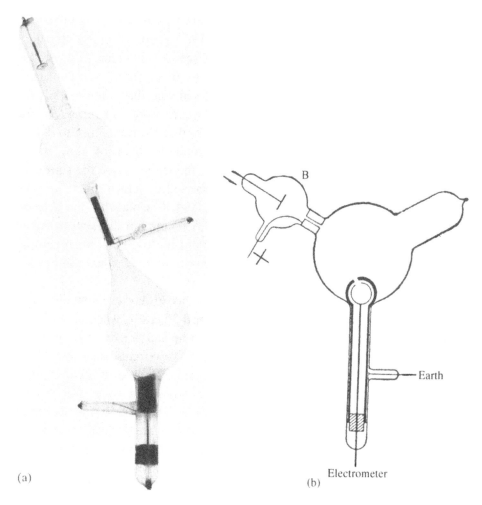

(a)   (b)

Fig. 1.1 Thomson's modified form of Perrin's apparatus. The equipment shown in part (a) is from the Cavendish Laboratory museum, and the diagram in part (b) is from Thomson's account of his discourse at the Royal Institution (Thomson 1897a).

Zeeman, who had just published his account of how the spectral lines of sodium are changed by a magnetic field, had derived a similar value for m/e.

When, in October, JJ published a fuller account (Thomson 1897b), he remarked that there was a long history, extending back to Prout early in the 19th century, of the idea that all elements were composed of the same primordial matter; in his view this was not hydrogen (as Prout had conjectured) but something very much smaller, which could be detached by strong electric fields such as were to be found in a discharge tube. He questioned whether the corpuscles carried an ionic charge or something larger (as he was inclined to think at that time), a problem his continental contemporaries thought hardly worth dwelling on. They, and for that matter Stoney, could surely not have written, as JJ did in his habitual Maxwellian mode,

In the molecule of HCl, for example, I picture the components of the hydrogen atoms as held together by a great number of tubes of electrostatic force; the components of the chlorine atoms are similarly held together, while only one stray tube binds the hydrogen atom to the chlorine atom.

This fancy, however seriously considered at the time, evaporated as soon as the first intimation of the true value of *e* became known. But there are other matters to bring forward before turning to this question.

In the same paper of October 1897, JJ first described his measurements on electric and magnetic deflection of electrons with the apparatus of Fig. 1.2. It was no light matter to reduce the gas pressure enough to eliminate the conductivity that had misled Hertz, but this first account slides over the problem arising from the very slow pumps that were available and the outgassing of the unbaked glassware. JJ explains in his reminiscences (Thomson 1936, p. 334) that he (it was, more likely, his assistant Everett) pumped and

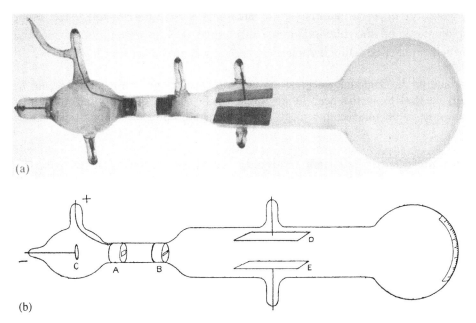

(a)

(b)

Fig. 1.2 Apparatus used for electric and magnetic deflection of cathode rays. The equipment shown in part (a) is from the Cavendish Laboratory museum, and part (b) is from Thomson (1897b). The rays start from the cathode, C, are accelerated towards the anode, A, and collimated to a thin beam by the slits A and B. The glass of the bulb on the right fluoresces when the beam strikes it.

sealed the tube, and then ran the discharge for several days before re-pumping. He could then arrange that the electric field E in the length l between the plates D and E, and the magnetic field B (which for some unexplained reason he took to be uniform over the same length l) gave the same beam deflection θ, from which $m/e = B^2 l / E\theta$. With the same gases as before – air, H_2 and CO_2 – he obtained once again consistent values of m/e, in SI units 1.3×10^{-11} kg/C, but by this time he had revised his original Perrin-type value from 1.6×10^{-11} down to 0.41×10^{-11}, 0.52×10^{-11} and 0.9×10^{-11} with three versions of the equipment, and he thought the last was probably the best of the three. The accepted modern value is 0.57×10^{-11} kg/C, and it is not clear how a two-fold error could have arisen in his second experiment. At the time, though, it did not matter in the least whether the electron was 800 or 1800 times lighter than a hydrogen atom; the smallness was all, as well as the discovery that the same result was obtained with different gases and different electrodes. Whatever its precise mass, the electron was a constituent of all matter.

The first published commentary on JJ's discovery was given by Fitzgerald (1897) only three weeks after the Royal Institution discourse. Fitzgerald remarked that if the formation of cathode rays was the outcome of a wholesale disruption of an atom into its corpuscles, one might expect their reconstruction to result in different atoms – the alchemist's dream. He suggested other explanations of the phenomenon, but hoped JJ's interpretation would be verified, and added with characteristic prescience and becoming loyalty,

It would be the beginning of great advances in science, and the results it would be likely to lead to in the near future might easily eclipse most of the other great discoveries of the nineteenth century, and be a magnificent contribution to the Jubilee year.

We must not allow such admirable sentiments to distract us from the discoveries being made in parallel on the mainland of Europe. Pieter Zeeman's (1897) observation of what is now called the Zeeman effect, and which JJ mentioned in his discourse, was not as clear-cut as is often thought. He did not find that the sodium orange lines were split, but only that a magnetic field applied to the flame source caused the lines to be broadened. Lorentz's expectation of three lines when viewed at right angles to the magnetic field, and two along the field, led Zeeman to drill holes through his magnet pole pieces and observe with a quarter-wave plate and polariser, which would allow only one sense of circularly polarised light to pass. On changing from one sense to the other, Zeeman noticed a shift in the apparent position of the line in the spectroscope, confirming for him that there were two unresolved lines of opposite circular polarisation. Lorentz showed him how to set up the theory of the effect, and Zeeman concluded that e/m was (in SI units) about 10^{11} C/kg. Later he refined his estimate to 1.6×10^{11} C/kg, closer than JJ's to the accepted value. He made no comment on the magnitude of e/m, or on the likelihood of particles very much smaller than an atom being involved. In the following year, Preston (1898) managed to resolve the magnetically split lines, as did Cornu (1898) and Michelson (1898), and they found not three but four or even six components. If there had not been other demonstrations of the electron's existence, Lorentz's explanation would have been very suspect. It was only much later, with the hypothesis of the spinning electron and the developments arising from Bohr's atomic theory, that these complexities of the Zeeman effect were sorted out.

I think it not unfair to deny Zeeman priority in discovering the electron, if only because he did not concern himself with free particles, but with the internal constitution of the atom. A stronger claim may be put forward for Emil Wiechert (1897), who gave a full account of his work to the Königsberg Physikalisch-ökonomische Gesellschaft on January 7, 1897. Probably JJ was unaware of it when he made his first announcement, but he referred to it later, with reservations about its accuracy (Thomson 1903, p. 102); in fact, Wiechert was considerably closer to modern values of e/m and m than was JJ. In his first experiments, Wiechert measured the magnetic deflection of cathode rays, but instead of electric deflection he relied on the potential difference V between anode and cathode to give eV as the upper limit of the kinetic energy of the particles. He next borrowed an idea of Des Coudres to measure the particle velocity directly. A collimated beam of cathode rays was deflected transversely by high-frequency coils oscillating in phase, but separated from one another along the direction of the beam. If the beam takes half a cycle to cover the distance between them, the deflection by the first coil will be annulled by the second, and the beam at the detector will be found to be still sharp and undisplaced. This procedure obviates the assumption about kinetic energy and, together with magnetic deflection, gives e/m. Wiechert was more definite than was JJ at that time about the value of e, which he took to be one 'electron', i.e. the same magnitude as for a hydrogen atom, though he allowed, with little conviction, that it may be an integral multiple. His final conclusion was that the mass of the particle was between 2000 and 4000 times less than that of hydrogen. He carried this excellent work no further, but went on to make his reputation in seismology (Bullen 1976).

At the same time, Walter Kaufmann (1897) was experimenting very much along the same lines as Wiechert, and was puzzled to discover how large e/m was, and especially that it was independent of the gas in the tube and the metal of the electrodes. He did not speculate on the mechanisms involved, but went off in a different direction, using β-rays from radioactive sources to see if the mass depended on velocity, as had been predicted by Heaviside and would later be incorporated into the theory of relativity. The effect was undeniable, but his accuracy was not quite good enough to decide between Heaviside's and other rival theories.

What elevates JJ's contribution over the rest is not that it was more accurate but that he went on to show that the same electron is found in different gases and metals; moreover, he and his colleagues began the determination of the electric charge e, which in Millikan's hands attained high precision, and was relied on for nearly 20 years in fixing the values of fundamental constants. This story must be our next concern, but it is worth noting why Stoney (1881) had earlier proposed much too small a value for e. The faraday, the charge on one gramme of hydrogen ions, was well known, and in Stoney's (and Helmholtz's) view was equal to Ne, where N is Avogadro's number, the number of atoms in a mole. Stoney's estimation of e assumed a value for N in 1881 that was 30 times too large; by 1900 JJ and others were no further astray than 30%, while Millikan, who started working with oil drops in 1909, attained a consistency approaching one part in a thousand, and was only let down by an error in the viscosity of air, which is not easily measured with sufficient accuracy.

1.2 The determination of e

By good fortune, the University of Cambridge decided in 1895 to admit graduates from elsewhere as research students. In the Cavendish Laboratory there is a complete sequence of photographs of research workers from 1897 on, and one is impressed by the growth of their numbers; even more, by the subsequent distinction of so many of the early ones. C. T. R. Wilson, a Cambridge graduate, had been inspired by the coronas and glories he saw in the clouds from the summit of Ben Nevis, and began in 1895 to study the condensation of moist air. Ernest Rutherford arrived from New Zealand, and John Townsend from Dublin, at the end of the same year; Harold Wilson came from Leeds in 1897 and Owen Richardson arrived in 1900, the only one of the last four to have taken his undergraduate courses at Cambridge. There were others, of course (including the Frenchman Paul Langevin and the American John Zeleny), but those singled out for mention were especially close to JJ's work on the electron. Three, with JJ as a fourth, won Nobel prizes.

Presumably, it was JJ's residual doubt of the identity of the ionic and electronic charges, as well as the uncertain value of Avogadro's number, that gave impetus to the considerable effort in the Cavendish, after 1897, to determine e directly. The story has been told in some detail by Millikan (1935, chaps. 3–5) and need not be repeated in full. It begins with C. T. R. Wilson, whose first cloud chambers, like the relic shown in Fig. 1.3, were made up by himself and had none of the machine-finish of the final version used for photographing particle tracks. It was the early examples that played a central part here; there were many of them, for the glassware was short-lived, and Wilson was patiently scrupulous in his experimental mastery (Blackett 1960). He found that an isolated volume of saturated damp air could be caused to condense as mist or rain when adiabatically expanded, and that after several repetitions the expansion needed to initiate condensation settled to a highly

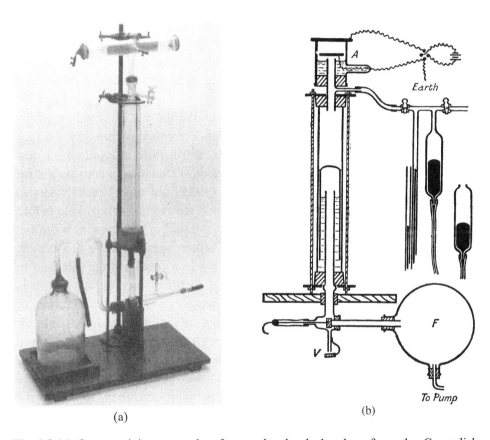

(a)　　　　　　　　　　(b)

Fig. 1.3 (a) One surviving example of an early cloud chamber, from the Cavendish Laboratory museum; (b) a diagram (C. T. R. Wilson 1900) of a similar chamber. The test-tube-shaped piston can be set at any desired height by means of the mercury reservoir on the right, and restored to atmospheric pressure by opening the air seal V. It is suddenly drawn down, by closing V and pulling back the seal to F, so that the air above suffers a known expansion. The experimental chamber A is initially filled with saturated water-vapour. The water sealant for the piston is no longer in the museum exhibit.

consistent value. By this time the original dust particles and other nuclei had been eliminated. A volume expansion of 25% cooled the air to the point of being 4.2 times supersaturated (the vapour pressure was 4.2 times the equilibrium pressure over a flat surface of water) and raindrops were formed; at an expansion of 28%, and a supersaturation of 7.9, a fine mist appeared. The demonstration by JJ that X-rays caused air to conduct electricity led Wilson to study their effect, and that of ultra-violet light, in the cloud chamber. A supersaturation of 4.2 was now enough to produce a fine mist, but with less expansion nothing happened, whether or not the air was irradiated. Wilson was satisfied that with a supersaturation of 4.2 water would condense on any nuclei present, and the copious ions produced by irradiation were responsible for the mist. But the cleanest air still contained a small number of nuclei which would attract all the excess vapour and form raindrops. Only in 1911 was it shown (Hess 1911) that cosmic rays are the principal cause of the residual ionisation. With supersaturation in excess of 7.9, the spontaneous condensation is rapid enough to need no accidental nuclei.

At the time that these measurements stimulated the use of cloud chambers for measuring e, little was known of the process of condensation, but this lack of knowledge did not matter. It is, however, worthwhile interpolating a few remarks on the theory which occupied a number of statistical mechanicians for quite a while before Becker and Döring (1935; see also Frenkel (1946)) published the first generally acceptable account. Kelvin had shown long before that the vapour pressure in equilibrium with a liquid drop increases when the drop is smaller. This means that at a given vapour pressure there is a critical size above which a drop will tend to grow, while a smaller drop will tend to evaporate – that is to say, the equilibrium of a critical drop is unstable. It is important to note the word *tend*; growth involves the chance accretion of vapour molecules, evaporation the chance loss, so that it is not impossible for too small a drop to collect extra molecules by random impact and grow to a size where further growth is more likely than not. Becker and Döring solved the problem of how long one might expect to wait before a critical drop grew spontaneously, and their formula shows a remarkable sensitivity to the degree of supersaturation, which determines the critical size. For a not-unrealistic choice of parameters, one may expect to wait 10^{41} years with two-fold supersaturation, 6 days for four-fold, and less than one second for six-fold. This is not too far from Wilson's discovery of spontaneous misting in clean air at 7.9 times supersaturation. Kelvin's argument, which related the increased vapour pressure of a drop to the excess internal pressure caused by surface tension, is readily extended (Thomson 1893a) to a charged drop; the electric field, pulling on the surface, opposes the action of surface tension and encourages the growth of a very small drop to the size at which it is an efficient nucleus for condensation. The effect of X-rays is thus explained, and of course the same reasoning underlies the use of the cloud chamber for visualising the ionisation due to particle tracks.

For 12 years after the discovery of the electron, attempts to measure its charge were based on Wilson's cloud chamber, with refinements in the technique used to determine the droplet size. Inevitably with a cloud of drops, the rate of fall under gravity, which was then seen to be as good a method as any, was made inexact by the spread of drop sizes. Townsend, C. T. R. Wilson and JJ made progress, but a significant advance came when H. A. Wilson (1903), following a suggestion of Townsend's, applied a vertical electric field acting upwards on negative droplets. He found evidence that the cloud became stratified as different groups of drops sank at different speeds, in a manner consistent with their carrying small multiples of a basic charge. By measuring the rate of fall with and without the field, Wilson could deduce the charge carried by the drops in any one group. His value of *e* was too small by about 35%, but this was of little consequence when Millikan took over the idea. For some time he was as troubled as the Cambridge workers by the tendency of drops to evaporate, but in 1909, returning by train to Chicago, he suddenly thought of using oil drops instead of water (Millikan 1951). He began to develop the idea immediately, and in the following years produced a succession of papers detailing his increasingly improved measurements. He could keep a single drop suspended in the electric field for hours, or time its rise and fall through many cycles of the applied field. Occasionally the charge would change as an ion struck the drop, and he showed with great precision that the change was always one unit of the basic charge, rarely a multiple. There might be hundreds of changes in the course of a few hours' observations. The demonstration of the quantisation of charge involved one assumption only, that the velocity of a given drop was proportional to the force on it.

This fundamental demonstration was challenged by the Austrian Ehrenhaft, whose experimental procedures were, in Millikan's view, all too susceptible to random disturbance. Convinced of the falsity of Ehrenhaft's attacks, he emphasised that he had found no exceptions to precise quantisation of charge. In recent years, long after Millikan's death, his notebooks were sedulously scrutinised, and it was suggested that he had been less than perfectly candid, in that he occasionally failed to include in his tables results which seemed anomalous. One cannot believe that he would have taken such pains in presenting his data if he had entertained the smallest doubt of their correctness, nor have the many repetitions by others given support to Ehrenhaft. Probably the report by Fairbank's group (La Rue *et al.* 1981) of evidence from a different type of experiment that metal particles could be charged in non-integral multiples of *e*, combined with the postulate of quarks with charge $(1/3)e$, was responsible for the reappraisal. No one, however, who compares Millikan's exhaustive descriptions with Fairbank's exiguous accounts would seriously decide in favour of the latter, especially as Fairbank later confessed to Franklin (1986) that he had suspicions about his apparatus.

This digression on a topic that has excited the attention of several philosophers of science, and which has demonstrated the pitfalls that beset the most careful of experimenters, has taken us away from the question of the exact value of *e*. The weight of a drop is found from its speed of fall under gravity, and thence the charge on it from the electric field that serves to balance the weight and hold the drop at a fixed height. Until Millikan began using oil drops, the validity of Stokes's law was assumed, the force on a sphere of radius *a*, moving with velocity *v* through a fluid of viscosity η, being $6\pi\eta av$. Millikan soon found, however, that the value obtained for *e* varied systematically with the size of drop and with the gas pressure, which he was able to change by a large factor. The resulting curve could be extrapolated to give the limiting value at high pressure, when the mean free path of the gas molecules was much less than *a*, and Stokes's law could be relied on. Eventually he felt confident that the measurements were capable of an accuracy of about one part in a thousand, and the most serious remaining difficulty was in the determination of η. When he believed this had been overcome, he gave as his final value of *e* $(4.770 \pm 0.005) \times 10^{-10}$ esu, or 1.591×10^{-19} C, and for a long time this was taken as the best available. It was an extraordinary improvement over the Cambridge values of some 10 years before, and though ultimately it has been changed to $1.602\,177\,33 \times 10^{-19}$ C, after dedicated work by many skilled experimenters using entirely different phenomena, this does not detract from Millikan's spectacular achievement. It was not his oil-drop measurements that were to blame for the error of nearly 1% but, after all, the value of η. And we must not forget that this was by no means his only major success, though his work on Brownian motion, the photoelectric effect and cosmic rays lies outside this story.

1.3 Electrons in metals

It was hard to understand conduction in metals before the discovery of quantum mechanics. How could there be so many charged particles moving so freely through the lattice? Yet this was not the prime worry for JJ and other Maxwellians. Electrical resistance, as Buchwald (1985a, b) has discussed in detail, was something of a mystery to those who held that every significant action took place in the aether. It was a big step from the fields around a moving charge to the fields around a conductor, as one realises when reading JJ's discussion (Thomson 1893b, p. 36). An uncharged straight wire has no resultant electric field around it, but something must be happening when a current passes. There must be electric tubes of force that cancel by pointing in opposite directions, and whose differential motion is responsible for the magnetic field. So far, so good – but the modern reader who is prepared to share this conception would not assume the tubes to lie *parallel* to the conductor, and move radially. We who believe in free electrons in a fixed positive lattice would be happier extending JJ's picture of the single charge so that the tubes of force from the electrons combine to run radially from the wire, and move along with the mean electron flow, while the compensating tubes from the positive lattice remain at rest. Both pictures give the same magnetic field pattern, since B is produced at right angles to the electric tubes and their direction of flow. It demands patience and sympathy to follow JJ's argument, and the effort is nowadays unnecessary for any but enthusiasts for this branch of Maxwellian thought.

By the time of the international physics conference in Paris (Thomson 1900), JJ was ready to explain metallic conduction in terms that a modern reader can easily follow and assent to. His picture was similar to the independent development by Drude (1900), following Riecke's (1898) slightly earlier (and less elegant) presentation. It seems to have been generally accepted by those tackling the problem at this time, that metals had a sort of spongy structure through which the charged particles could pass fairly freely. Riecke, with ionic conductors in mind, thought there might be many kinds of particles, and in particular that they might be positively and negatively charged in order to explain why the Hall effect was positive in some metals and negative in others. Drude followed this train of thought, but suggested that the magnitude of e/m was the same for both kinds of particle; with this assumption he derived the Wiedemann–Franz (W–F) law, $K/\sigma T = L$ (a constant) that had been known for some decades to specify the relationship between electrical conductivity, σ, and thermal conductivity, K. JJ would clearly have preferred to entrust every process to electrons, especially as (in contrast to electrolytic conduction) there was no evidence of material transport when a current flows – a result easily explicable if the same electrons are involved in every metal. Nevertheless he recognised the difficulty of explaining the Hall effect, which bedevilled all thinking about metals until Peierls (1929) showed how quantum mechanics provided the solution.

The derivations of the W–F law by Drude and JJ contain a minor curiosity. While both quote Maxwell's expression for the thermal conductivity of a gas, which is based on an exponential distribution of free paths, they calculate the electrical conductivity on the assumption that all free paths are the same. This apparently innocuous simplification halves the value of σ, and leads to an expression for L that agrees rather well with experiment. When Lorentz took up the question – in around 1905, according to Bohr (1972) – he applied classical statistical methods consistently and obtained a markedly less satisfactory value of L, only two-thirds of Drude's. The discrepancy, which was removed 20 years later by Sommerfeld's quantum theory of metals, did not prevent the explanation of the W–F law from being regarded as a triumph of the free-electron theory, and a challenge to every one of the alternatives that were proposed.

Apart from this success, the theory was beset by serious difficulties. The high thermal conductivity, an essential ingredient of the W–F triumph, showed that the electron gas was responsive to temperature changes, yet made a negligible contribution to the thermal capacity. This indicated to some, such as Bridgman, that there were few free electrons, and they devised various unconvincing explanations of how the electrons could move far enough, without colliding, to give the observed conductivities. There was also Kamerlingh Onnes' (1911) discovery of superconductivity, which was seen, until the Meissner effect was revealed in 1933, as essentially the abrupt disappearance of electrical resistance at a certain critical temperature. It is clear, from papers written after 1911, that the hope of explaining superconductivity was the driving force for many unsuccessful models of metallic conduction. It was fortunate, perhaps, that Bohr's doctoral thesis (see Bohr (1972)) remained hidden in the obscurity of Danish, for he erected there a further obstacle to progress along conventional lines, in the form of a general demonstration that classical mechanics does not permit an assembly of charged particles to possess any magnetic properties. On the other hand, this was surely an important step for Bohr himself towards his realisation (Heilbron 1985) that classical mechanics needed some restraint, and that quantum theory could provide it.

At the fourth Solvay conference in 1924, Lorentz (1927), the chairman, expounded the limited successes and real difficulties of the still-classical theory, on which he had not been closely engaged since the early years of the century, and left to Bridgman the telling of attempts to improve it. There had been many, but Bridgman spoke only in terms of his own work, which received no further support during the long discussion that followed. It is hardly surprising that JJ's enthusiasm for the free-electron model had soon wavered. Shortly after his Paris lecture, he looked longingly at an old love of his, the pre-Arrhenius theory of electrolytic conduction developed by the Lithuanian Theodor von Grotthuss (1785–1822), and he tried to develop a model

(Thomson 1907, chap. 5) in which the electrons were bound together in chains, but could be transferred from one chain to another. Fortunately his interests were diverted into more profitable channels when he developed the technique of positive-ray parabolas. Out of this came the discovery that it was not only radioactive elements that had a range of isotopes, and a life's work for Aston in improving the mass-spectrograph.

But JJ could not abandon the problem, and later (Thomson 1915) he elaborated his model by introducing an analogy with Weiss's theory of ferromagnetism – that internal electric fields cause chains of atoms to acquire spontaneous polarisation at a critical temperature, below which electrons can be transferred along the chains without resistance (the details of this part of the argument are left to his reader's imagination). Only a little earlier, Lindemann (1915) had proposed that Coulomb interactions would cause the electrons to condense into a rigid lattice which could move bodily through the ionic lattice but which, like a Debye solid, would have little thermal capacity and still conduct heat well. Like JJ, he hoped, vainly, to find in this way an explanation of superconductivity. But we must leave the realms of wishful thinking and turn to a development of the earliest free-electron models, which, supported by experiment, had far-reaching consequences; this was the systematic elucidation of thermionic emission, in parallel with the invention of the radio valve.

1.4 Thermionics

Owen Richardson graduated at Cambridge in 1900 and began research under JJ's direction. In November, 1901, he read a paper to the Cambridge Philosophical Society on 'the negative radiation from hot platinum', and later published a comprehensive account (Richardson 1901, 1903). A later book (Richardson 1916) tells the history of the thermionic effect, which had been known for a long time, insofar as du Fay described, in 1733, how a hot body cannot hold an electric charge. JJ's own investigation (Thomson 1899) relied on his analysis of the cycloidal motion of a charged particle in crossed electric and magnetic fields. A hot carbon filament was placed *in vacuo* between parallel plates to produce E normal to the filament, and B was applied parallel to the filament. According to the theory, if all the emitted particles start at rest, none will reach a plate set at a distance greater than $2mE/eB^2$. The observed cut-off was not very sharp, but was sharp enough to show that the particles concerned were electrons. In the same set of experiments, JJ also showed that electrons were the particles emitted when metals were irradiated with ultraviolet light. He was well placed, therefore, to take part a year later in the discussion on metallic conduction.

Richardson's task was to study the electron emission systematically as a function of temperature, and his final account is exemplary in its care over detail (it won him the Nobel prize in 1928). He chose not carbon, but platinum; the variation with temperature of its resistance had been determined some time before by Callendar, another JJ disciple, so that this part of the work was ready to hand. As usual, obtaining and keeping a good vacuum was a tedious affair, and another question to be resolved was the voltage needed between the wire (as cathode) and the anode collector to ensure that any space-charge built up between cathode and anode was insufficient to drive electrons back to the wire. The current then remained the same with further increase of voltage. So long as the filament was thoroughly degassed by prolonged heating to white heat in a high vacuum, this saturation current i was well defined. It varied with temperature in a way that agreed well with Richardson's theoretical expectation: $i \propto T^{1/2} e^{-b/T}$, where b is a constant characteristic of the metal. Later, Richardson (1916) decided that a better formula was $i \propto T^2 e^{-b/T}$. Because the exponential varies so rapidly with temperature, b being many times greater than any attainable value of T, the difference between $T^{1/2}$ and T^2 is hard to discern, as Fig. 1.4 illustrates. Nevertheless, it is worth noting how the difference arises, for which purpose Dushman's (1923) thermodynamical argument, originally used by H. A. Wilson, is well suited.

The derivation is a simple extension of Clapeyron's expression for the variation with temperature of the vapour pressure of a condensed phase,

$$\mathrm{d}P/\mathrm{d}T = L/VT,$$

with the simplification that the volume per unit mass of the vapour phase, V, is much greater than that of the condensed phase. We may think of the electrons as evaporating from the solid, and needing latent heat l for each electron (i.e. $L = l/m$); in addition, the volume v per electron in the vapour phase ($V = v/m$) may be written as $k_B T/P$ on the assumption that the evaporated electrons, in equilibrium with the solid, are so sparse as to form a perfect gas. It remains to devise a plausible form of l. Richardson and Wilson assumed that electrons were held in the solid by a potential energy step, of magnitude ϕ, and that, in addition, some thermal energy would be needed to extract them. Since the thermal energy of a gas is proportional to temperature, the latent heat per electron takes the form $l = \phi + \alpha k_B T$, where α is a constant. Clapeyron's equation then has solution $P \propto T^\alpha e^{-\phi/k_B T}$.

Now, the pressure of a gas of n particles per unit volume is proportional to nT, while the flux of particles through an element of area is proportional to $nT^{1/2}$, since the mean velocity is proportional to $T^{1/2}$. If, therefore, the

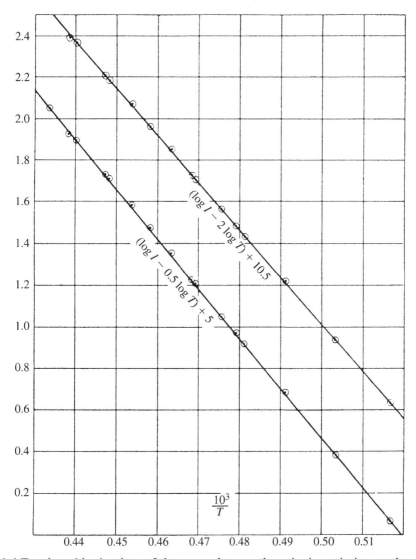

Fig. 1.4 Two logarithmic plots of the same data on thermionic emission to show that no choice between alternative forms, with pre-exponentials $T^{1/2}$ and T^2, can be made by this means (Dushman 1923).

potential difference is great enough to ensure that every particle in the outgoing flux is prevented from returning, the saturation current will be given by

$$i \propto T^{\alpha - 1/2}\, e^{-\phi/k_B T}.$$

In Richardson's first model, the electrons in the solid behave like a classical monatomic gas with specific heat, at constant volume, of $(3/2)k_B$ per particle. In the gas phase, but not in the solid, they are free to change their volume, and have specific heat $(5/2)k_B$ per particle. Hence, $\alpha = 5/2 - 3/2 = 1$, and

Richardson's original formula results. He derived it by a kinetic argument, but it may be remarked that it also follows immediately from Boltzmann's energy distribution; if the particles outside and inside both form a classical gas, the ratio of densities is $\mathrm{e}^{-\phi/k_\mathrm{B}T}$. His revised derivation, in which he was followed by Dushman, took note of the (as yet unexplained) experimental evidence that the electrons made no contribution to the specific heat of a metal, so that ϕ had to be supplemented by the whole $(5/2)k_\mathrm{B}T$ needed by the external electrons; hence the pre-exponential T^2.

By the time of Dushman's involvement in thermionic emission, many metals had been studied, and ϕ, the barrier height, determined for each. The lower the barrier, the easier to draw electrons from the metal at low temperatures; Wehnelt (1904) reported the technique of coating a filament with an oxide of strontium or barium, and thereby reducing the value of ϕ.

This may be taken as the start of the commercial interest in thermionics, which sprang directly from Marconi's success in transmitting wireless messages between ships and shore, and even across the Atlantic. Progress was slow at first (Fleming 1910, 1922; see also Tillman and Tucker (1978) and Ryder and Fink (1984)). Wehnelt realised that a hot filament and anode would constitute a one-way conductor, which he called a *valve*, and later in 1904 Ambrose Fleming showed how to use the new device, a *diode*, to rectify high-frequency oscillations. The following year, Lee de Forest took out a patent on a slightly different (in fact, better) arrangement of the rectifying circuit, which he called an *audion*. Fleming had no hesitation in accusing de Forest of plagiarism and infringement of his own patent, and battle was joined in the law courts and learned journals. It is not clear which, if either, was the winner, and in any case the matter soon became rather marginal, through two more inventions in 1906. Dunwoody developed the carborundum and cat's whisker rectifier, which replaced the old coherer and other primitive ways of detecting a weak radio signal; more importantly, de Forest added a control grid to his diode (he still called the tube an audion) and thus cleared the way for electronic circuitry. It seems that he was very much an inventor in the Edison mould, and had only limited appreciation of the processes at work in his audion.

Before 1912, very little was achieved, but then progress became rapid; there is no doubt that the First World War stimulated intense development of communications, with General Electric (Schenectady, NY) playing a leading part. It was here that Dushman and Irving Langmuir pushed the physics ahead, while E. H. Armstrong, at Columbia University and in the US Army Signal Corps, systematically developed the understanding of triodes in circuits. The thermionic diode now came into its own. The cat's whisker was relegated in the post-war years to the lash-ups of amateurs, of whom there were a great many once commercial broadcasting began in about 1920; only with the re-invention of waveguides and their application to radar in the Second World War did the crystal detector return as an essential circuit element. The whole development of thermionics and its use in research is a big story, which has not, as far as I know, been written up as fully as it deserves, and the short selection of references (Fleming 1910, 1922; see also Tillman and Tucker (1978) and Ryder and Fink (1984)) must suffice as an introduction. It would take us too far from the central theme to enlarge further.

It is, however, of interest to note when, and how, the successes of radio communication began to encourage scientists to use thermionic circuits as instruments to help their research into other fields. Not surprisingly, this had to wait until 1919 when the war-stimulated knowledge and commercial production were freed for civil applications. Without having searched the literature assiduously, I have failed to find anything earlier than a remarkable short paper by Barkhausen (1919), which reported two quite different observations made with the help of a multi-stage amplifier. One of these is well known, the clicks heard when an iron bar is magnetised, and the individual domains have their sense of magnetisation suddenly changed, to induce voltage pulses in a surrounding coil. The other observation was made by connecting the amplifier to a primitive aerial, when occasionally a descending tone was heard. Barkhausen gives a diagram of the waveform, and describes the sound (rather accurately) by the word *piou*. It is now known, particularly after Storey's (1953) thorough investigation, that these *whistlers* are originally pulses generated by lightning flashes on the other side of the world, which travel through the ionosphere with hardly any lateral divergence along lines of magnetic field, and which by dispersion are spread out so that the higher frequencies arrive first.

About three months after Barkhausen, Joachim (1919) described a study of dielectrics at a frequency of 100 MHz, and opened up the investigation of dielectric loss, which had been given a sound theoretical basis by Debye (Davies 1970). Until then, the only measurements were at low frequencies, but an enormous literature has since grown up with the availability of high frequencies generated by applying positive feedback to amplifying circuits.

These are two early examples of a development which has completely altered the character of physical research – and of much more than physics. To be sure, it was the ousting of the thermionic tube by the transistor, and especially the deposition of many circuits on a single silicon chip, that made affordable such a large range of powerful instruments as is to be found in every laboratory. It is hard to believe that, without electronics, physics could have supported more than a small fraction of its present research strength.

1.5 Stoney's dream

Since Johnstone Stoney, admittedly in a minor way, began this story with his invention of the word *electron*, he may be allowed a last appearance with another of his ideas, again not wholly irrelevant. In 1874 he addressed the British Association at their Belfast meeting, but did not then publish his remarks. He felt it to be undesirable (Stoney 1894) that all physical measurements should depend on an arbitrary choice of units, such as the kilogramme, the second and the metre, and proposed that they be replaced in due course by a set of units based on constants of nature. In the days of cgs units the electrical properties of free space required neither ε_0 nor μ_0, both of which were defined to be unity. Thus three basic units were needed, for which he suggested c, the ratio of the electrostatic to the electromagnetic units – or, as we should say, the velocity of light – the Newtonian constant of gravitation G, and the charge carried by an ion in electrolysis, which we now know as e. His conviction that this last was a fundamental constant was far from generally accepted, but he held it strongly, and much later protested (Stoney 1894) that Helmholtz was credited with making the suggestion, when he himself had done so several years earlier. It was from this firm belief that his proposal sprang to give a name to the fundamental charge.

It is not surprising that these ideas of Stoney's were ignored. The purpose of an agreed set of units is that it allows the results of different measurements to be compared as precisely as necessary, and this applies equally to the operations of science and law. An arbitrary unit must be something that can be copied and, if necessary, compared directly with the prime standard. If it is to be replaced by a fundamental physical unit, means must be available to translate this new unit into useful substandards. No one wants to carry an atomic clock with him, but the quartz crystal clock is nowadays cheap and good enough for almost everyone; above all, it is possible to pick up broadcast time signals generated by an atomic clock when high-precision calibration is essential. There is no practical barrier, therefore, against adopting the principle of Stoney's idea to define the unit of time. Again, the velocity of light can be measured with an accuracy that is only limited by the difficulty of measuring a length well enough; from the point of view of a physicist it makes sense to redefine the metre as the distance travelled by light in a specified time. In this case, as with the quartz clock, copies of the metre are readily available which are accurate enough for most purposes. If they are not, one must call on a standards institution to perform the required calibration.

We have, as a result, fundamental standards of length and time, arbitrary only in the sense that they depend on verbal agreement, not on the preservation of prime standards and the hope that they will not change with the passage of time. Only two-thirds of Stoney's dream has yet come to pass – there is still no fundamental standard of mass, and we must rely ultimately on what is locked up in a safe in the Bureau International des Poids et Mesures. It seems inconceivable that the Newtonian G should ever find a place in this system, nor can we expect to rely on e, which is far from being determined directly to something approaching one part per billion, a degree of precision that is possible with a balance. But if the dream should be fully realised one day, it is to be hoped that Stoney's champions (if any remain) will not have cause to declare that once again his originality has been forgotten.

2

The isolated electron

W. N. COTTINGHAM
University of Bristol

It has been convincingly shown, over the last 100 years, that matter in all its various forms is made up of just a few types of elementary particles, and the electron takes pride of place at the top of the list. The electron was the first elementary particle to be discovered, and indeed, apart from the photon, it was 40 years before the next was identified. However, it is not only the electron's antiquity that gives it the right to retain its first place: for humanity it is the most significant of particles. Not only is it the particle that we can manipulate to drive our modern electrical and electronic technology, but in science it is the particle that plays the main role in providing our understanding of the material world. For example, along with Planck's constant, the electron's mass, m, and electric charge, e, set the scales of energy and length in chemistry and materials science in general. A physicist would quote the chemical bond length in angstroms, $1 \text{ Å} = 10^{-10}$ m and energies in electron volts; $1 \text{ eV} = 1.602\,177\,33$ $(49) \times 10^{-19}$ J.

These units correspond closely with what can be called the natural units

$$\text{of length} = \frac{4\pi\varepsilon_0\hbar^2}{me^2} = 0.529\,177\,249\ (24)\ \text{Å}$$

$$\text{and of energy} = \frac{m}{2\hbar^2}\left(\frac{e^2}{4\pi\varepsilon_0}\right)^2 = 13.605\,698\,1\ (40)\ \text{eV}.$$

These values are quoted to the precision with which they have currently been experimentally determined (Particle Data Tables 1994). Section 2.2 tells something of the story of the high precision of the measurements, and section 2.3 gives a good reason for taking so much trouble.

The electromagnetic field around an electron has the restricted angular structure characteristic of a spin-1/2 particle. Apart from its total charge, it can have no moments more complicated than dipole. I will denote the magnetic dipole moment by $\boldsymbol{\mu}_e$; it is a vector with direction opposite to that of the electron spin. Besides e and m, μ_e has also been determined with great precision. Moreover, we have a theory, basically quantum electrodynamics, which provides a very precise relationship between e, m and μ_e; the astonishing success of this theory in relating these three quantities is the subject of section 2.2. Section 2.3 deals with the secret that the electron still keeps to itself, its electric dipole moment $\boldsymbol{\eta_e}$. In spite of intense efforts, measurements are still consistent with the electron having no electric dipole moment at all, but there is no reason, from our current understanding, to believe η_e to be identically zero. Indeed, the electron may still have a surprise in store for us.

2.1 The electron mass and the electron charge

The story of J. J. Thomson's contribution to establishing the existence of the electron has been told in chapter 1. Several physicists contributed with somewhat different experiments and with somewhat different interpretations. Extracting numbers from his own measurements of 1897, Thomson found the charge to be negative, $-e$,* and $m/e \approx 0.4$, 0.5 and 0.9×10^{-11} kg/C, the mass to charge ratio of cathode rays from three different cathode ray tubes. Together with his measurement of the unit of charge on an ion (Thomson 1898) $e = 2.2 \times 10^{-19}$ C, we can infer

$$m \approx 1.4 \pm 0.5 \times 10^{-30} \text{ kg}; \quad e \approx 2.2 \times 10^{-19} \text{ C}.$$

Both estimates are within a factor of two of today's accepted values.

Twenty years later, accuracy had improved to better than 1%. There was much independent work on the measurement of e/m; K. Woltz (1909) gives a comprehensive list which includes his own measurement of $e/m = 1.764(3) \times 10^{11}$ C/kg, but in particular, by 1916, Millikan, by his careful and inspired experimental work, had made precision measurements of both e and of Planck's constant $\hbar = h/2\pi$. His experiments with oil drops (Millikan 1917) established the fact that, on the atomic scale, electric charge comes in integral multiples of e, and his experiments with the photoelectric effect (Millikan 1916) verified Einstein's assertion on the quantum nature of the absorption of electromagnetic radiation. He found:

$$e = 1.592 \times 10^{-19} \text{ C}; \quad \hbar = 1.054 \times 10^{-34} \text{ J s}.$$

With the measurements of Woltz these give

$$m = 9.025 \times 10^{-31} \text{ kg}.$$

---

* The sign convention was established in the 18th century. For example, electric charge is exchanged if a glass rod is wiped with a silk handkerchief. The charge on the glass was defined to be positive, that on the silk negative. Today we know that, in the process of wiping, electrons migrate from the surface of the glass onto the silk.

Today we know all three fundamental constants to better than one part in a million (see, for example, the Cohen and Taylor (1986) adjustment of the fundamental constants). For the properties of the electron, a good part of this knowledge is due to the efforts of H. G. Dehmelt and co-workers, of the University of Washington, and it is on this work that I will focus particular attention. Just as Millikan was able to follow the motion of a single oil drop in varying electric fields over several hours, now it is possible to record, over similar time scales, the oscillatory motion of a single electron in the electro-magnetic field of a Penning trap. The advantage we have over Millikan is our electronic technology that can be used to measure frequencies with very great precision. It is with the oscillatory frequencies of electrons (and also protons and ions) that the Washington group have been concerned.

The basic measurements in these experiments are the two frequencies of motion of a low-energy electron in a uniform magnetic field, $\boldsymbol{B}$. An electron moving with velocity $\boldsymbol{u}$ in the field experiences a net force $-e\boldsymbol{u} \times \boldsymbol{B}$ and a torque $\boldsymbol{\mu}_e \times \boldsymbol{B}$. Setting aside any relativistic corrections, the force results in circular motion about the field direction with the cyclotron frequency ω_C, and the torque results in a precession of the electron spin with the Larmor frequency ω_L, where

$$\omega_C = \frac{eB}{m}; \quad \omega_L = \frac{2\mu_e B}{\hbar}. \tag{1}$$

Both of these formulae can be inferred from classical mechanics. Consider the magnetic field to be in the z direction, $\boldsymbol{B} = (0, 0, B)$, and the position of the electron to be (x, y, z). Newton's law, force equals mass times acceleration, gives $m(\mathrm{d}^2 z/\mathrm{d}t^2) = 0$, which implies a constant drift velocity along z;

$$\frac{\mathrm{d}^2 x}{\mathrm{d}t^2} = -\omega_C \frac{\mathrm{d}y}{\mathrm{d}t}, \tag{2a}$$

and

$$\frac{\mathrm{d}^2 y}{\mathrm{d}t^2} = \omega_C \frac{\mathrm{d}x}{\mathrm{d}t}. \tag{2b}$$

The classical motion in the x, y plane can be thought of as the motion of a complex number $Z = x + \mathrm{i}y$ in an Argand diagram. If we add eq. (2a) to i times eq. (2b) we obtain

$$\frac{\mathrm{d}^2 Z}{\mathrm{d}t^2} = \mathrm{i}\omega_C \frac{\mathrm{d}Z}{\mathrm{d}t},$$

which has the general solution

$$Z = R\mathrm{e}^{\mathrm{i}\omega_C t} + C, \tag{3}$$

where R and C are complex numbers. The solutions show that the motion in the x, y plane is in a circle of arbitrary radius $|R|$ about an arbitrary centre C and all the circles are executed with the cyclotron frequency.

Considering the spin precession, classically the rate of change of spin angular momentum equals the applied torque. The spin angular momentum is a vector, say $\mathbf{s} = (s_x, s_y, s_z)$ of magnitude $\hbar/2$ aligned with the magnetic moment so that we can write $\boldsymbol{\mu}_\mathbf{e} = (2\mu_e/\hbar)\mathbf{s}$.

The torque $\boldsymbol{\mu}_\mathbf{e} \times \mathbf{B} = \omega_L(s_y, -s_x, 0)$ and the classical equations are

$$\frac{\mathrm{d}s_x}{\mathrm{d}t} = \omega_L s_y; \quad \frac{\mathrm{d}s_y}{\mathrm{d}t} = -\omega_L s_x; \quad \frac{\mathrm{d}s_z}{\mathrm{d}t} = 0,$$

with the solution

$$s_x = s_1 \sin(\omega_L t); \quad s_y = s_2 \cos(\omega_L t); \quad s_z = s_3;$$

s_1, s_2 and s_3 are constants. These solutions describe the precession of the spin with the Larmor frequency.

The ratio ω_L/ω_C (which can be very well determined) is independent of the magnetic field (which in practice can not) and gives a formula for determining the magnetic moment

$$\mu_e = \mu_B \omega_L/\omega_C, \tag{4}$$

where I have introduced the natural unit of magnetic moment, the Bohr magneton

$$\mu_B = e\hbar/2m. \tag{5}$$

For the practical determination of these frequencies, the electron has to be confined in as near isolation as possible. This is done in a Penning trap, a device for holding an electron in an electromagnetic field. The principles were invented 50 years ago by F. M. Penning (1937).

An ideal trap consists of an electric quadrupole field, for example as represented by the electrostatic potential $V = (V_0/l^2)(x^2 + y^2 - 2z^2)$, and a uniform field $\mathbf{B}$, parallel to the z axis. V_0/l^2 is a constant, $V_0 > 0$ is a voltage and l is a length. Fig. 2.1 illustrates the lines of electric force experienced by an electron. It can be seen that the electric field constrains the electron to be near the x, y plane, and a large enough magnetic field will constrain the electron to the neighbourhood of the z axis, hence confinement. Again, classical mechanics gives the relevant frequencies. The force on the electron is $-e(-\nabla V + \mathbf{u} \times \mathbf{B})$ and Newton's law now gives the three equations

$$\frac{\mathrm{d}^2 x}{\mathrm{d}t^2} = -\omega_C \frac{\mathrm{d}y}{\mathrm{d}t} + \frac{2eV_0}{ml^2} x,$$

$$\frac{\mathrm{d}^2 y}{\mathrm{d}t^2} = \omega_C \frac{\mathrm{d}x}{\mathrm{d}t} + \frac{2eV_0}{ml^2} y,$$

$$\frac{\mathrm{d}^2 z}{\mathrm{d}t^2} = -\frac{4eV_0}{ml^2} z.$$

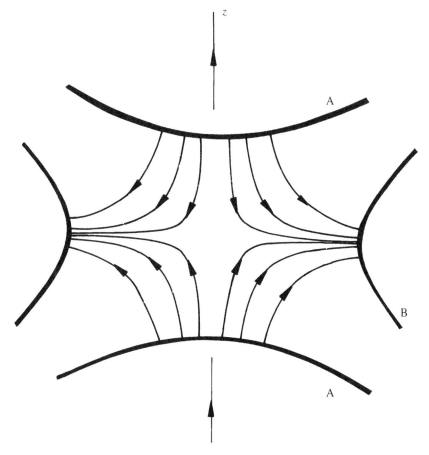

Fig. 2.1 A schematic diagram of a Penning trap. It has cylindrical symmetry about the z axis. The inner surface of the end cap electrodes, labelled A, are machined in copper into the forms $x^2 + y^2 - 2z^2 = -1^2$ and the ring electrode, labelled B, into the form $x^2 + y^2 - 2z^2 = +1^2$. Some lines of electric force on the electron are also indicated.

The motion perpendicular to the x, y plane is simple harmonic with frequency

$$\omega_Z = \frac{2}{l}\left(\frac{eV_0}{m}\right)^{1/2}.$$

In the plane we form the complex quantity $Z = x + iy$ and obtain

$$\frac{d^2 Z}{dt^2} = i\omega_C \frac{dZ}{dt} + \frac{\omega_Z^2}{2} Z.$$

This equation has solutions of the form $Z = R\,e^{i\omega t}$ with

$$\omega^2 - \omega_C\omega + \frac{\omega_Z^2}{2} = 0.$$

There are two possible frequencies, a faster one with frequency ω_1 and a slower with frequency ω_2:

$$\omega_1 = \frac{\omega_C}{2}\left[1 + \left(1 - 2\left(\frac{\omega_Z}{\omega_C}\right)^2\right)^{1/2}\right]; \quad \omega_2 = \frac{\omega_C}{2}\left[1 - \left(1 - 2\left(\frac{\omega_Z}{\omega_C}\right)^2\right)^{1/2}\right].$$

(6)

The classical motion in the plane can be thought of as the motion of the complex number $Z = x + iy$ in an Argand diagram. There are two circular motions, apart from phase factors, $Z_1 = R_1\,e^{i\omega_1 t}$ and $Z_2 = R_2\,e^{i\omega_2 t}$; R_1 and R_2 are the radii. The general motion is given by

$$Z = Z_1 + Z_2.$$

With no electric field, $\omega_Z = 0$, $\omega_1 = \omega_C$, $\omega_2 = 0$, and we recover eq. (3) with $Z_2 = C$. In practice, $\omega_2 \ll \omega_1$. With no electric field, C, the centre of the cyclotron circular orbit, could be anywhere in the x, y plane; the electric field causes the centre to rotate slowly with frequency ω_2 about the point $x = y = 0$. The total energy of motion in the x, y plane can be calculated to be

$$\text{energy} = \frac{m}{2}(\omega_1 - \omega_2)(\omega_1 R_1^2 - \omega_2 R_2^2).$$

Note that the energy associated with circle 2 is negative and decreases without bound as R_2 increases.

It is also an interesting exercise to solve the quantum mechanical problem of motion, which can be done in terms of elementary functions (Sokolov and Pavlenko 1967). It is perhaps not surprising that the three harmonic motions are quantized somewhat like harmonic oscillators, but as with the classical motion the energy associated with ω_2 is negative. The quantized energies are characterized by three independent integers, n_1, n_2 and n_Z that are positive or zero; there is also the spin index $s = \pm 1/2$:

$$E(n_1, n_2, n_Z, s) = \hbar[(n_1 + 1/2)\omega_1 - (n_2 + 1/2)\omega_2 + (n_Z + 1/2)\omega_Z + s\omega_L].$$

(7)

The frequencies are the classical frequencies. These are the energies in an ideal Penning trap. Large values of the integers correspond to large classical orbits, and the size of a practical trap implies upper limits on them. Comparing the classical with the quantum energy, the radii of the classical orbits are

$$R_1 = (n_1 + 1/2)^{1/2}d; \quad R_2 = (n_2 + 1/2)^{1/2}d,$$

where the length $d = [2\hbar/m(\omega_1 - \omega_2)]^{1/2} \sim 0.3\ \mu\text{m}$. Also, in practice it is necessary to apply corrections for deviations of the fields from perfect electric quadrupole and for magnetic inhomogeneities, but skilful design and data analysis makes corrections small and under good control.

Details of the trap design can be found in Wineland *et al.* (1973), Wineland and Dehmelt (1975) and Van Dyck *et al.* (1976a, b, 1977). The traps have dimensions of a few millimetres and are operated at a few kelvin. They are evacuated to pressures of 10^{-14} Torr, at which pressure the electron storage time is days. Magnetic fields are of the order of one tesla, and the electric fields are weak, V_0/l^2 being of the order of 10 V/cm^2. Typical frequencies are

$$\frac{1}{2\pi}\omega_Z \approx 5 \times 10^7 \text{ Hz}; \quad \frac{1}{2\pi}\omega_C \approx 5 \times 10^{10} \text{ Hz},$$

and these result in

$$\omega_1 \approx \omega_C; \quad \omega_2/\omega_1 \approx 5 \times 10^{-7}.$$

As advertised, ω_2 is much smaller than ω_1.

After evacuation, electrons are introduced into the trap by ionizing some of the molecules of the residual gas using a 1 keV electron beam. (The positive ions are automatically removed since their orbits are not confined.)

Communication with the electrons is established by making the trap part of a radio frequency electric circuit with variable frequency ω. The trap when empty is a capacitor; when it contains free electrons it becomes a more active circuit element with various resonant frequencies ω_R. These frequencies are such that $\hbar\omega_R$ equals the quantum energy level differences, hence ω_1, ω_2, ω_Z and ω_L are all resonance frequencies. The width of the resonant frequencies, which ultimately determines the accuracy with which they can be measured, is not due to the trap offering a resistance but is largely caused by deviations of the internal potential from pure quadrupole; the sharp frequencies of an ideal trap become a band of frequencies.

The operation of the circuit with a large amplitude on resonance is used to drive out unwanted electrons. At low amplitudes and with the required number of electrons the radio frequency output across the trap is the measured signal. At appropriate resonance frequencies the amplitude of the signal gives a measure of the number of free electrons; small numbers can be counted. As a function of frequency, peaks in the signal give the resonant frequencies, from which one can deduce the cyclotron and Larmor frequencies.

The cyclotron and Larmor frequencies involve the B field, which is not very well known, but B is eliminated if one considers frequency ratios. One important piece of work done by the Washington group was also to liberate protons in the same trap (but on different runs with V_0 reversed). They then determined the ratio of the proton to the electron mass through the relationship

$$\frac{m_p}{m} = \frac{\omega_C \text{ (for the electron)}}{\omega_C \text{ (for the proton)}}.$$

The value (Van Dyck *et al.* 1986) which is used in the 1986 adjustment of the fundamental constants is $m_p/m = 1836.152\,701\,(37)$. A more recent value (Farnham *et al.* 1995), obtained by comparing the electron cyclotron frequency with that of C^{6+} ions gave $m_p/m = 1836.152\,666\,5(40)$. This number gives the electron mass in units of the proton mass to nine significant figures. To put the electron mass in SI units requires m_p in kilogrammes. The measurement of the proton mass is a different story; nuclear masses are well determined, but not so well as this ratio. The present determination is

$$m = 9.109\,389\,7(54) \times 10^{-31} \text{ kg}.$$

2.2 The electron magnetic moment and the electron charge

One of the triumphs of the Dirac theory of the electron, as we shall see in chapter 3, was that it predicted that its magnetic moment should equal the Bohr magneton. As eq. (4) implies, the cyclotron and Larmor frequencies should then be the same, and, to a very good approximation, they are the same. The difference between them can be expressed in terms of a dimensionless parameter a, which can be called the anomaly

$$\frac{\mu_e}{\mu_B} = \frac{\omega_L}{\omega_C} = 1 + a, \tag{8}$$

and from the Washington experiments

$$a = 0.001\,159\,652\,188\,4(43)$$

(Van Dyck *et al.* 1987). Not only is the anomaly small, it is also known with some precision. Precision is important here because the standard model of particle physics, basically that part of it which is quantum electrodynamics, accurately predicts the anomaly. In fact, the anomaly provides the theory with its most challenging test. The remarkable success of the theory in passing this test is a cause for confidence in the renormalization techniques of quantum electrodynamics and in the more general strategy of demanding that a basic theory, such as the standard model of particle physics, be a renormalizable field theory. This strategy has served science well over the last 50 years.

Quantum electrodynamics is a theory with only three parameters: m, e and $\hbar$. The fine-structure constant α is the only combination of them that is dimensionless: $\alpha = e^2/4\pi\varepsilon_0\hbar c \approx 1/137$ (c is fixed, it defines the unit of length, and ε_0 is fixed by the definition of the Coulomb; they are not additional parameters). The only combination which has the dimensions of magnetic moment is the Bohr magneton. Hence, quantum electrodynamics predicts the electron magnetic moment to be a scale factor times μ_B. The scale factor is $(1 + a)$, and the anomaly a is a function only of α.

The calculation of this function is a formidable task, and so far only the first four terms of a power series expansion in α have been estimated. In fact, to obtain the last significant figure of the experimentally determined anomaly also requires contributions from outside quantum electrodynamics, in particular from contributions due to the muon. Including the muon introduces another dimensionless number $(m/m_\mu)^2 \approx 10^{-5}$, but within the standard model the muon contributions to the first few terms of the power series have been calculated so that today one can write

$$a = C_1\left(\frac{\alpha}{\pi}\right) + C_2\left(\frac{\alpha}{\pi}\right)^2 + C_3\left(\frac{\alpha}{\pi}\right)^3 + \ldots + \delta_a, \tag{9}$$

where $\delta_a = 0.000\,000\,000\,001\,69(4)$ is the estimated contribution from all other sources. At the quoted level of this tiny correction only the light hadrons contribute. The C values are given by

$$C_1 = 0.5,$$

$$C_2 = \left(\frac{197}{144} + \frac{\pi^2}{12}(1 - 6\ln(2)) + \frac{3}{4}\zeta(3)\right) + \delta C_2 = -0.328\,478\,444,$$

$$C_3 = 1.176\,11\,(42); \quad C_4 = -1.434\,(138),$$

where

$$\delta C_2 = 0.000\,000\,521 \text{ is the contribution of the muon,}$$

and

$$\zeta(3) = 1.202\,056\,903\,2\ldots$$

is the Riemann zeta function of 3. By contemplating the expressions for C_1 and C_2, it can be appreciated that the calculations for the higher terms in the expansion become rapidly more difficult. There are no expressions for C_3 and C_4 in terms of ancient functions, and the estimates given are the results of numerical computation of integrals.

C_1 was first calculated by Schwinger (1948), and C_2 was calculated by Sommerfield (1957) and Peterman (1957). C_3 and C_4 were calculated by Kinoshita and Lindquist (1990). These values are the result of a large body of numerical work published in several papers cited in this reference. The anomaly results from the interactions of the electron with itself through its own electromagnetic field. Calculations employ a technique introduced by Feynman (1949). He systematized the calculations, and his diagrams give some visual indication of the nature of a particular contribution. With each diagram there is associated an integral, generally multi-dimensional. C_1 is associated with the diagram of Fig. 2.2(a), C_2 with the electrodynamic diagrams of Fig. 2.2(b). The muon also contributes to C_2 through quantum fluctuations, which create muon pairs. The diagram is also shown in Fig. 2.2(b). It is through similar

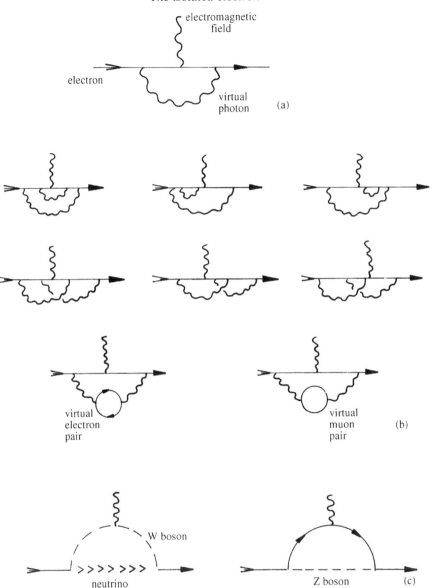

Fig. 2.2 Feynman diagrams indicating the nature of some quantum corrections to the interaction of an electron with an electromagnetic field.

processes, highly suppressed, that mesons and nucleons contribute. The heavy gauge bosons of the weak interaction contribute even to C_1 with a term

$$\delta C_1 = \left(\frac{m}{m_W}\right)^2 \left[\frac{1}{3}\sin^2\theta_W - \frac{1}{6} + \frac{1}{4\sin^2\theta_W}\right] \approx 3 \times 10^{-11} \qquad (10)$$

(Itzykson and Zuber 1980). The Feynman graphs are shown in Fig. 2.2(c). m_W is the W boson mass and θ_W is the Weinberg angle. The contribution of δC_1 to the dipole moment is negligible at present levels of accuracy, but the expression does illustrate the small size of heavy-mass-particle contributions. To compare the theory with the experiment, we need a value for α. Two independent and consistent values are (Cohen and Taylor 1986);

$$\alpha^{-1} = 137.035\,997\,9(32) \text{ from the quantized Hall resistance,}$$

$$\alpha^{-1} = 137.036\,003\,0(260) \text{ from muonium hyperfine structure.}$$

Taking the better determined Hall resistance value leads to

$$a \text{ (theory)} = 0.001\,159\,652\,140\,0\,(41 + 53 + 271)$$

and we recall the experimental value of a to be

$$a \text{ (experiment)} = 0.001\,159\,652\,188\,4\,(43).$$

The agreement between theory and experiment is to better than nine places of decimals, and, with this value for α, the theory is within 1.7 standard deviations of the experimental measurement. As for the errors given in the theoretical value of a, the 41 is from the uncertainty in C_4, the 53 is from the uncertainty in C_3 and the 271 is from the quoted error in the quantized Hall resistance. It is also of interest to note that if the theory at its present level is accepted then the electron anomaly can be used to give an estimate of α with a better precision than that from the quantized Hall resistance. Matching the theory with the measurement,

$$\alpha^{-1} = 137.035\,992\,22\,(51 + 111).$$

The error of 51 is from the measurement and the 111 is from the calculation. From the measurements of $\hbar$, we infer

$$e = (4\pi\varepsilon_0\hbar c\alpha)^{1/2} = 1.602\,177\,31\,(49) \times 10^{-19}\text{C}.$$

2.3 The electron electric dipole moment

The electric dipole moment is a most elusive quantity. Measurements to date are consistent with zero. Particle Data Tables (1994) essentially give an upper bound on the magnitude. The electric dipole moment is a vector aligned with the direction of spin; denoting the size by η_e, experiment gives

$$\eta_e = (-0.3 \pm 0.8) \times 10^{-28} e \text{ metre} \tag{11}$$

$$= (-1.5 \pm 4) \times 10^{-16} \mu_B/c.$$

Taking $\mu_B/c = e\hbar/2mc$ to be the natural unit of electric dipole moment, the factor of 10^{-16} is a considerable tribute to experimental skill.

The methods adopted for searching for electric dipole moments to some extent parallel those for the measurement of magnetic moments: they can be considered to be the measurement of the precession frequency of the electric dipole in a large static electric field. An isolated electron, however, poses a particular problem in that, being charged, it tends to accelerate out of any confining region. Most experiments, in fact, investigate atomic electric dipole moments and interpret the results in terms of electronic moments.

The consideration of atomic dipole moments has produced one of the nice surprises of theoretical physics. At first, the situation seemed bleak. Schiff (1963) had shown that, within the context of the Schrödinger equation, intrinsic dipole moments of the electrons would distort their wave functions in such a way as to produce a spatial moment to cancel the intrinsic moments. To the lowest order in η_e (which we have seen to be very small), the total atomic dipole moment should be zero. Schiff's theorem certainly applies to light atoms in which the electrons are non-relativistic, but in the inner regions of heavy atoms the electrons become relativistic and Dirac's equation, not the Schrödinger equation, applies. In fact, for heavy atoms such as caesium and thallium, the conclusions of the theorem are completely reversed. The distortions induced in these heavy atoms magnify the intrinsic dipole moments (Sandars 1965). The results quoted in eq. (11) are interpretations from measurements on atoms of thallium (Abdullah *et al.* 1990), for which the magnification factor is of the order of 600.

The thallium states that were investigated have spin $1\hbar$. An electric dipole moment, say η, will be aligned with the spin and, by analogy with the magnetic case, will precess in an electric field E with frequency $\Delta\omega = \eta E/\hbar$. The experiments are conducted with a static magnetic field parallel to the electric field so that the total precession frequency is $\omega = \omega_0 + \Delta\omega$, where $\omega_0 = \mu B/\hbar \approx 7.5 \times 10^5/\text{s}$ (μ is the atomic magnetic moment). The electric fields employed and the measured bounds on η imply that $\Delta\omega < 10^{-3}/\text{s}$. The experiments involve the use of lasers and radio frequency magnetic fields both to prepare particular quantum states in atomic beams and to analyse the states after they have evolved over a period of time. The small frequency shift $\Delta\omega$ imposes a small change in the final state over what it would be in the absence of the electric field. It is very interesting to understand the sophisticated techniques that make measurements of the small changes possible, but the reader is referred to the original papers.

Turning to theory, the electric dipole moment is small because it is a property that is induced on the electron by its weak interactions, and this can be deduced from considerations of symmetry. The strong and the electromagnetic interactions preserve the laws of physics under both the parity transformation (reflection of space co-ordinates) and under the time reversal transformation (reversal of the arrow of time). Not so the weak interactions: the weak interaction respects neither symmetry. A finite electric dipole moment on an electron breaks both parity and time reversal symmetry, so for two good reasons it must be a property of the weak interaction. The symmetries are broken because an electric dipole moment is an ordinary vector which changes sign under space reflection but not under time reversal; the electron spin, like all angular momenta, changes sign under time reversal ($L = r \times p$ and $p = m(\mathrm{d}r/\mathrm{d}t)$ changes sign), but, being a vector product, it does not change sign under the parity operation. The orientation of the electric dipole moment with respect to the spin therefore changes sign under either operation. This is to be contrasted with the magnetic dipole moment, which is anti-aligned with the spin in both left and right handed co-ordinate systems, and if the arrow of time is reversed. Therefore it is the weak interaction that is responsible for the electric dipole moment.

The weak interaction sector of the standard model of particle physics provides an elegant description of how parity is violated in nature, but this is not the case with the violation of time reversal symmetry. Also, many independent experiments verify the standard model of parity violation, but there is only one confirmed example, in the decay of K mesons, that clearly demonstrates time reversal violation. Although the observations of time reversal asymmetry in the behaviour of K mesons are consistent with the standard model, the standard model is by no means wholly confirmed. Different interpretations abound, and the question of the origin of time reversal asymmetry provokes much speculation. More experimental information would be most welcome; hence the interest in electric dipole moments.

One certain fact is that processes which violate time reversal symmetry are more suppressed than the ordinary weak interaction. The known smallness of the electron electric dipole moment is an example of the suppression. We saw in eqs. (9) and (10) that the weak interaction makes a contribution to the magnetic dipole moment $\delta C_1(\alpha/\pi)\mu_B \approx 10^{-13}\mu_B$. If time reversal symmetry were violated with the full strength of the weak interaction, one might expect an electric dipole moment of order $10^{-13}\,\mu_B/c$, three orders of magnitude bigger than the present experimental limit. In fact, the standard model predicts an electric dipole moment 11 orders of magnitude smaller than the present experimental limit. It is the contrived way that time reversal asymmetry is built into the standard model that is responsible for this very small number. In the standard model, only the weak interactions of quarks are asymmetric in time, and the manifestation of asymmetry requires the participation of all three quark generations. This results in the dipole moment only showing itself in the seventh order of perturbation theory. A calculation can be found in Hoogeveen

(1990). The largest contributions are associated with ten Feynman graphs, each one similar to that of Fig. 2.3, which in total result in an expression

$$\eta_e = \alpha^3 F \left(\frac{m}{M_W} \right)^2 \left(\frac{M_t^2 M_b^2 M_c M_s}{M_W^6} \right)^2 \frac{\mu_B}{c} \sim 3 \times 10^{-27} \frac{\mu_B}{c}, \tag{12}$$

where M_W is the W boson mass, and M_t, M_b, M_c and M_s are respectively the top, bottom, charmed and strange quark masses (see also Jarlskog (1985)). F is basically a combination of multi-dimensional integrals and elements of the Kobayashi–Maskawa coupling matrix. After numerical computation, Hoogeveen finds F to be about 60. However, because this value is so small, it must be concluded that, according to the standard model, it would be hopeless to try to measure the electric dipole moment.

A review of theoretical work on the electric dipole moment in various extensions of the standard model can be found in Bernreuther and Suzuki (1991). Many of these extensions, including super-symmetric models, left–right symmetric models and extended Higgs models, induce an electric dipole moment on the electron even at the level of Feynman triangle graphs. The predicted moments are correspondingly much larger than that of the standard model. The graphs are like those of Fig. 2.2(c) but with the particles on the internal lines being new particles conjectured in the particular model. Based upon these theoretical considerations, it is fair to say that reasonable speculation suggests that the electron might well have an electric dipole moment not

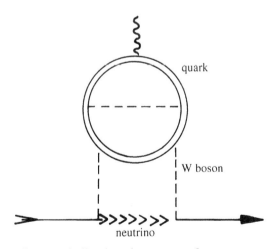

Fig. 2.3 A Feynman diagram indicating the nature of a quantum correction from the standard model of particle physics that induces an electric dipole moment on the electron. The doubly inscribed circle indicates a quark that changes its flavour at every interaction vertex with the W boson.

much below the present experimental limit. If it has, and if it was measured now, then the electron, the first of all the particles, would be the first to demonstrate clearly that the standard model it started is not yet the final word on the theory of matter and radiation.

3

The relativistic electron

D. I. OLIVE

University of Wales, Swansea

3.1 Introduction

During the early 1920s, when the principles of quantum mechanics were being developed and elucidated, it was clear that a crucial test was to be a successful derivation of the energy spectrum of the hydrogen atom. This spectrum was well established experimentally yet inexplicable according to classical mechanics. The partial explanation afforded by the Bohr model suggested that a fuller understanding would follow from a proper quantum mechanics. The challenge was to explain Bohr's *ad hoc* rules in terms of more acceptable principles and to obtain a more detailed and precise picture.

In the classical picture, the hydrogen atom consisted of an electron in orbit around the hydrogen nucleus, which consisted of a single proton. The principle of the conservation of energy combined with the known values of the energy indicated that the electron moved sufficiently slowly that the effects of Einstein's theory of special relativity could be safely ignored. Since quantum theory was also expected to respect the conservation of energy, the unimportance of relativity in this context could be expected to carry over.

Wolfgang Pauli (1926a) was the first to achieve the goal of a satisfactory treatment of the hydrogen atom, and he used Heisenberg's matrix mechanics. Soon after, Erwin Schrödinger (1926a–d) developed a different approach to the same question, which had the advantage of using a formulation that felt more comfortable to the physicists of the time. This was because it revealed a link to Maxwell's equations in the sense that Schrödinger also dealt with differential equations to be satisfied in space. However, rather than the electric and magnetic field strengths, the solutions to his new equations were provided by the wave function of the electron. Despite the differences between the two methods of Pauli and Schrödinger, the results agreed perfectly, and were regarded as very satisfactory because they also provided the basis for the

39

understanding of atoms that are more complicated in that they possess more electrons in orbit around a heavier central nucleus.

The picture found was that the electrons added in heavier and more complicated atoms successively filled up 'shells' of increasing energy before moving on to populate the next shell. The importance of all this is that it determines the chemical properties of the elements made up in aggregating these atoms. For example, atoms in which the last shell is just full, just underfull or just overfull each display similar chemical properties. This is reflected in the pattern of the periodic table, constructed by chemists in the 19th century as a means of systematising possible chemical reactions between different elements. This recipe, once mysterious, was explained by the revolutionary new quantum mechanics.

However, it has to be admitted that this degree of understanding required two further ingredients from quantum theory that had to be added 'by hand'. The first is the fact that the electron carries an intrinsic spin of one-half in suitable units and, as a consequence, possesses two spin states. The second is that the electron obeys Fermi–Dirac statistics. This means that the wave function describing the behaviour of a number of electrons is antisymmetric with respect to the interchange of electrons by an odd permutation. As a consequence, the electrons obey what has come to be known as the Pauli exclusion principle.

These two new ingredients were simply accepted as unexplained facts which were begging for further justification. There was another niggle, and this is what particularly concerned Paul Dirac. He asked whether the principles of quantum mechanics could be compatible with the principles of special relativity, so far ignored. One possible scenario is that this unification could be difficult, and achievable only at a price. The identification of this price then becomes the interesting issue, since it could be that the price exacted by the possibility of a relativistic quantum theory buys much more predictive power. This is exactly what Dirac (1928) found. His results led to a natural incorporation of the electron spin already mentioned into the relativistic theory. Later, others correlated this with the Fermi–Dirac statistics. Most unexpectedly, it led to the prediction of a new particle, now called the positron. At first this was an embarrassment for Dirac, but it became his crowning achievement when it was discovered experimentally. This was perhaps the most celebrated coup in the history of theoretical physics, and rapidly earned Dirac the award of a Nobel prize, followed by other honours.

The equation that Dirac found was called by him the 'relativistic equation of the electron', but it is now known to the world as the Dirac equation. In this chapter we explain the reasoning behind it, together with some of its many consequences and ramifications. In particular, these include the inexorable march towards the formulation of relativistic quantum field theories, which furnish the basis for all modern fundamental theories of elementary particles and their interactions.

3.2 The Dirac equation and the spin of the electron

The behaviour of the wave function associated with a slowly moving electron is governed by the Schrödinger equation. As Schrödinger saw it, this equation can be inferred by considering the expression for the energy E in terms of the momentum p by making the replacements

$$E \to i\hbar \frac{\partial}{\partial t}, \quad p \to -i\hbar\nabla \tag{1}$$

when acting on the wave function. Thus, if the electron is free, that is free of all external forces, so that

$$E = \frac{1}{2m}p^2, \tag{2}$$

then the Schrödinger equation reads

$$i\hbar \frac{\partial \psi}{\partial t} = -\frac{\hbar^2}{2m}\nabla^2\psi. \tag{3}$$

We shall concentrate on this freely moving situation before including the effects of an electromagnetic field acting on the electron. This second step would be needed to formulate the behaviour of the electron within a hydrogen atom, for example. The reason for this strategy is that first we want to study the unification of quantum principles with special relativity in the simplest possible situation.

Einstein had discovered that when a free particle moves quickly, the principle of special relativity requires that the above expression (2) for the energy of a free particle be replaced by

$$E = c(m^2c^2 + p^2)^{1/2}. \tag{4}$$

To a good approximation, this reduces to $E = mc^2 + p^2/2m$ when the magnitude of the momentum, $|p|$, is much smaller than the mass m multiplied by the speed of light, c. This, of course, agrees with (2) in normal circumstances in which mass is neither lost nor created.

There is a neat way of rewriting this energy momentum relation (4) which makes clearer the connections with the relativity principle and the geometry of space time. This involves the introduction of the energy momentum four-vector

$$p^\mu \equiv (p^0, p^1, p^2, p^3) = \left(\frac{E}{c}, p\right). \tag{5}$$

The scalar product between any two four-vectors is defined by the metric $g_{\mu\nu}$, which, as a 4 by 4 matrix, is diagonal, with entries $1, -1, -1, -1$ on the diagonal. This scalar product should then be Lorentz invariant, that is independent of the frame of reference. In particular,

$$p^2 \equiv \left(\frac{E}{c}\right)^2 - \boldsymbol{p}^2 = m^2 c^2, \tag{6}$$

which is equivalent to (4) upon rearrangement, except for the subtle difference that a negative root in (4) is also allowed. This will be important later.

The result (6) can be deduced from the expression for the four-momentum

$$p^\mu = m \frac{\mathrm{d}x^\mu}{\mathrm{d}\tau}, \tag{7}$$

where x^μ is the space time coordinate of the particle expressed as a four-vector

$$x^\mu = (ct, \boldsymbol{x}), \tag{8}$$

and τ is its proper time, defined infinitesimally by

$$c^2 \, \mathrm{d}\tau^2 = (\mathrm{d}x)^2 \equiv c^2 \, \mathrm{d}t^2 - \mathrm{d}\boldsymbol{x}^2. \tag{9}$$

Notice that

$$\partial_\mu \equiv \frac{\partial}{\partial x^\mu} = \left(\frac{1}{c}\frac{\partial}{\partial t}, \nabla\right) \tag{10}$$

is a four-vector with a lowered suffix, in contrast to the four-vectors met previously, which all have upper suffices. Raising the suffix by means of $g^{\mu\nu}$, the inverse matrix to the metric $g_{\mu\nu}$ previously described, yields

$$\partial^\mu \equiv \frac{\partial}{\partial x_\mu} = \left(\frac{1}{c}\frac{\partial}{\partial t}, -\nabla\right). \tag{11}$$

Thus, the replacement (1) can be unified into a single Lorentz covariant equation

$$p^\mu \rightarrow i\hbar\partial^\mu. \tag{12}$$

This shows that the principles of quantum theory embodied in the replacements (1) are indeed consistent with relativity. This is because the apparently contrary signs in the replacement (1) have become neatly aligned in (12), thanks to the magic of Einstein's special relativity.

There is now an obvious way of constructing a relativistically invariant Schrödinger equation for the wave function of a free particle. This is to insert the replacement (12) into the energy momentum relation in the form (6) when acting on the wave function. The result is

$$\partial^2 \psi \equiv \frac{1}{c^2}\frac{\partial^2 \psi}{\partial t^2} - \nabla^2 \psi = -\frac{m^2 c^2}{\hbar^2}\psi. \tag{13}$$

Indeed, Schrödinger himself had done precisely this prior to his famous paper.

He had even gone on to incorporate the electromagnetic interactions. Unfortunately, the results for the hydrogen spectrum that he found were disappointing, and he therefore discarded this analysis in favour of the one based on the non-relativistic equation (3), which gave better results, despite the approximation.

As explained above, in the introduction, Dirac felt that the determination of a satisfactory relativistic wave equation was of prime importance for reasons of principle (Dirac 1977). As he related in this retrospective article, he was aware of eq. (13), by then known as the Klein–Gordon equation, because of its rediscovery by those authors (Klein 1926; Gordon 1926). Like Schrödinger, he rejected it, but on different, more aesthetic, grounds.

Dirac's objection to it was that it was quadratic and not linear in the time derivative. He knew that this linearity was a characteristic feature of all successfully treated quantum systems and felt that it should persist. It guaranteed the conservation of total probability obtained by summing over positive contributions. This probability interpretation had been developed some time after Schrödinger's work and could not be applied to the Klein–Gordon equation (13) as it stood. Linearity in the time derivative could be achieved by inserting the replacements (1) in the relativistic expression (4) for the energy, but the price would be an ugly square root of a differential operator that treated space differently from time.

Dirac (1928) argued that it should be possible to eliminate the square root and formulate a differential equation linear in all four space time derivatives. To do this he noted that he could linearise the energy momentum expression (4)

$$c(m^2c^2 + p^2)^{1/2} = (\alpha_m mc + \boldsymbol{\alpha} \cdot \boldsymbol{p})c, \tag{14}$$

by introducing coefficients α_m and $\boldsymbol{\alpha}$. Although these commute with all the original degrees of freedom, they have to satisfy the anticommutation relations

$$\{\alpha^i, \alpha^j\} \equiv \alpha^i\alpha^j + \alpha^j\alpha^i = 2\delta^{ij}, \quad \alpha_m^2 = 1, \quad \{\alpha_m, \alpha^i\} = 0. \tag{15}$$

Then, as a consequence of putting together (4) and (14) with the replacements (1), he arrived at the relativistic equation of the electron, now universally known as the Dirac equation,

$$i\hbar\frac{\partial\psi}{\partial t} = i\hbar c\boldsymbol{\alpha} \cdot \nabla\psi + \alpha_m mc^2\psi. \tag{16}$$

Because of the way this was derived, the wave function ψ automatically satisfies the Klein–Gordon equation (13).

The fact that the coefficients α_m and $\boldsymbol{\alpha}$ anticommute, (15), means that they cannot be numbers but must describe some new degree of freedom in the physical system associated with the electron. Actually the anticommutation relations satisfied by the three quantities α^1, α^2 and α^3 are very reminiscent of the relations satisfied by the three Pauli matrices σ^1, σ^2 and σ^3, introduced by Pauli (1926b, 1927) in an *ad hoc* way to describe the spin of the electron that had been inferred from experimental data. Since the electron has two spin states, the three angular momentum matrices should each have two rows and two columns and should, of course, satisfy the angular momentum commutation relations that all angular momenta, such as the orbital contribution, obey. This determines the three matrices up to similarity, and the conventional choice is to take them equal to $\hbar/2$ times the three Pauli matrices given by

$$\boldsymbol{\sigma} = (\sigma^1, \sigma^2, \sigma^3) = \left(\begin{pmatrix} 0 & 1 \\ 1 & 0 \end{pmatrix}, \begin{pmatrix} 0 & -i \\ i & 0 \end{pmatrix}, \begin{pmatrix} 1 & 0 \\ 0 & -1 \end{pmatrix} \right). \tag{17}$$

Thus, there is the promise of a sort of explanation of the electron spin, resulting from the synthesis of relativity with quantum theory. This is essentially correct, as we shall see, but there are some further subtleties.

To proceed further, we need to investigate the conservation laws following from the new equation. Let A be a physical observable, such as, for example, the momentum or angular momentum of the electron. Then the expectation value of A with respect to a given wave function ψ is defined by

$$\langle A \rangle_\psi = \int d^3x\, \psi^\dagger A \psi. \tag{18}$$

A is conserved if its expectation value is constant in time for all wave functions satisfying the wave equation (16). Since we are working in the 'Schrödinger picture', it is the wave function ψ, rather than A, which varies with time. Consequently, the time derivative of the expectation value is given by

$$\frac{d}{dt} \langle A \rangle_\psi = \langle [A, H] \rangle_\psi / i\hbar. \tag{19}$$

Here, H is the expression on the right hand side of (14). We have used the wave equation (16), expressing the time dependence, together with a similar equation satisfied by $\psi^\dagger$ given that H is hermitian. As the momenta are automatically hermitian, this means that we are assuming that all four quantities, α_m and $\boldsymbol{\alpha}$, are individually hermitian. The reason for this is that if we choose A to be the unit operator, eq. (19) immediately implies that $\langle 1 \rangle_\psi$ is conserved. This is the conservation of probability that we sought.

If we now consider the momentum $\boldsymbol{p}$ of the electron, we see that, as it commutes with H, eq. (14), it is indeed conserved, as expected. In saying this we have used the canonical commutation relations related to the replacements (1)

$$[x^i, p^j] = i\hbar \delta^{ij}, \quad [x^i, x^j] = 0 = [p^i, p^j], \quad i, j = 1, 2, 3. \tag{20}$$

Since the electron is moving freely, its total angular momentum must be conserved, just as its linear momentum is. The orbital contribution to this is given by the conventional expression $x \times p$. As the canonical commutators, (20), imply

$$[x \times p, H] = i\hbar c\alpha \times p, \tag{21}$$

we see that the orbital contribution is not conserved. Hence there must be another contribution to the total angular momentum of the electron besides the orbital one. Indeed, if we consider $\alpha \times \alpha$, we find

$$[\alpha \times \alpha, H] = c[\alpha \times \alpha, \alpha \cdot p] = 4c\alpha \times p.$$

This means that it is the whole of the expression $x \times p - i\hbar\alpha \times \alpha/4$ which is conserved, since it is this which commutes with H, the expression on the right hand side of (14). Thus, besides the orbital contribution, the total angular momentum of the electron contains an extra contribution, $-i\hbar\alpha \times \alpha/4$ with components $-(i\hbar/2)(\alpha_2\alpha_3, \alpha_3\alpha_1, \alpha_1\alpha_2)$. Notice that the three components of this vector do satisfy the angular momentum commutation relations as a consequence of the anticommutation relations (16). Further, when $\hbar/2$ is divided out, these three quantities mutually anticommute and have square one, just like the Pauli matrices. Thus we can equate them formally and say that the electron has angular momentum

$$x \times p + \hbar\boldsymbol{\sigma}/2.$$

In this way, we can say that the fact that the electron carries an internal spin of one-half is incorporated in the relativistic wave equation.

3.3 The Dirac equation and the Clifford algebra

Let us now make this clearer by recasting the Dirac equation in a way that is more familiar nowadays. In this we follow Pauli (1936), and introduce 'gamma' matrices

$$\gamma^0 = \alpha_m, \quad \boldsymbol{\gamma} = \alpha_m\boldsymbol{\alpha}, \tag{22}$$

so that the anticommutation relations become unified in a more natural way as

$$\{\gamma^\mu, \gamma^\nu\} = 2g^{\mu\nu}, \tag{23}$$

where $g^{\mu\nu}$ is the inverse matrix to the metric tensor, actually equal to it, because of its numerical form $\mathrm{diag}(1, -1, -1, -1)$. Premultiplying the Dirac equation (16) by γ^0, we obtain the alternative form,

$$i\gamma^\mu\partial_\mu\psi = \frac{mc}{\hbar}\psi. \tag{24}$$

This looks more relativistic, and it is easier to check that it implies the Klein–Gordon equation. Another attraction is that the anticommutation relations (23) were already known in the mathematics literature as a 'Clifford algebra', after the 19th century English mathematician. This algebra (Clifford 1882) has many interesting properties in its own right, as we shall see.

First, let us rewrite the electron spin in terms of the gamma matrices:

$$\boldsymbol{\sigma} = -\frac{i}{2}\boldsymbol{\alpha} \times \boldsymbol{\alpha} = \frac{i}{2}\boldsymbol{\gamma} \times \boldsymbol{\gamma} = i(\gamma^2\gamma^3, \gamma^3\gamma^1, \gamma^1\gamma^2).$$

Next let us introduce the quantity

$$\gamma^5 = \gamma^0\gamma^1\gamma^2\gamma^3. \tag{25}$$

We immediately see that this has the properties

$$(\gamma^5)^2 = -1; \quad \gamma^5\gamma^\mu + \gamma^\mu\gamma^5 = 0. \tag{26}$$

Furthermore, using γ^5, we can rewrite the spin as

$$\boldsymbol{\sigma} = i\gamma^5\gamma^0\boldsymbol{\gamma} = i\gamma^5\boldsymbol{\alpha}. \tag{27}$$

This means that we can rewrite

$$H/c \equiv \boldsymbol{\alpha} \cdot \boldsymbol{p} + mca_m = i\gamma^5\boldsymbol{\sigma} \cdot \boldsymbol{p} + \gamma^0 mc. \tag{28}$$

The projection of the total angular momentum of the electron onto its direction of motion, given by $\boldsymbol{p}$, eliminates the orbital part, leaving what is known as the helicity, $\boldsymbol{\sigma} \cdot \boldsymbol{p}/|\boldsymbol{p}|$. That this is conserved is particularly clear from the last expression for H. This fact makes it easier to codify data from scattering experiments involving electrons and other elementary particles, such as protons, carrying spin one-half.

The three quantities γ^0, $i\gamma^5$ and $\gamma^0\gamma^5$ all commute with the three spin components σ^1, σ^2 and σ^3, and themselves satisfy the properties characteristic of Pauli matrices. They can therefore be thought of as a third set of Pauli matrices, ρ^1, ρ^2 and ρ^3, say. Alternatively they could be ρ^2, ρ^3 and ρ^1, or ρ^3, ρ^1 and ρ^2. Since ρ^3 is conventionally the one that is diagonal, the choice depends on which of the three it is, $\gamma^0\gamma^5$, $i\gamma^5$ or γ^0, that we wish to diagonalise. All this means that we have a construction of the quantities α or γ^μ in terms of a product of two independent sets of Pauli matrices. These quantities are there-fore matrices with four rows and four columns, not two. The wave function ψ is therefore a column vector with four entries, each themselves wave functions. This doubling of the number of components necessary simply to describe the electron spin means that there is a further degree of freedom to be understood. Considering zero momentum in (28), we have $H = \gamma^0 mc^2$. As γ^0 has eigen-values ± 1, we see that we are forced by quantum mechanics to consider both signs of the root in the energy expression (4). This is not so surprising. In the classical theory it is possible to forbid the negative root, but not in a quantum theory where discrete jumps in energy readily occur. Nevertheless, there is a serious problem, as the energy expression (4) with the negative sign can now take values arbitrarily large and negative. This is difficult to interpret physically unless we postulate the existence of a new particle, as we shall see later.

Before returning to this point, and to the question of interactions between the electron and the electromagnetic field, we shall make a small mathematical diversion in order to discuss a few of the properties of the Clifford algebra (23).

With an algebra such as this, or such as the angular momentum algebra (which is an example of a Lie algebra), it is possible to represent the elements in terms of square matrices with entries that are either complex or real numbers. The Pauli matrices were a simple example involving only two rows and two columns (17). Given two such representations, it is possible to form a third by considering larger matrices which are block diagonal, with two blocks given in turn by the original matrices. A representation which can be obtained in this way is said to be *reducible* and one that cannot, *irreducible*. Clearly it is interesting to list the irreducible ones, as they are the building blocks for the others. One way of stating the familiar classification of quantum angular momenta is to say that the irreducible representations are labelled by the spin quantum number J, which can take the values $0, 1/2, 1, 3/2, 2, \ldots$, and so on. The spin J representation involves matrices with $2J + 1$ rows and columns. Thus the angular momentum algebra (like the other more complicated Lie algebras that come up in physics) has an infinite number of irreducible representations. The Clifford algebra (23) satisfied by the gamma matrices is quite different. It has only one irreducible representation, the one that we have already found, with four rows and four columns.

In saying this we should make clear that two matrices, D and D', representing a given algebra are counted as the same, and said to be equivalent, if they are related by

$$D' = ADA^{-1},$$

where A is a non-singular matrix. Since the Clifford algebra has only the one irreducible representation, any two choices of irreducible gamma matrices must be so related. This point was very much emphasised and exploited by Pauli (1936).

We are now in a position to treat two interesting questions, concerning the possibility of real solutions to the Dirac equation, and its Lorentz covariance.

Complex conjugating the Dirac equation (24), we see that the complex conjugate wave function ψ^* satisfies the same equation with γ^μ replaced by $-(\gamma^\mu)^*$. As the latter satisfies the same Clifford algebra, it must furnish an equivalent representation, so that there must exist a non-singular matrix C such that

$$-(\gamma^\mu)^* = C\gamma^\mu C^{-1}.$$

It can be shown that C is either symmetric or antisymmetric. If it is symmetric, then, by a redefinition, it can be taken to be the unit matrix. In this case, the gamma matrices are purely imaginary, and it is possible for the Dirac equation to have real solutions. So far, everything we have said about the Clifford algebra holds in any space time of even dimension D, say, however many time and space components there are. The unique irreducible representation of the Clifford algebra consists of matrices with $2^{D/2}$ rows and columns. On the other hand, whether C is symmetric or antisymmetric depends very much on the nature of the space time, and becomes important in unified theories using extra dimensions of space time. In the space time of the real world, with one time and three space coordinates, C is always symmetric. This is demonstrated by the fact that we can write down the following purely imaginary representation of the Clifford algebra:

$$i(\gamma^0, \gamma^1, \gamma^2, \gamma^3) =$$

$$\begin{pmatrix} 0 & 0 & 0 & 1 \\ 0 & 0 & -1 & 0 \\ 0 & 1 & 0 & 0 \\ -1 & 0 & 0 & 0 \end{pmatrix}, \begin{pmatrix} 0 & 0 & 0 & 1 \\ 0 & 0 & 1 & 0 \\ 0 & 1 & 0 & 0 \\ 1 & 0 & 0 & 0 \end{pmatrix}, \begin{pmatrix} 0 & 0 & 1 & 0 \\ 0 & 0 & 0 & -1 \\ 1 & 0 & 0 & 0 \\ 0 & -1 & 0 & 0 \end{pmatrix}, \begin{pmatrix} 1 & 0 & 0 & 0 \\ 0 & 1 & 0 & 0 \\ 0 & 0 & -1 & 0 \\ 0 & 0 & 0 & -1 \end{pmatrix}.$$

$$(29)$$

With this choice, the Dirac equation is a real equation and can have real solutions.

The second question concerns the Lorentz covariance of the Dirac equation. The conservation of angular momentum that has already been demonstrated means that it possesses a rotational covariance in any frame of reference. The contribution of the spin to the angular momentum was given by $(i\hbar/4)\gamma \times \gamma$. This can be extended to give the generators for a general Lorentz transformation specified by the antisymmetric tensor

$$M^{\mu\nu} = \frac{i\hbar}{4}[\gamma^\mu, \gamma^\nu]. \qquad (30)$$

We shall not present all the details of the proof here, but shall confine ourselves to some comments. It is easy to check from the Clifford algebra (23) that

$$[M^{\mu\nu}, \gamma^\lambda] = i\hbar(\gamma^\mu g^{\nu\lambda} - \gamma^\nu g^{\mu\lambda}),$$

which means that γ^λ transforms as a four-vector. From this it is easy to check that $M^{\mu\nu}$ satisfies the commutation relations of the Lie algebra of the group of Lorentz transformations.

It follows that the $M^{\mu\nu}$ furnish a representation of the Lorentz group in terms of matrices with four rows and four columns. However, since γ^5, (25), anticommutes with all the γ^{μ}, it commutes with all the $M^{\mu\nu}$. This statement can be interpreted in two ways. It means that γ^5 is Lorentz invariant, thereby explaining its special significance. Secondly, it means that, although the four-dimensional matrices furnish an irreducible representation of the Clifford algebra (23), they furnish a representation of the Lorentz algebra which is reducible, rather than irreducible. This is because there is a lemma due to Schur that states that any matrix commuting with all the generators of an irreducible representation must be a multiple of the unit matrix. γ^5 cannot be such a multiple as it would then commute rather than anticommute with the γ^{μ}. As $i\gamma^5$ has square one, it has two eigenvalues, ± 1. Viewed in terms of the two corresponding eigenspaces, $M^{\mu\nu}$ has block diagonal form, and so is built out of two representations, each involving matrices with two rows and two columns. These are the two spinor representations of the Lorentz group, and they are indeed inequivalent and irreducible.

If instead we had considered rotations in D dimensions (with D even), we would have found a similar pattern, with two inequivalent irreducible spinor representations of dimension $2^{(D/2)-1}$.

3.4 The gauge principle and the electromagnetic properties of the electron

Although, as explained above, the free Dirac equation (24) may have real solutions, the wave function of the electron is a complex valued function (or, more precisely, it consists of a four component column vector with complex entries). This complex nature reflects the fact that the electron carries non-zero electric charge, q, say. The electron is therefore subject to electromagnetic forces and is itself a source of the field. Classically this is familiar, as we regard the electron as a point particle subject to the Lorentz force exerted by the ambient electromagnetic field, itself determined by Maxwell's equations.

It was Hermann Weyl (1929) who first understood the direct connection between the complex nature of the electron wave functions and its electromagnetic properties. This is embodied in what has become known as the *gauge principle*. Nowadays, this is seen as one of the most important geometric principles for determining both the nature and the effects of the fundamental forces. It has been extended to include the weak forces responsible for radioactive beta decays and the like. It is also thought to apply to the strong forces experienced by quarks within the nuclear particles. Despite its power and versatility, it has to be admitted that the basis of this principle is still imperfectly understood. Its vindication is furnished by its sheer elegance, augmented by the simple fact that it actually works, as testified by all the data supplied by the giant particle accelerators.

The term 'gauge' is actually a misnomer, a relic from a previous incarnation of Weyl's idea. Originally, before the formulation of quantum theory was found, Weyl sought a geometrical interpretation of Maxwell's electromagnetic theory, based on a unification with Einstein's general relativity theory of gravity, then very new, by considering local changes of scale, or gauge. This idea had to be abandoned as it turned out to have unacceptable physical consequences. Later, after the advent of quantum theory, Weyl was able to revive it in a modified form within the new context to very good effect.

When considering the probability interpretation of wave functions, it is often convenient to normalise them. This leaves an overall constant phase that is arbitrary. Indeed, physical quantities, such as expectation values, (18), are also independent of this overall phase and so display a symmetry, with respect to change of phase.

The gauge principle recognises this but goes further: it posits that all physical quantities be unaffected by changes in the phase of the wave function that may vary from point to point in space and in time. Such alterations can be written as

$$\psi(x, t) \rightarrow \exp\left(\frac{iq\chi(x, t)}{\hbar c}\right)\psi(x, t), \tag{31}$$

where $\chi(x, t)$ is the function specifying the alteration in phase. The space time derivative of ψ also enters physical quantities and, of course, the equations of motion. In order to accord with the gauge principle, it ought to transform in the same way as ψ, as given above. This is clearly impossible if χ is not a constant, unless we modify the theory by introducing a new quantity, the four-vector function, A_μ, and replace the space time derivative ∂_μ by what is called the 'covariant derivative'

$$\partial_\mu \rightarrow \mathscr{D}_\mu \equiv \partial_\mu + \frac{iq}{\hbar c} A_\mu, \tag{32}$$

whenever it appears acting on a wave function for a particle of charge q.

Then it is easy to check that $\mathscr{D}_\mu\psi$ transforms exactly like ψ, providing the four-potential A_μ simultaneously alters as

$$A_\mu \rightarrow A_\mu + \partial_\mu\chi.$$

Notice that this alteration in the four-potential is independent of the charge q (and $\hbar$). Furthermore, it implies that the 'curl' of A_μ,

$$F_{\mu\nu} \equiv \partial_\mu A_\nu - \partial_\nu A_\mu, \tag{33}$$

is unaltered by the gauge transformation.

As $F_{\mu\nu}$, (33), is antisymmetric in its two indices, it has six independent components, that is the same as the number for the electric and magnetic fields taken together. Indeed, it is possible to interpret $F_{\mu\nu}$ as the electromagnetic field strength tensor that arises naturally in relativistic treatments of electromagnetism. By its definition, it automatically satisfies the homogeneous Maxwell equations. Thus the emergence of the electromagnetic field strengths in the electron theory becomes a natural consequence of the gauge principle.

It is possible to combine successive gauge transformations of the form (31), with functions χ and χ', to obtain a third with a function χ''. These are related by

$$e^{-iq\chi''/\hbar c} = e^{-iq\chi/\hbar c} \, e^{-iq\chi'/\hbar c}.$$

It is also possible to 'undo' a gauge transformation specified by $e^{-iq\chi/\hbar c}$ by applying $e^{iq\chi/\hbar c}$. It follows that, if we focus attention on one particular space time point, the gauge transformations form a group, namely that made up of phase factors. The order in which two factors are multiplied is irrelevant, and this group is said to be abelian. It is denoted by $U(1)$, indicating that it is formed by unitary matrices consisting of a single row and column.

Thus, according to this point of view, Maxwell theory emerges as what is known as the $U(1)$ gauge theory. It is possible to extend the idea and consider square matrices with more rows and columns acting on column vector wave functions. As the order in which these matrices are multiplied usually matters now, the new groups they form are no longer necessarily abelian. It turns out that the other elementary particle interactions are governed by gauge theories of this type. For example, the group $U(2)$, made up of unitary matrices with two rows and two columns, leads to the electroweak theory of Weinberg and Salam, whose predictions were verified at CERN, earning Carlo Rubbia a Nobel prize, as well as the founders.

Notice that the $U(1)$ gauge transformation multiplies each of the four components of the Dirac wave function column vector by the same phase.

We are now in a position to apply the replacement (32) to both the Dirac equation (24) and the Klein–Gordon equation (13) and compare the resultant predictions concerning the effect of the electromagnetic field on the particles described by the two types of wave function. The assumption tacitly made, that this procedure yields the correct results, is essentially one of simplicity. It is possible to envisage more complicated possibilities that still respect the gauge principle and also reduce to the accepted laws in the classical limit. Nevertheless, this assumption of simplicity was stunningly successful for the electron, being vindicated by its observed properties.

The insertion of the replacement (32) into the Dirac equation (24) and the Klein–Gordon equation (13) yields, respectively,

$$i\gamma \cdot \mathscr{L}\psi = i\gamma \cdot \left(\partial + i\frac{q}{\hbar c}A\right)\psi = \frac{mc}{\hbar}\psi,$$

and

$$\mathscr{L}^2\psi = \left(\partial^\mu + \mathrm{i}\frac{q}{\hbar c}A^\mu\right)\left(\partial_\mu + \mathrm{i}\frac{q}{\hbar c}A_\mu\right)\psi = \left(\frac{mc}{\hbar}\right)^2\psi.$$

In contrast to the situation when the vector potential A^μ vanishes, the first equation does not imply the second. Instead, because of the identity

$$\gamma^\mu\mathscr{L}_\mu\gamma^\nu\mathscr{L}_\nu = \tfrac{1}{2}[\gamma^\mu\gamma^\nu + \gamma^\nu\gamma^\mu + \gamma^\mu\gamma^\nu - \gamma^\nu\gamma^\mu]\mathscr{L}_\mu\mathscr{L}_\nu$$

$$= \mathscr{L}^2 + \tfrac{1}{2}[\gamma^\mu, \gamma^\nu][\mathscr{L}_\mu, \mathscr{L}_\nu]$$

$$= \mathscr{L}^2 + \frac{\mathrm{i}q}{4\hbar c}[\gamma^\mu, \gamma^\nu]F_{\mu\nu}$$

$$= \mathscr{L}^2 + \frac{q}{\hbar^2 c}M^{\mu\nu}F_{\mu\nu},$$

we find an extra term in which the following expression acts on ψ:

$$\frac{q}{\hbar^2 c}M^{\mu\nu}F_{\mu\nu} = \frac{q}{\hbar c}(\gamma^5 \boldsymbol{E} + \boldsymbol{B})\cdot\boldsymbol{\sigma}.$$

We have substituted our expression for the gamma matrices to obtain this. The γ^5 term mixes the large and small components of the wave function and therefore has a small effect for slowly moving electrons. The term involving the magnetic field $\boldsymbol{B}$ is responsible for a coupling between the electron spin and the magnetic field. This means that the electron spin produces a magnetic moment whose precise value can be evaluated by considering a slowly moving electron (Itzykson and Zuber 1980). This value is

$$\boldsymbol{\mu}_\mathrm{e} = \frac{q}{mc}\left(\frac{\hbar\boldsymbol{\sigma}}{2}\right).$$

This is to be compared with the magnetic moment produced by the orbital angular momentum **L**, namely $(q/2mc)$**L**. The ratios between the two sorts of angular momentum and the magnetic moment they produce differ by a factor of two. This unexpected factor agrees precisely with the experimental data already available at the time of Dirac's analysis, and so provided an important vindication of his argument.

3.5 Fermi–Dirac statistics and second quantisation

To gain a first insight into the question of whether the electron obeys Fermi–Dirac or Bose–Einstein statistics, it is worth enquiring whether the Dirac equation follows from an action principle. In other words, does there exist a real expression, called the action, with the property that it is stationary with respect to any small variations of the wave function whenever the wave function satisfies the Dirac equation (and not otherwise)? This question is less innocuous than it seems. It seems to be a fact of nature that all fundamental equations do follow from such a principle. The underlying reason is not entirely clear, but the existence of an action certainly provides a step towards quantisation following the ideas of Richard Feynman. In this case, since it is the wave function itself that would be quantised, in addition to the electron degrees of freedom, the procedure would correspond to what is known as 'second quantisation'. We shall consider this later.

If we denote $\bar{\psi} = \psi^{\dagger}\gamma^0$, then the expression

$$\mathscr{L} = c(i\bar{\psi}\gamma \cdot \partial\psi + \frac{mc}{\hbar}\bar{\psi}\psi) \tag{34}$$

is real. Furthermore, it is a Lorentz scalar function of position in space time, x^{μ}.

We shall temporarily consider small variations $\delta\psi$ and $\delta\psi^{\dagger}$ of the wave function and its conjugate which are not necessarily related by complex conjugation but are independent. Then the corresponding variation of $\mathscr{L}$ is given by

$$\delta\mathscr{L} = c\delta\psi^{\dagger}\gamma^0\left(i\gamma \cdot \partial\psi + \frac{mc}{\hbar}\psi\right) + c\left(i\gamma \cdot \partial\psi + \frac{mc}{\hbar}\psi\right)^{\dagger}\gamma^0\delta\psi,$$

plus terms that are total derivatives of the space time coordinates. This shows that the integral of $\mathscr{L}$ over space time is stationary with respect to these general variations whenever ψ satisfies the Dirac equation. This integral is therefore a prime candidate for what was described as the action, above.

But there is a serious problem with the above expression, (34), for $\mathscr{L}$, the potential Lagrangian density. This is seen most clearly in the choice (29) mentioned above in which the gamma matrices are purely imaginary. Then the Dirac equation possesses real solutions, and we can consider the possibility that the wave function ψ is real, even though we know this cannot apply to the electron when the effect of its electric charge is taken into account. This is just a temporary stratagem to reduce the following argument to its simplest terms.

The natural expectation, assumed tacitly until the 1970s, when the theory of supersymmetry was developed, was that the four components of the Dirac wave function consisted of c-number functions, as in the conventional treatment of non-relativistic wave functions. However, this leads to a fundamental difficulty in the definition of the relativistic Lagrangian density $\mathscr{L}$ (34), namely that it vanishes identically (up to a total derivative). For example, the mass term $\bar{\psi}\psi = \sum_{i,j=1}^{4} \psi_i(\gamma^0)_{ij}\psi_j$ vanishes since γ^0, being hermitian and imaginary, is an antisymmetric matrix, as is witnessed by (29). Likewise, the term in $\mathscr{L}$ involving the time derivative, namely,

$$i \sum_{i=1}^{4} \psi_i \frac{\partial \psi^i}{\partial t} = i \frac{\partial}{\partial t} \left(\frac{\psi^2}{2} \right),$$

and so vanishes up to a total derivative with respect to time. Similar arguments apply to the other terms, and altogether this means that the expression for the action is actually an integral over the surface at infinity. This vanishes if the wave function decays asymptotically. Clearly it is fruitless to vary a quantity which vanishes identically. It is difficult to think of any alternative expression which would suffice as an action. Instead, we must accept the correctness of the action being the integral of the Lagrangian density (34) over space time, but agree that it is wrong to suppose that the wave function possesses c-number components. Rather we must assume that its components consist of anti-commuting functions. More precisely, they satisfy

$$\psi_i(x)\psi_j(y) + \psi_j(y)\psi_i(x) = 0 \tag{35}$$

and are said to be 'Grassmann' functions. In particular, this implies that $\psi_i(x)^2 = 0$, which seems odd but is nevertheless consistent. Then it is easy to see that none of the terms in $\mathscr{L}$, (34), vanish any more, even when the wave function is real. As a consequence, we now have a more satisfactory action principle. Notice that, so far, we are still treating the wave function classically, as Planck's constant is absent from (35).

The startling conclusion is that the existence of an action principle for the Dirac equation requires the wave function to be treated as anticommuting, rather than commuting. It is reassuring to apply the same argument to the action for the Klein–Gordon equation, and find that its wave functions must be commuting rather than anticommuting.

These two results are a classical analogue of a result that has become known as the connection between spin and statistics. The argument shows that the essential ingredients are relativity and the validity of an action principle. In order to obtain the full result with the implication that spin-1/2 particles obey Fermi–Dirac statistics and hence the Pauli exclusion principle so crucial in atomic structure, we need a further step. This is the procedure of second quantisation, whereby we treat the components of the wave function as quantum objects (q-numbers) and find a connection between the field and particle concepts.

Before considering this, it is worth noting a new difficulty: this is that, if the Dirac wave function is real and Grassmann, then the probability density, $\sum_{i=1}^{4} \psi_i^2$, vanishes. There is no simultaneous resolution of this and the previous difficulty without second quantisation. As a result, that procedure becomes imperative for a consistent and satisfactory treatment of the Dirac equation. In other words, the Dirac equation contains within it the seeds of quantum field theory, that is a theory of an indefinite number of particles, not just one.

A cornerstone of quantum mechanics is the procedure of canonical quantisation, linking quantum mechanics to the Hamiltonian version of classical mechanics and its concomitant action principles. The canonical conjugate 'momentum' of a variable is the response to varying the Lagrangian (the integral of the Lagrangian density over 3-space) with respect to the coordinate. Hence, from (34), the canonical conjugate to the Dirac wave function $\psi(x)$ is $i\psi^\dagger(x)$, where we have reverted to treating the wave function as complex.

Because, in the classical limit in which Planck's constant vanishes, we must recover the Grassmann or anticommuting property rather than the c-number property, we are forced to replace the canonical commutator by an anticommutator. Thus,

$$\psi(x)\psi^\dagger(y) + \psi^\dagger(y)\psi(x) = \hbar I_4 \delta(x - y) \tag{36a}$$

at equal times for the space time points x and y. Here I_4 is the unit matrix in the four-dimensional spinor space. In addition,

$$\psi(x)\psi(y) + \psi(y)\psi(x) = 0, \tag{36b}$$

also at equal times. Notice that the sign is fixed in the first of these relations as both sides have to be positive. It is at this stage that ψ has become what is known as a 'quantum field'. Moreover, for reasons that become clear below, it is a 'fermionic quantum field'.

The next step is the construction of a Hamiltonian with the property that the Heisenberg equations of motion for the Dirac ψ, calculated using the canonical anticommutation relations (36), reduce to the Dirac equation (16). There is a quirk in applying the standard procedures to the present situation which was resolved by Dirac's theory of constraints. We shall avoid a digression into this theory by following the historical treatment and simply making a straightforward guess as to the answer:

$$H = c \int d^3x \left(i\bar\psi \sum_{i=1}^{3} \gamma_i \cdot \nabla_i \psi + \frac{mc}{\hbar} \bar\psi\psi \right). \tag{37}$$

Actually, this expression will have to be revised in the light of information still to come, but it is easy to check, at least formally, that it works, in the sense that

$$[\psi(x), H] = i\hbar c\boldsymbol{\alpha} \cdot \nabla\psi + \alpha_m \frac{mc^2}{\hbar}\psi. \tag{38}$$

Now consider a complete set of c-number solutions to the Dirac equation which are also energy eigenfunctions. Call them $f_i(x)$ if the energy E_i of the solution is positive, and $g_i(x)$ if the energy $-E_i$ of the solution is negative. Notice that, with this convention, E_i is always positive and that we are going to suppose that the index i is discrete in order to simplify the formalism. The expression for the probability density resulting from (18) can be used to define a scalar product between two wave functions f and f':

$$(f, f') \equiv \int d^3x f^* f'. \tag{39}$$

The virtue of this is that it is conserved, that is, independent of the time chosen for the integration over 3-space, whenever the two wave functions satisfy the Dirac equation. We shall suppose that we can choose an orthonormal basis for these solutions:

$$(f_i, f_j) = \delta_{ij} = (g_i, g_j), \quad (f_i, g_j) = 0. \tag{40}$$

Then the quantum field $\psi(x)$, satisfying the Dirac equation and the canonical anticommutation relation (36) can be expanded,

$$\psi(x) = \sum_i f_i(x) B_i + g_i(x) C_i^*, \tag{41}$$

where the coefficients in the expansion,

$$B_i = (f_i, \psi), \quad C_i^* = (g_i, \psi), \tag{42}$$

are constants, independent of time, by the previous remarks. From the second of the canonical anticommutation relations, (36b), they satisfy

$$\{B_i, B_j\} = \{C_i, C_j\} = \{B_i, C_j\} = 0 \tag{43a}$$

and similarly for their conjugates. From (36a),

$$\{B_i, B_j^*\} = \hbar \delta_{ij} = \{C_i, C_j^*\}, \quad \{B_i, C_j^*\} = 0. \tag{43b}$$

Thus we have constructed a fermionic annihilation–creation operator pair for each element of our basis of wave functions. It is natural to enquire what these destroy and create in physical terms. From the Heisenberg equation of motion above, and the conservation of the scalar product, we find

$$[H, B_i] = -E_i B_i, \quad [H, C_i^*] = E_i C_i^*. \tag{44}$$

This means that B_i^* and C_i^* both create positive energy E_i, while B_i and C_i destroy it. There is another conserved quantity in the theory that we can use to distinguish the two operators creating the same energy. This is the electric charge Q obtained by integrating the time component of the conserved electric current $j_\mu(x) = q\bar{\psi}(x)\gamma_\mu\psi(x)/\hbar$, namely

$$Q = (q/\hbar) \int d^3x \psi^\dagger(x)\psi(x). \tag{45}$$

Again this is constant in time but the definition has to be provisional, as was that of H. We find from the canonical anticommutation relations (36) that

$$[Q, B_i] = -qB_i, \quad [Q, B_i^*] = qB_i^*, \quad [Q, C_i^*] = -qC_i^*, \quad [Q, C_i] = qC_i.$$
$$(46)$$

So B_i^* creates electric charge q while C_i^* creates $-q$. The interpretation of (44) and (46) is that B_i^* creates a state of an electron, with energy E_i and charge q. As C_i^* creates energy E_i and charge $-q$, we have to interpret this as corresponding to a new particle with opposite charge to the electron. This is the positron, the antiparticle to the electron observed first in 1932 by studying tracks made in a Wilson cloud chamber by cosmic rays (Anderson, 1933). Although Dirac was one of the founders of the method of second quantisation, the argument that led him to predict the existence of the positron was different, involving what he called the 'hole theory'. Nowadays, this is regarded as a pictorial version of the second quantisation argument. The hole theory argument relied heavily on the exclusion principle satisfied by the electrons, whereas the second quantisation argument can be extended to scalar, Klein–Gordon particles which do have antiparticles too but do not satisfy the exclusion principle. Nevertheless, the result was a triumphant vindication of Dirac's interpretation of the hitherto bewildering negative energy solution of the Dirac equation.

Given this interpretation, it remains to describe the quantum mechanical vector space in which the quantum field ψ lives. We define a vacuum $|0\rangle$ annihilated by all the destruction operators,

$$B_i|0\rangle = 0, \quad C_i|0\rangle = 0. \tag{47}$$

This is unique. States describing a single electron are obtained by applying the appropriate creation operator, to obtain $B_i^*|0\rangle$, and similarly for states with a single positron, given by $C_i^*|0\rangle$. States describing two electrons are obtained by applying two creation operators, $B_i^* B_j^*|0\rangle$. Notice that, according to (43a), this is antisymmetric with respect to interchange of the labels i and j. This is the Fermi–Dirac statistics satisfied by the electron. When i and j are equal, the state vanishes. This is the Pauli exclusion principle. The process can be extended to build up states describing any number of electrons and positrons, always with Fermi–Dirac statistics. The space found in this way is known as a Fock space and possesses a natural positive definite scalar product, just as the principles of quantum mechanics would require.

But there is one remaining difficulty. This concerns the expressions for energy and electric charge, (37) and (45) which fail to vanish as they should when acting on the vacuum. However, it is easy to see that the following revised expressions do vanish on the vacuum and, moreover, preserve the commutation relations (45) and (46) with the annihilation and creation operators:

$$H = \sum_i E_i(B_i^* B_i + C_i^* C_i)/\hbar, \quad Q = q \sum_i (B_i^* B_i - C_i^* C_i)/\hbar. \qquad (48)$$

Because of the structure of the Fock space, it follows by a physically intuitive version of Schur's lemma that the two expressions for energy and charge can only differ from each other by a *c*-number, interpreted as the false, unrevised numerical value of the energy or charge of the vacuum. Thus the revision of the energy and charge expressions is a simple redefinition of the vacuum energy and charge, and hence relatively innocuous, it would seem. It is known, technically, as normal ordering. The only problem is that the constants concerned are infinite, so that the redefinition involves an infinite subtraction. This procedure is the simplest instance of what is known as a renormalisation procedure. Unfortunately, such procedures are endemic in quantum field theory unless something sophisticated, such as supersymmetry, is invoked. Notice the irony of the radical change in the physical interpretation of $\psi^\dagger \psi$ that this revision entails. Originally, when ψ was treated as a *c*-number wave function, its positivity was sacrosanct as it was interpreted as a probability density for the electron distribution. Now that ψ is treated as a fermionic quantum field, $\psi^\dagger \psi$ has to be normal ordered and is no longer positive. It is interpreted as being proportional to the density of electric charge with the negative contributions coming from the positrons.

3.6 Current perspectives

We have taken the story of the Dirac equation and the electron as far as we can within a limited compass. We shall close by reviewing it in the light of modern knowledge, both experimental and theoretical.

Since the time that the positron made its debut on the experimental stage, the discovery of further new elementary particles has continued unabated until the present day. Ever more powerful particle accelerators can produce ever heavier particles, often with higher values of their spin. Everything we know confirms that, like the electron, all these particles possess antiparticles of the same mass but opposite charge. Furthermore, they all obey the 'connection between spin and statistics'. This means that if they possess spin with a half integral value, like the electron, they obey Fermi–Dirac statistics. On the other hand, if they possess spins with integral values, they obey Bose–Einstein statistics. Examples of this latter are the scalar Klein–Gordon particles with zero spin, and the photon, the quantum associated with Maxwell's equations, which possesses unit spin. Indeed, this is the principle underlying the action of the laser, which is now familiar to us in so many ways.

From this viewpoint, there is nothing special about the electron; it is just one species amongst the many inhabitants of a whole zoo of particles. Besides the electron, there are other important particles carrying spin one-half, for example the proton and neutron, the constituents of the nuclei of atoms. The proton carries the same electric charge as the positron, but the experimental value of its magnetic moment is not given by the Bohr magneton as the simple Dirac equation would predict. It is understood that there is no contradiction with general principles as it is possible to add extra terms to the equation, destroying its simplicity, but nothing more.

As we have outlined, the study of the Dirac equation has unveiled the key features of relativistic quantum field theory. These results have since been consolidated with proofs which are ever more comprehensive and sophisticated yet, at the same time, more economical of assumptions. This understanding can be used to seek theories giving realistic descriptions of nature as revealed by the particle accelerator data. Grand unified theories are model theories attempting to find simplifying patterns amongst the particle zoo, and the forces they exert on each other, by making use of the generalisation of the gauge principle mentioned earlier. These theories restore the very special role played by particles carrying spin one-half. Indeed the electron is associated with its neutrino and what are called the 'up' and 'down' quarks, the constituents of the proton and neutron, all of which carry spin one-half. Altogether there are exactly three generations of leptons and quarks like this, corresponding to the electron, the muon and the tau lepton, and all other particles are seen as conglomerates of these in some way or other.

Since these model theories overlook gravity, more elaborate constructions are required for this, and they usually embody a new principle, 'supersymmetry'. Although unproven experimentally it guarantees a number of features which are well nigh irresistible to the theorist. This is the beginning of a new story, beyond the scope of the present volume. The point that is relevant is that the argument reveals a new role for Dirac's ideas. Quantities called 'supercharges' are the new ingredients, and augment the Poincaré algebra expressing the space time symmetry. For a long time it was thought that these could only be Lorentz scalars, with well understood effects, until it was realised that there was just one other, very interesting possibility: that they could be Dirac spinors.

4

The electron glue

B. L. GYORFFY

University of Bristol

4.1 The nature of the problem

In this chapter we shall deal with condensed matter, which consists of electrons and positively charged nuclei. Our focus will be on the role of the electrons as the bonding agent which keeps the nuclei together. One of the principal aims of such considerations is an understanding of the equilibrium structure (phase diagram) of infinite (bulk) materials.

The above picture of solids was contemporaneous with the view of atoms as mini solar systems, with the heavy nuclei playing the role of the sun and the electrons that of the planets, and it was developed along similar lines (Lorentz 1909). However, in classical electrodynamics this model is plagued by the same lack of stability against radiation-induced collapse as overtakes a classical model for atoms (Lieb 1976) and hence it must be treated fully quantum mechanically from the start. But, as in atomic physics, the nuclei can be regarded as positive point charges (Ze), and hence a complete theory of our model is quantum electrodynamics (QED) (Schweber 1994). Of course, this is not meant to be a useful remark, in the sense of offering an easy insight into why certain solids form and others do not, but a reminder that, in principle, there are no adjustable parameters in the theory. As will be readily appreciated, such a reminder is called for from three different points of view. First, it provides a justification for the often used phrase 'first-principles calculation'. In this connection we wish to stress that this phrase does not mean that the calculation is exact in any sense, but merely that there is an exact theory which is approximated without introducing adjustable parameters. Secondly, it should occasion some astonishment that such relatively simple systems, consisting of only two kinds of charges interacting via Maxwell's equations, can describe such an incredible variety of materials as we find in nature. Thirdly, it may be interesting to point out occasionally that the QED relevant to solids is that for an infinite condensed system. While QED for a few electrons and photons, as in atoms, is well understood (Schweber 1994), for dense systems many of its features remain unexplored.

Fortunately, most of the subtler features of QED play only a minor role in binding the nuclei into solids. Thus, we shall neglect vacuum fluctuations and radiative corrections (Schweber 1994) altogether and mention velocity dependent (relativistic) forces only in passing. In short, we shall study systems of point charges interacting via the instantaneous Coulomb forces. Thus, any theory of the system at hand will be based on the Hamiltonian

$$\hat{H} = H_{\text{ions}}(\{R_i\}) + H_{\text{electrons}}(\{r_\alpha\}) + H_{\text{int}}(\{R_i\}, \{r_\alpha\}), \tag{1}$$

where $\{R_i\}$ stands for the set of position coordinates of the nuclei labelled by $i = 1, \ldots, N$, $\{r_\alpha\}$ is the corresponding set for the electrons with labels $\alpha = 1, \ldots, ZN$, and the individual contributions H_{ions}, $H_{\text{electrons}}$, H_{int} take the following standard form:

$$
\begin{aligned}
H_{\text{ions}} &= -\frac{\hbar^2}{2} \sum_i \frac{1}{M_i} \nabla_i^2 + \frac{1}{2} \sum_{i,j} \frac{Z_i Z_j e^2}{|R_i - R_j|}, \\
H_{\text{electrons}} &= -\frac{\hbar^2}{2m} \sum_\alpha \nabla_\alpha^2 + \frac{1}{2} \sum_{\alpha,\beta} \frac{e^2}{|r_\alpha - r_\beta|}, \\
H_{\text{int}} &= -\sum_{\alpha,j} \frac{Ze^2}{|r_\alpha - R_j|}.
\end{aligned}
\tag{2}
$$

Note that the nuclei are specified by their masses $\{M_i\}$ and atomic numbers $\{Z_i\}$. A pictorial representation of the above system is given in Fig. 4.1. Evidently, the electronic mass m is much smaller than the ionic masses $\{M_i\}$. This simplifies matters enormously. Indeed, as was stressed by Migdal (1958), the small parameters $(m/M_i)^{1/2}$ can be used as small parameters of a

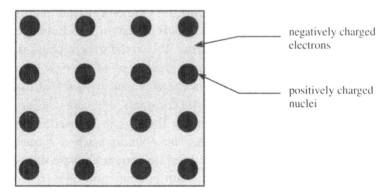

negatively charged electrons

positively charged nuclei

Fig. 4.1 Schematic structure of matter: the light shaded area represents the electron 'glue'.

systematic perturbation theory starting with infinitely heavy ions as the unperturbed system. This zeroth-order (Born–Oppenheimer) approximation will suffice for most of the discussions in the present chapter. An exception will be our brief account of lattice dynamics and the electron–phonon interaction.

The principal gain in assuming the ions to be infinitely heavy is that, as a consequence, the position vectors of the nuclei $\{\boldsymbol{R}_i\}$ can be regarded as classical variables. Thus, in this limit we are faced with a quantum many-body problem for the electron *glue* only. Of course, we shall regard the nuclei as inert, namely as external charges without spin providing the electrons with a fixed external potential, while solving the many-electron problem for a prescribed configuration $\{\boldsymbol{R}_i\}$. We will then assume that the ground, or equilibrium, state of the full system corresponds to that configuration $\{\boldsymbol{R}_i^0\}$ for which the ground-state, or free, energy is minimum. Evidently, without the electrons the set $\{\boldsymbol{R}_i^0\}$ would be such that the nuclei are infinitely far apart, and it forms a dense set of points, corresponding to matter as we find it in nature, only in the presence of the electron glue. That such a collection of charges form a stable system whose energy is an extensive quantity is a non-trivial consequence of quantum mechanics. An elegant proof of this was given by Lieb (1976). Thus, the task of a theory for the structure of matter is to find all the stable, equilibrium configurations $\{\boldsymbol{R}_i^0\}$ allowed by this glue.

Since what is at stake is nothing less than an understanding of the structure of condensed matter, it is not surprising that there are many different approaches to the problem at hand, generated by physicists, chemists and metallurgists. To provide a context for the electron-based theory at the centre of attention in this chapter, I interrupt this narrative by recalling, briefly, some of these.

The historically first, and most immediately useful, description of matter is in terms of frankly phenomenological theories. A canonical example of these is the visco-elastic theory of solids and liquids (Chaikin and Lubensky 1995). Other examples are the Ginzberg–Landau–Wilson theories of phase transitions (Chaikin and Lubensky 1995). From our present point of view, the relevant feature of these is that each material is characterised by a set of parameters, such as the elastic, diffusion and dielectric constants, etc: constants which have to be determined experimentally before the theory can be used to predict the outcome of some further experiments. Thus, within such a framework the answer to a question such as 'Why does a certain structural phase transition take place?' might be that it is because a certain combination of elastic constants changes sign as the temperature is lowered or the pressure is increased. Such theories may be exact, in a certain well-defined sense, and may be regarded, in spite of reductionist objections, as fundamental, but they may leave many questions unanswered.

Alternative descriptions are provided by microscopic models. An example of this is one which attributes the total energy of a solid or liquid to a sum of pairwise interactions between atoms. Namely, the potential energy of the system is assumed to be given by

$$V(\{\boldsymbol{R}_i\}) = \frac{1}{2} \sum_{i,j} v_{\text{eff}}^{(2)}(\boldsymbol{R}_i - \boldsymbol{R}_j), \tag{3}$$

and the functional form of the effective pair potential $v_{\text{eff}}^{(2)}(\boldsymbol{R})$ is specified by some phenomenological recipe. Evidently, such models give a more detailed physical picture of what happens when a solid forms. For instance, they may explain that a binary alloy orders because unlike atoms attract each other more than like atoms. Indeed, much modern simulation of material properties is based on such microscopic pair potential models (Ducastelle 1991).

Although the above, conceptually straightforward, approach does not even mention electrons, it offers an attractive strategy for describing matter in electronic terms. Namely, we may associate each group of Z electrons in our model, depicted in Fig. 4.1, with one nucleus and anticipate that they form weakly interacting neutral atoms. Then, we proceed to calculate the interaction energy of an isolated pair of such atoms, using the full machinery of quantum mechanics and the model of point charges interacting via electromagnetic fields, as a function of their separation $\boldsymbol{R}_i - \boldsymbol{R}_j$.

Following G. N. Lewis's suggestion in 1916 that the chemical bond between atoms consists of a pair of electrons held jointly by the two nuclei, Heitler and London (1927) were the first to calculate such interaction potentials, in 1927. As was fully acknowledged by Pauling and Wilson (1935) a few years later, this work opened up a new epoch in chemistry of molecular structure and valence theory.

Nowadays, such calculations of $v_{\text{eff}}^{(2)}(\boldsymbol{R}_i - \boldsymbol{R}_j)$ can be implemented numerically to high accuracy and very rapidly. Also, subsequent classical molecular dynamics calculations can readily deal with 10^4–10^5 atoms. Nor is one strictly limited to pairwise contributions only in eq. (3). With some extra effort, three-body, four-body or even higher order interactions can be included. Thus, one might think that, in this way, we can treat quantitatively systems large enough to be considered macroscopic, starting from first principles and without introducing adjustable parameters.

Unfortunately, the electron glue cannot be eliminated from the problem by the above, admittedly attractive, device. The fact of the matter turns out to be that, in general, one must stick to the model of point charges and treat all the electrons on an equal footing. In principle, this is because the electrons form a highly quantum mechanical, degenerate, Fermi liquid, and each of the bonding electrons, in particular, interacts with all the nuclei. In practice, the situation is more controversial. Nevertheless, even in this case there is much evidence against universally applicable 'atomic potential' models (Gyorffy 1993).

The state of the argument may be summarised as follows. Whilst there are many successful calculations of phase diagrams based on effective few-body potentials on the one hand and first-principles total energy calculations for infinite periodic systems on the other, they are not convincing because, in the successful cases, such classical energy models have to fit only small variations in energy between rather similar nuclear configurations. More decisive is the very elegant, systematic work of Heine *et al.* (1991). They studied ten dramatically different structures of Al atoms and investigated the binding energy per atom, $U(c)$, as a function of the coordination number c. Their results, obtained in a way to be described later, are depicted in Fig. 4.2. The results fit

$$U_{\text{electron}}(c) = ac - bc^{1/2} \tag{4}$$

where the coefficients a and b, obviously, depend on the fact that Al atoms were used. On the other hand, straightforward arguments suggest that a many-atom expansion yields

$$U_{\text{many-atom}}(c) = \sum_{n=0} a_n c^n, \tag{5}$$

where the coefficient of the c^{n-1} term, a_{n-1}, is mainly determined by the n-atom interaction. Evidently the function $U_{\text{electron}}(c) = ac - bc^{1/2}$ in eq. (4)

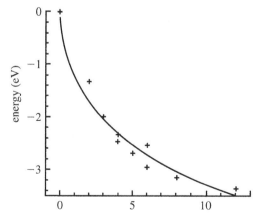

Fig. 4.2 The ground-state energy per atom versus the coordination numbers for Al atoms (Heine *et al.* 1991).

cannot be represented by the power series of eq. (5). In fact, Heine *et al.* (1991) have shown that if one wishes to represent $U_{\text{electron}}(c)$ by $U_{\text{many-atom}}(c)$, the coefficients a_n increase with increasing n. That is to say the many-atom expansion diverges. Thus we may conclude that both in principle and in practice the 'effective pair-potential' models do not do justice to the complexity of the problem of cohesion.

Having established the legitimacy of the electronic point of view in describing the binding of atomic nuclei in condensed matter, I wish to continue by recalling some simple empirical correlations which directly implicate the electrons in the cohesion of atoms in all condensed matter.

The Mendeleev table of elements represents an empirical correlation between the properties of atoms. In short, nearby atoms have similar properties. One of the principal successes of the positively charged nuclei and negatively charged electrons model of atoms plus quantum mechanics is the explanation of these correlations. Clearly, the dependence of properties on the number of electrons, Z, is a consequence of the model, and the periods are a quantum mechanical effect. Naturally, one would expect similar regularities in solids and liquids.

The valence electron per atom ratio (e/a) was identified as an important factor in determining the structure of a binary compound early on by Hume-Rothery (1936). Over the past 50 years or so, other factors, such as atomic size and electronegativity differences, were also found to be useful in predicting structure types. However, the most successful correlations, involving some 8000 compounds (Villars *et al.* 1989) are those of structure maps based on the Mendeleev number M associated with each element by Pettifor (1988). Pettifor defined M by the string through the modified Periodic Table shown in Fig. 4.3, and proceeded to represent all AB binary compounds as points on a plot of M_A against M_B. The resulting structure map is shown in Fig. 4.4. Remarkably, the points corresponding to the same structure types are near each other and form separate clusters on this map. Also, missing compounds are easily identified and their structure type can be guessed. Moreover, similar good separation occurs for other stoichiometries A_nB_m.

Whilst such maps are of practical use in searches for the right alloy for a given application (Villars *et al.* 1989), in the present context their significance is that they are challenges a first-principles theory must meet. From the way the Mendeleev string winds its way through the Periodic Table, we can conclude that in determining the crystal structure of a particular alloy both the number of valence electrons per atom and quantum mechanics (their orbital character) are at play (Pettifor 1988).

Let us now return to our principal task of studying an infinite array of nuclei

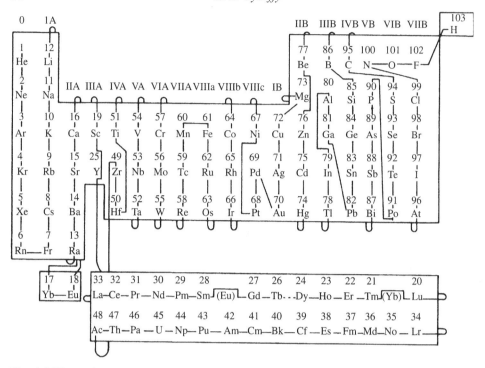

Fig. 4.3 The string woven through the Periodic Table defines the Mendeleev number M_α for each element α as its rank along the string (Pettifor 1988). From Sutton (1993) by permission of Oxford University Press.

bound together by electrons. There are two general approaches to this problem. The first harks back to chemistry, or atomic physics, and attempts to make use of the fact that we understand isolated atoms rather well. The hallmark of this approach in the representation of the one-electron states for the whole solid in terms of atomic-like orbitals centred on the individual atoms (Heitler and London, 1927). It is often referred to as the tight-binding method (TBM) (Mott and Jones 1936). While historically it has provided us with the insightful language of describing cohesion in solids as the formation of chemical bonds, I will describe it, briefly, only to contrast it with the more modern, represent-ation-independent second approach of density functional theory. This will be the principal topic of the next section.

Whatever method or approximation one uses to describe the electrons of an infinite solid, one must choose a representation for the one-electron wave functions. Because of the translational symmetry of the problem, these will have to satisfy the Bloch theorem. In the case of the tight-binding method, such Bloch functions, corresponding to a given wave vector k and band index ν, are given by

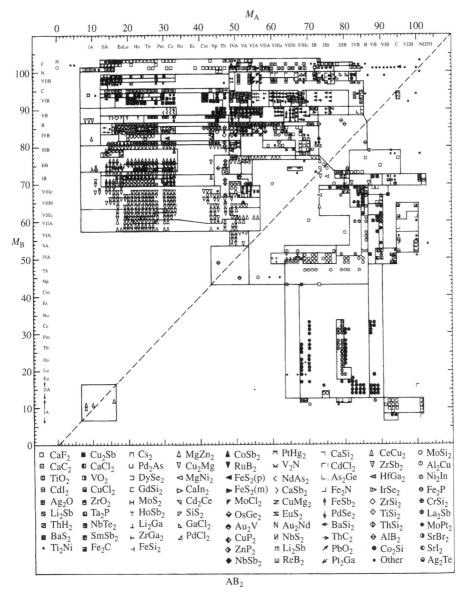

Fig. 4.4 The AB_2 – binary compound structure map (Pettifor 1988). From Sutton (1993) by permission of Oxford University Press.

$$\psi_{k,\nu}(\mathbf{r}) = \sum_{i,\mu} a_\mu(\mathbf{k},\nu)\varphi_\mu(\mathbf{r}-\mathbf{R}_i)\,e^{i\mathbf{k}\cdot\mathbf{R}_i}, \tag{6}$$

where the orbitals $\varphi_\mu(\mathbf{r}-\mathbf{R}_i)$ designated by a generalised quantum number μ and centred on the nuclear positions $\mathbf{R}_i$ are the results of an atomic-like calculation. Thus, the first task of theory is to determine the coefficients $a_\mu(\mathbf{k},\nu)$

and the corresponding energy eigenvalues $\varepsilon_{k,\nu}$. They are found as eigenvectors and eigenvalues, respectively, of a Hamiltonian matrix which depends on the hopping integrals $t_{\mu\mu'}(i,j)$. These are the probability amplitudes that an electron at the site i, in the state $\varphi_\mu(r - R_i)$, makes a hop to the site j into the state $\varphi_{\mu'}(r - R_j)$, and they can be calculated as matrix elements of the crystal potential, seen by the electrons, with respect to the above orbitals.

One of the great advantages of this approach is its physical transparency. The bands of the infinite solid, described by $\psi_{k,\nu}(r)$ and $\varepsilon_{k,\nu}$, arise from the atomic orbitals. Because the corresponding hopping integrals are small, the low-lying atomic core states are hardly modified by the formation of the solid, while the highest-filled states in the atom become the valence and conduction bands, whose position in energy and width are determined by the atomic energies ε_μ corresponding to the orbitals φ_μ and the hopping integrals $t_{\mu\mu'}(i,j)$, respectively. Thus it is possible, and useful (Mott and Jones 1936), to talk of s–p bonded metals, d- and f-band metals or semiconductors, whose directed bonds are sp^3 hybrids. Indeed, the power of the above conceptual framework enabled Wigner to draw an amazingly accurate qualitative picture of cohesion in solids, all through the Periodic Table in his famous article which launched the *Solid State Physics* series edited by F. Seitz (Wigner and Seitz 1955).

Unfortunately, for accurate quantum mechanical calculations, the tight-binding method is rather inconvenient. The principal trouble is that the useful, physically meaningful, orbitals of $\{\varphi\}$ are very far from forming a complete set. Thus, in a demanding calculation, where mathematics takes over from physical intuition, one can very easily end up with 30 to 40 orbitals per site instead of the sensible $1s + 3p + 5d = 9$ to reach convergence. Moreover, while the extra orbitals make the calculation mathematically more rigorous, they no longer have the physical significance of the so-called 'minimal basis-set'. In short, one comes to the conclusion that in large scale numerical calculation, where high accuracy is important, the choice of the representation of the wave function should be governed by mathematical convenience rather than an appeal to physical intuition trained on the model of weakly interacting atoms.

Having said that, the tight-binding approach remains the most effective road to semi-phenomenological model Hamiltonians such as the Hubbard model (Hubbard 1963) or the impurity and the periodic Anderson models (Anderson 1961) referred to in chapter 8. Such Hamiltonians parameterise the free particle motion from atom to atom by phenomenological hopping integrals $t_{\mu\mu'}(i,j)$ and the electron–electron interaction by interaction constants $U_{\mu\mu'}(i,j)$. Unfortunately, such models are useful only in circumstances when a few parameters are sufficient to describe the phenomena in question. This is frequently the situation in the cases of effective Hamiltonians which are designed to describe only the low energy physics near the Fermi energy (see chapter 5). However, to describe the system through the full physical range between 10^{-2} eV to 10^2 eV, the number of adjustable parameters becomes so large that such models lose all their predictive powers. It thus appears that the 'electron glue' is such a complex quantum many-body system that it cannot be usefully parameterised. By contrast, the interacting lattice vibrations, the phonons, of a rigid body form a simple system. Namely, their dispersion relations can be fitted by a set of phenomenological force constants of modest size. The electronic energy bands $\varepsilon_{k,\nu}$ (the spaghetti), on the other hand, defy such easy representation.

Thus, we conclude that, whilst pair potentials and effective electronic Hamiltonians are frequently useful descriptions of the electronic binding in solids, they are no substitute for first-principles calculations based on the Hamiltonian given in eq. (2). Namely, there are many interesting questions, such as the nature of metallic magnetism, charge density waves etc., which involve phenomena on many energy scales, and which can be answered only by reliable, materials specific, parameter free calculations. In the next section we shall turn to the basic principles of these.

4.2 Density functional theory

4.2.1 The basic theorem

Of course, the fact that the simple qualitative models of the previous section have given way to quantitative first-principles calculations is largely to do with the advent of modern computers. But that is only half of the story. The other half has to do with density functional theory. One way or another, canonical many-body theory forces one to work with explicitly antisymmetrised wave functions, namely Slater determinants, and these are particularly recalcitrant to numerical computations. For instance, modern configuration interaction (CI) calculations use wave functions which are linear superpositions of 10^9 Slater determinants and are still only for relatively light atoms like Xe (Cowan 1981). Moreover, such calculations scale as $\sim N^7$ with the number of atoms N. Whilst canonical perturbation theory (Fetter and Walecka 1971), based on the use of creation and annihilation operators and Feynman diagrams, treats infinite systems from the start, it also resists numerical implementation. Density functional theory makes headway because it deals with the complexities introduced by the permutation symmetry in a way that is radically different from the other methods. Namely, it eliminates them by a surprising, and extremely general, theorem (Hohenberg and Kohn 1964). It is easy to dislike this theorem for it is based on a *reductio ad absurdum* argument and it lacks any intuitive appeal. But it is hard to deny its power: it reduces the general many-body problem to a self-consistent one-electron problem without any approximation. As will be seen presently, this renders any approximation scheme within this framework readily tractable.

Of course, the density functional strategy is not entirely new. Indeed it was originated by Thomas and Fermi (Fermi 1927; Thomas 1927). But the fact that there is a version of these arguments which is not an approximation but an exact result is startling.

The actual proof of the theorem would be out of place in such a discursive treatment of the subject as this. However, for clarity I wish now to recall its simplest version. It refers to a non-spin-polarised, non-relativistic electron system at zero temperature, $T = 0$, in the presence of an external potential $V^{\text{ext}}(r)$. According to the Hohenberg–Kohn theorem (Hohenberg and Kohn 1964), there is a unique energy functional $E[n(r)]$ of the charge density $n(r)$, whose minimum $E_0[n_0]$ is the ground-state energy of the system; and the charge density $n_0(r)$, for which $E[n]$ takes on its minimum value, E_0, is the ground-state charge density. Moreover, the energy functional can be shown to be of the form

$$E[n(r)] = \int dr\, n(r) V^{\text{ext}}(r) + F[n(r)], \qquad (7)$$

where $F[n]$ is a universal functional in the sense that it does not depend on the external potential $V^{\text{ext}}(r)$. In fact, $F[n(r)]$ is completely determined by the Coulomb interaction between the electrons. Evidently, the ground-state energy E_0 and the corresponding charge density $n_0(r)$ are to be found by minimising $E[n]$ with respect to arbitrary variations in $n(r)$ with the constraint that the total number of particles

$$N = \int dr\, n(r) \qquad (8)$$

is fixed. This is achieved, using the method of Lagrange multipliers, by minimising $E[n] - \mu \int dr\, n(r)$, where the Lagrange multiplier μ is the chemical potential. To make the above theory useful one must construct the functional $F[n]$ explicitly. A partial solution of this problem is to write

$$F[n] = F_0[n] + e^2 \int dr' \int dr \frac{n(r)n(r')}{|r - r'|} + E_{\text{xc}}[n], \qquad (9)$$

where $F_0[n]$ is the functional for free electrons, the next term is the fairly obvious Coulomb contribution and the last one is, by definition, the as-yet unknown, exchange-correlation functional. Knowing the formal solution for the non-interacting many-electron problem in an external potential $V^{ext}(r)$ implies that we can construct the non-interacting functional $F_0[n]$. Surprisingly, following the same arguments for the fully interacting case leads to a very transparent procedure for finding the minimum of $E[n]$. In short, the charge density is represented by the sum

$$n(r) = \sum_{\nu} |\psi_{\nu}(r)|^2 f(\varepsilon_{\nu}), \tag{10}$$

where $f(\varepsilon_{\nu}) = 1$ for $\varepsilon_{\nu} < \mu$ and $f(\varepsilon_{\nu}) = 0$ for $\varepsilon_{\nu} > \mu$; the wave functions $\psi_{\nu}(r)$ form an orthogonal complete set, and they are the solution of the Kohn–Sham equations (Kohn and Sham 1965) with eigenvalues ε_{ν}. These are Schrödinger-like equations

$$(-\nabla^2 + V_{eff}(r; [n]))\psi_{\nu}(r) = \varepsilon_{\nu}\psi_{\nu}(r), \tag{11}$$

where the effective potential $V_{eff}(r; [n])$ is a function of r and a functional of the charge density $n(r)$. It is given by

$$V_{eff}(r; [n]) = V^{ext}(r) + e^2 \int dr' \frac{n(r')}{|r - r'|} + V^{xc}(r; [n]), \tag{12}$$

where the exchange-correlation contribution is

$$V^{xc}(r; [n]) = \left(\frac{\delta^2 E_{xc}}{\delta n(r)\delta n(r')} \right). \tag{13}$$

Note that eq. (11) has the form of a self-consistent field, independent particle Schrödinger equation. To solve it one must start with an initial guess, $n_{in}(r)$, for the charge distribution, calculate the effective potential $V_{eff}(r; [n])$, solve the 'Schrödinger' equation (11), recalculate $n(r)$ using eq. (10) and compare the result $n_{fi}(r)$ with the initial guess $n_{in}(r)$. If $n_{fi}(r) \cong n_{in}(r)$ to within a prescribed tolerance, then the problem is solved and $n_{fi}(r)$ is the ground-state charge distribution. If the agreement is not sufficiently close, a new guess is made and the procedure is repeated until convergence. Once the ground-state charge density $n_0(r)$ has been determined, the ground-state energy is calculated using the following formula (Kohn and Sham 1965):

$$E_0 = \sum_{\nu} \varepsilon_{\nu} f(\varepsilon_{\nu}) - e^2 \int dr \int dr' \frac{n_0(r)n_0(r')}{|r - r'|} + E_{xc}[n_0]. \tag{14}$$

As stressed earlier, the straightforward minimisation procedure specified by eqs. (10)–(14) gives a full formal solution to the many-electron problem at hand. Of course, this solution is only useful if we know the exchange correlation potential functional $V^{xc}(r; [n])$ defined in eq. (13). Unfortunately, there is no systematic, well controlled procedure for constructing successively better approximations to the exact $V^{xc}(r; [n])$. Nevertheless, there are well tried schemes (Dreizler and Gross 1990), and in what follows all discussions will be based on these. However, the power of the method does not lie with the ease with which $V^{xc}(r; [n])$ can be constructed, but rather with the readiness with which the procedure can be implemented numerically for a given approximate $V^{xc}(r; [n])$. Over the past 30 years, a whole arsenal of highly efficient band theory methods have been devised for solving precisely the above problem. These, together with the explosive growth of computer power, completely transformed the scientific ambience in which Wigner could write (Wigner and Seitz 1955):

If one had a great calculating machine, one might apply it to the problem of solving the Schrödinger equation for each metal and obtain thereby the interesting physical quantities, such as the cohesive energy, the lattice constant, and similar parameters. It is not clear, however, that a great deal would be gained by this. Presumably the results would agree with the experimentally determined quantities and nothing vastly new would be learned from the calculation. It would be preferable instead to have a vivid picture of the behaviour of the wave functions, a simple description of the essence of the factors which determine cohesion and an understanding of the origins of variation in properties from metal to metal.

Of course, Slater and a few others (Slater 1975) always objected to this point of view on the grounds that one does not know what to highlight with a 'vivid picture' until the principal factors have been identified by quantitative calculations which treat all factors on an equal footing. But, not until successful first-principles calculations, based on the formally solid foundations of density functional theory, began to appear in the 1970s did a large community of theorists accept Slater's views as a new paradigm. As an illustration of the general acceptance of this change I note that in the 1967 (third) edition of *Introduction to Solid State Physics* Kittel quotes the above passage with approval, but that in the fifth edition, of 1976, it no longer appears. In the next section I will argue more forcefully that, far from being unnecessary, first-principles calculations are one of the principal ways forward in condensed matter physics. However, before turning to the highlights of results from such calculations, I complete this section by giving a brief account of the most frequently used approximation for $V^{xc}(r; [n])$.

4.2.2 The local density approximation (LDA)

The local density approximation (LDA) was the first guess made by Kohn and Sham (1965) at what the exchange-correlation potential functional $V^{xc}(r; [n])$ might be like. It has considerable physical appeal, and it has proved a reliable guide to bonding in solids. The defining simple idea of the LDA is the assumption that the potential $V^{xc}(r; [n])$ depends only on the charge density at r:

$$V^{xc}(r; [n]) \cong V^{xc}_{LDA}(r; n(r)). \tag{15}$$

Because $V^{xc}_{LDA}(r; n(r))$ must be a universal function of $n(r)$, independent of the external potential $V^{ext}(r)$, the above simple statement immediately translates into a practicable scheme for first-principles calculations. The point is that there is a potential $V^{ext}(r)$ for which the many-electron problem can be solved more or less exactly, and hence the function $V^{xc}(r; n(r))$ can be determined and used for all other external potentials. Of course, this potential is the constant corresponding to an infinite homogeneous positive background of charge. Thus, by solving the problem of an infinite, dense system of interacting electrons with a uniform positively charged, neutralising background, namely that of the famous 'jellium' model, we can calculate an approximate functional $V^{xc}_{LDA}(r; n(r))$ to be used in the density functional theory. This procedure is called the LDA.

Given the theme of this book, it behoves us to pause and discuss the 'jellium' in a bit more detail than is strictly necessary from the point of view of the LDA. Interestingly, perhaps the first to use this model was J. J. Thomson. After the discovery of the electron, but before Rutherford's experiments, Thomson modelled neutral atoms by a set of negatively charged point electrons moving in a finite sphere of homogeneously distributed positive charges (Thomson 1907, chaps. 6 and 7). Although it yielded encouraging results, which would have improved with quantum mechanics, Rutherford's discovery of the nucleus rendered it irrelevant. Nevertheless, as if to vindicate his thinking, there is at present a strong revival of interest in such models in connection with quantum dots (Heitmann and Kotthaus 1993).

Evidently, infinite jellium is the generalisation of J. J. Thomson's model of atoms (Thomson 1907, chaps. 6 and 7) for metals. Having smeared out the positive nuclear charges into a neutralising background, the ion–ion and ion–electron interactions have been eliminated from the problem, and hence it becomes a pure electron–electron many-body problem. As such it is of considerable general interest. The early work on the subject is comprehensively reviewed in the beautiful book by Pines and Nozières (1966). In addition to the treatment of approximate microscopic calculations, it includes a full discussion of the phenomenological Landau–Silin theory, which regards the electron system as a charged Fermi liquid. As illustrated in chapter 5, in spite of its phenomenological nature, this remains a very fundamental point of view. By contrast, modern microscopic work on jellium tends to concentrate on exact numerical solutions using quantum Monte Carlo methods (Ceperley 1978; Ceperley and Alder 1980; Silvestrelli *et al.* 1993).

From the point of view of the LDA, the relevant information to be extracted from the above calculations is the ground-state energy per particle as a function of the density n_0, in the form

$$\varepsilon(n_0) = \varepsilon^{\mathrm{HF}}(n_0) + \varepsilon^{\mathrm{xc}}(n_0) \tag{16}$$

where $\varepsilon^{\mathrm{HF}}(n_0)$ is the Hartree–Fock contribution and $\varepsilon^{\mathrm{xc}}(n_0)$ is, by defini-tion, the exchange-correlation energy per particle. Evidently, for jellium, the ground-state charge density number is a constant and $\varepsilon^{\mathrm{xc}}(n_0)$ is a smooth function of n_0. Thus, the universal LDA exchange potential functional is given by

$$V_{\mathrm{LDA}}^{\mathrm{xc}}(r; n(r)) = \left(\frac{\mathrm{d}\varepsilon^{\mathrm{xc}}(n_0)}{\mathrm{d}n_0} \right)_{n_0 = n(r)}.$$

In short, an electron at r feels an exchange-correlation potential appropriate to an infinite jellium with density equal to the local density $n_0(r)$.

Whilst the above statement does not lack physical appeal, it does not prepare us for the remarkable successes of the LDA to be discussed in the next section. One has only to recall the very rapid changes of the charge density in a solid to be surprised that such a local scheme as the LDA is anything but a rough guide. Fortunately, careful analysis reveals that its accuracy is governed by more subtle effects than the slow variation of $n(r)$. In effect, there are two general reasons why the LDA works as well as it does. One is to do with the fact that, unlike in the Hartree–Fock theory, the LDA exchange-correlation hole always contains exactly one 'electron'. Thus the electron and the positively charged background taken together are neutral, as they should be. Of course, being spherically symmetric, the shape of the hole is wrong, particularly near a surface or an atomic nucleus. However, this fact does not do as much harm as one might expect because $V_{\mathrm{LDA}}^{\mathrm{xc}}(r; n(r))$ turns out to depend on only an average over the exchange-correlation hole and hence it is rather insensitive to errors in its shape. This is the second reason behind the success of the LDA (Perdew 1995). While these remarks go some way towards explaining why and how it works, they also highlight the fact that, however physically appealing, it is an '*ad hoc*' approximation. In other words, it is not one step in a sequence of well controlled approximations which, when carried to convergence, lead to the exact answer. For instance, one might have thought that since it is obviously exact in the constant density limit, which corresponds to jellium, it can be systematically improved by including gradients and higher derivatives of $n(r)$. Unfortunately, the failure of the gradient expansion to bring decisive improvement (Perdew 1995) seems to indicate that this is not the case. In fact, as yet, there is no generally accepted next move beyond the LDA.

The above blemishes notwithstanding, the virtues of the LDA, from the point of view of implementation, are easy to see. It does not contain any adjustable parameters and can be readily deployed under any circumstances. Indeed it has been applied in studies of atoms and molecules as well as in condensed matter physics. Moreover, in the latter case not only bulk materials have been treated but also point defects, like impurities and vacancies (Dederichs *et al.* 1989), and infinite defects, like surfaces and interfaces (Lang 1973). The point to appreciate is that the self-consistent, local one-electron Kohn–Sham equations (eqs. (10)–(11)) are incomparably easier to solve numerically than any other many-body procedure. Moreover, because these are Euler–Lagrange equations of a variational problem, the calculation of the ground-state energy is rather forgiving as far as numerical accuracy is concerned. It is probably fair to say that the whole density functional approach to the many-electron problem would have remained a formal curiosity if the LDA, in spite of its shaky foundation, did not turn out to be such a useful and generally reliable tool.

As mentioned earlier in this section the above discussion of density functional theory and the LDA refers to the simplest case, where the ground state is a spin singlet. The most useful generalisation of this is to the spin-polarised case. Evidently, this is the relevant generalisation for describing metallic magnetism, and as such it requires some comments.

Following the same line of argument as above, it is straightforward to show that a spin-polarised density functional theory must feature a spin dependent exchange-correlation potential which is the functional of both the charge and the magnetisation densities, $n(r)$ and $m(r)$, respectively. It also follows that in spin-polarised local density approximation (SPLDA) this potential $V_\sigma(r; [n(r), m(r)])$ is to be generated from the exchange correlation energy of spin-polarised jellium, which is a function of both n_0 and m_0, the ground-state charge density n_0 and magnetisation density in the z-direction m_0, respectively. The difficulty I have referred to above concerns the fact that the ground state of jellium is not magnetic at metallic densities.

This is a well known but still interesting story, the end of which is yet to come. It begins with Bloch's observation (Bloch 1929) that in the Hartree–Fock approximation the ground state of jellium is magnetic provided the density is low enough ($r_s > 7.5$). This was a spectacular discovery because it demonstrated that the Pauli exclusion principle, together with Coulomb repulsion, can give rise to a magnetic ground state without any overtly magnetic force. Briefly, electrons with parallel spins are kept apart by the exclusion principle and therefore will pay a smaller Coulomb penalty than antiparallel electrons, which frequently come close to each other. The striking thing about this mechanism is that it does not require the presence of atoms with local moments and hence it is an attractive candidate for being the prototype of metallic magnetism. Unfortunately, a few years after Bloch's discovery, Wigner (1938) showed that electron–electron correlations keep electrons with anti-parallel spins sufficiently apart such that there is not enough gain in lining up their spins to form a magnetic ground state. Modern quantum Monte Carlo calculations confirm Wigner's arguments (Ceperley 1978) for metallic densities. Since it is precisely such computations that provide $\varepsilon_{xc}(n_0, m_0)$ on which the SPLDA calculations are based, it is, at first sight, puzzling how magnetisation can arise in a solid when there was none in jellium. Fortunately, this dilemma is based on a misunderstanding: $\varepsilon^{xc}(n_0, m_0)$ is not the exchange-correlation contribution to the ground-state energy of jellium with a fixed n_0; that quantity is $\varepsilon^{xc}(n_0, m_0 = 0)$, but rather that for jellium, which is constrained to have a magnetisation m_0. In practice, this constraint is enforced by a constant external, 'spin only' magnetic field, and thus $\varepsilon^{xc}(n_0, m_0)$ is extracted from calculations on jellium with constraints to be magnetic. The surprising fact is that, for some crystals with highly inhomogeneous charge density $n(r)$, the self-consistent solution of the Kohn–Sham equation, with the SPLDA exchange-correlation potentials, turns out not to be $m_0(r) = 0$ but some finite $m_0(r)$. In fact, as will be illustrated in the next section, the SPLDA is a re-markably accurate predictor of magnetic ground states all through the periodic table.

At this stage of the argument, it is difficult to resist the temptation to compare the above circumstances with that prevailing with respect to super-conductivity. Although it is not easy to imagine repulsive electrons in jellium being superconducting, Kohn and Luttinger (1964) have argued that correlations may conspire to promote pairing and therefore superconductivity. But recently it has been shown that, due to the long range of the Coulomb repulsion between the electrons, this possibility can be discounted (Alexandrov and Golubov 1992). However, not all is lost, for correlations may lead to super-conductivity in an inhomogeneous system like a crystal. In fact, the density functional theory for superconductivity has already been developed (Dreizler and Gross 1990). Not surprisingly the central feature of the theory is the exchange-correlation energy $\varepsilon^{xc}[n, \chi]$, which, in addition to the charge density $n(r)$, is a functional of the pairing amplitude $\chi(r, r')$. This quantity is the wave function of the Cooper pairs, and as such it is the order parameter of the problem. Thus it is the analogue of the magnetisation density $m(r)$ in spin-polarised density functional theory. Clearly, the central problem of the theory is to find a generalisation of the LDA and the SPLDA for $\varepsilon^{xc}[n, \chi]$. Intriguingly, this may be done either by invoking the electron–phonon interaction, which is the cause of pairing in conventional superconductors, or by exploring more exotic electron correlation mechanisms with relevance to the high T_C cuprates. The first efforts along these lines have appeared recently, and the prospects for progress are encouraging (Gross *et al.* 1995).

Having mentioned the LDA and the concomitant jellium model for magnetism and superconductivity, we close this discussion with two more generalisations, which deal with finite temperature and relativistic effects.

The finite temperature generalisation of the density functional theory by Mermin (1965) appeared shortly after the pioneering work of Hohenberg and Kohn (1964) and Kohn and Sham (1965). Conceptually, the change is that instead of the ground state the theory is designed to describe the equilibrium state. Namely it deals with the grand potential functional, $\Omega[n(r)]$, of the equilibrium density $n(r; T)$, which is, obviously, a function of the temperature T. Technically, the main new feature is that the step function $f(\varepsilon)$ in eq. (10) now becomes the usual Fermi function $f(\varepsilon) = (\exp(\beta(\varepsilon - \mu)) - 1)^{-1}$, and in the LDA the exchange-correlation contribution to the grand potential functional $\Omega_{\text{LDA}}^{\text{xc}}(n, T)$ is to be calculated by solving for the equilibrium state of jellium at finite temperatures. Clearly, such considerations become important when the thermal wavelength of the electrons, $\lambda = \hbar/(2mk_B T)^{1/2}$ becomes comparable with r_s, the average distance between electrons. While r_s, corresponding to the average density in a metal, $r_s = (3/4\pi n)^{1/3}$ is much smaller than λ at any relevant temperatures, locally the density does become smaller, and correspondingly the effective r_s increases. Although such temperature dependence has been taken into account in a number of calculations (Staunton *et al.* 1985), the full implications have not been explored.

We have focused above on the fact that when applying the LDA, inevitably one encounters regions in a solid where $n(r)$ is small and, hence, locally r_s is large. As stressed, this circumstance may lead to temperature dependent effects. We now turn to the need to generalise the LDA on account of the fact that locally $n(r)$ can become very large and hence the effective r_s becomes very small. Interestingly, this limit has to be dealt with using relativistic quantum mechanics. Clearly, relativistic effects are important when $\beta \equiv v_F/c$ is of the order of unity. In this limit, r_s becomes comparable to the Compton wavelength $\lambda_C = \hbar/mc$. To be precise, recall that for jellium in the high density limit the Fermi velocity $v_F = (1/m)\hbar k_F \sim 1/70.2 r_s$. Thus, a degenerate Fermi system will always become relativistic if it becomes dense enough. Indeed, $\beta \sim 1$ near the nuclei in a mercury crystal. This is another surprising consequence of the exclusion principle.

The relativistic formulation of density functional theory is due to Rajagopal

(Rajagopal 1978; Dreizler and Gross 1990). As might be expected, the ground-state energy or the grand potential is found to be a functional of the current and charge density four-vector $(\vec{j}, icn)$, and the analogue of the Kohn–Sham equation is a self-consistent Dirac equation. Once again there is an LDA, albeit with some complications, and to implement it one must have the energy or the grand potential for relativistic and spin-polarised jellium as a function of the current and charge density four-vector. Naturally, they are determined not by using the Coulomb potential but by the fully quantised Maxwell's equations. Namely, they include the interaction between electrons arising from the exchange of photons. Unfortunately this classic many-body problem is not as well understood as the non-relativistic jellium even at zero temperature (Wilson and Gyorffy 1995). Nevertheless, relativistic LDA calculations based on approximate relativistic jellium data are currently making considerable impact on our understanding of magnetic anisotropies (Wilson *et al* 1991) and circular X-ray dichroism (Ebert 1996).

To summarise: density functional theory based on the LDA is a powerful and very general method for dealing with almost all questions that can be asked within quantum electrodynamics, including radiative corrections (Engel *et al.* 1995) not considered here, and can be implemented for atoms, molecules and solids without adjustable parameters. Its shortcoming is that it is not a well controlled approximation and there is no systematic way to study corrections to it. In particular, the LDA as it stands is powerless in the face of the challenges represented by the group of problems discussed in connection with highly correlated electron systems elsewhere in this book (e.g. in chapter 8). Nevertheless, as will be illustrated in the next section, the LDA is highly effective in describing the binding of atoms into molecules and solids.

4.3 Bonding by electrons in the local density approximation

4.3.1 Molecules

Although our principal interest is the 'electron glue' in solids, it is useful to start with a brief mention of LDA calculations for molecules. In the beginning, such work was done on simple molecules, and the aim was to establish the credentials of the LDA by comparing its results with those of conventional quantum chemistry calculations and experiments. One of the most influential early papers was that of Gunnarson *et al.* (1977). Their results for the binding energy, atomic separation and vibrational frequencies of the diatomic boron, carbon, nitrogen, oxygen and fluorine molecules, respectively, are shown in Fig. 4.5. Evidently the agreement with experiments is quite satisfactory.

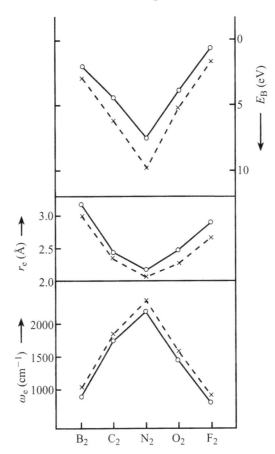

Fig. 4.5 Binding energies (E_B), equilibrium separation (r_e) and vibrational frequencies (ω_e) for the ground states of first-row diatomic molecules. The open circles are calculated values and the crosses are experimental values. (Jones and Gunnarson 1989.)

Reassuringly, the LDA results are much better than those obtained by using the Hartree–Fock approximation but not as good as achieved by the best configuration interaction (CI) calculations.

As a footnote to the history of the density functional theory, it is interesting to recall that initially the chemistry community was rather hostile to the whole enterprise. They were suspicious of the basic theorem and regarded the LDA as an *ad hoc* procedure no better than the X–α method of Slater (1975). Surprisingly, even Slater himself, who has done more than anyone else to solve the many-electron problems in atoms, molecules and solids quantitatively, was not enthusiastic (Slater 1975). However, after the above paper by Gunnarson *et al.* (1977) and under the pressure of its sequel (Jones and Gunnarson 1989)

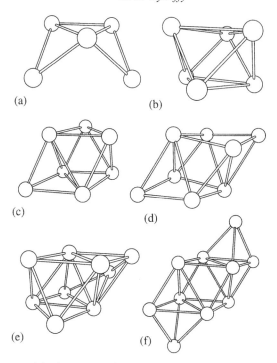

Fig. 4.6 The most stable isomers of Al_5 to Al_{10} (parts (a) to (f), respectively) as determined by local density approximation (LDA) calculations combined with simulated annealing (Jones 1991).

attitudes began to change. In fact, by 1989 such a distinguished chemist as Robert G. Parr had written a book (Parr and Yang 1989) with the evident purpose of advocating density functional theory and the LDA for use in chemistry. Nowadays, the method is not so much tested as used to solve long standing problems. To mention but one we show in Fig. 4.6 the most stable geometries determined by a combined LDA and molecular dynamics procedure for a series of isomers Al_5 to Al_{10}. Note that in these calculations a large number of local configurational minima have been found and the corresponding energies were determined sufficiently accurately that the one with the lowest energy, the global minimum, could be selected. Evidently the number of local minima increases rapidly with N, the number of atoms, and, even for the smallish 10–20 atom clusters considered above, the LDA plus molecular dynamics calculations have no first-principles competitors.

4.3.2 Semiconductors

For matter containing light elements the plane wave expansion of the pseudopotential method is a very efficient approach to solving the Kohn–Sham equations (Srivastava and Weaire 1987). Therefore, problems involving carbon, silicon, gallium arsenide or sodium chloride are usually dealt with in this way. To illustrate the results of such calculations we show, in Fig. 4.7, the ground-state energy of silicon for various crystal structures as functions of the volume. Clearly, for a given symmetry the equilibrium density, or, what amounts to the same thing, the appropriate lattice parameter a, is determined by locating the minimum of the corresponding energy versus volume curve. Then the question is: which one of the local minima is the lowest? The total energy at the equilibrium volume is best quoted as the cohesive energy, ε_c, which is the difference between the total energy per atom in the bulk and the energy of an isolated atom. Another useful quantity that can be extracted from these curves is the bulk modulus B, which is the second derivative of the ground-state energy E_0 with respect to the volume at the equilibrium volume V_{eq}. In Table 4.1 a, B and cohesive energy ε_c, calculated for a selection of materials within the LDA, are compared with experiments. These results are very satisfactory indeed and lend considerable credence to the whole project of first-principles calculations in general and the use of LDA in particular.

One of the great virtues of a reliable, quantitative, first-principles theory is that one can use it to study materials not found in nature or matter under

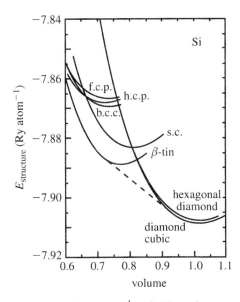

Fig. 4.7 The structural energy (in Ry atom^{-1}) of silicon in seven crystal structures as a function of volume normalised to the observed equilibrium volume. The crystal structures considered are: diamond cubic, hexagonal diamond (wurtzite), β-tin, simple cubic (s.c.), face-centred cubic (f.c.c), body-centred cubic (b.c.c.) and hexagonal close-packed (h.c.p.) (Yin and Cohen 1982).

Table 4.1 *Calculated and experimental (in parentheses) values for the lattice parameter* a, *the bulk modulus* B *and the cohesive energy* ε_c *for a selection of crystalline semiconductors (Srivastava and Weaire 1987).*

Material	a (Å)	B (Mbar)	ε_c (eV)
C	3.60 (3.57)	4.41 (4.43)	7.57 (7.35)
Si	5.45 (5.43)	0.98 (0.99)	4.67 (4.63)
GaAs	5.57 (5.65)	0.73 (0.75)	
GaP	5.34 (5.45)	0.897 (0.887)	
NaCl	5.56 (5.60)	0.284 (0.266)	
Al	4.01 (4.02)	0.715 (0.722)	3.646 (3.401)
V	2.97 (3.03)	2.0 (1.62)	5.83 (5.30)

conditions not accessible to experiments. An example of the former case is the search, by Liu and Cohen (1989), for low compressibility solids. Intriguingly, preliminary calculations suggest that structures held together by carbon–nitrogen bonds may have a higher bulk modulus, i.e. lower compressibility, than diamond. As an illustration of how LDA calculations for matter under extreme conditions can be useful, one may refer to studies of Fe–O systems at pressures and temperatures appropriate to those at the Earth's core (Ringwood 1977). From the point of view of contending models, whether Fe–O is a solid or a liquid, or whether it phase separates or orders at the relevant temperatures and pressures are key questions, and yet laboratory experiments cannot be performed to answer them. By contrast, solid progress is being made on the basis of first-principles calculations (Ringwood 1977).

The above studies are for bulk (infinite), crystalline solids. For such systems the spatial periodicity simplifies the task of solving the Kohn–Sham equation enormously. When this translational symmetry is broken, the computations become more arduous, although the basic LDA procedure remains the same. Thus, much effort is currently going into solving problems without full crystalline symmetry.

An example of this concerns the spatial arrangements of 'atoms' at surfaces. Whilst not as venerable as crystallography itself, thanks to low energy electron diffraction and surface X-ray diffraction, surface crystallography is today a well developed subject (MacLaren *et al.* 1987). By now, thousands of surface structures have been determined, and they usually differ intriguingly from what would result by cutting the bulk material in two. Evidently, the understanding of these surface structures and their evolution with pressure and temperature is the foundation of 'surface science', and as such is the subject of much scientific and technological interest. As an example, where LDA-based first-principles calculations have played a major role in establishing an unexpected surface structure and in identifying the driving force behind it, one might point to the Takayanagi reconstruction on a Si(111) surface (see Table 4.2; Stich *et al.* 1992). The complex structure is displayed in Fig. 4.8.

Other important cases where some of the crystal symmetry is removed but the density functional strategy, within the LDA, can be accurately implemented are: periodically deformed solids which represent vibrational modes of equilibrium lattices, crystals with dislocations or grain boundaries and interfaces of

Table 4.2 *First-principles local density approximation (LDA) calculations for variously reconstructed surfaces.*

Clearly, the changes in the surface energy are saturating towards the 7×7 structure (Stich *et al.* 1992).

	3×3	5×5	7×7
Energy per unit cell (eV)	10.765	29.205	56.509
Energy per surface atom (eV)	1.196	1.168	1.153

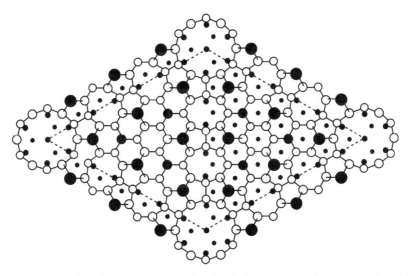

Fig. 4.8 Top view of a dimer-atom-stacking-fault (DAS) model proposed by Takayanagi and colleagues for the Si(111) – 7×7 surface reconstruction. The 7×7 unit cell is bounded by a dashed line. Atoms at increasing distances from the surface are indicated by circles of decreasing size. The large solid circles denote the 12 adatoms. The smaller solid circles represent the rest atoms. The faulted half of the unit cell is on the left. Small open circles denote dimers, while small solid circles and dots represent atoms in the unreconstructed lattice (Brommer *et al.* 1992).

artificially layered materials. With each step towards complexity, the importance of first-principles calculations increases. The point is not that there are no simple phenomenological models which give 'vivid pictures' of the basic physics of such defects, but rather that the validity of such models is difficult to establish experimentally. Thus, an appeal to first-principles calculations when eliminating irrelevant variables or unimportant features of the problem is often more effective.

Before closing our brief discussion of the successes of the LDA for semiconductors, one must pause to comment on its failure to predict accurately the energy gap separating the ground state from the excited states. As is well known, the LDA yields a gap which is roughly a factor of two smaller than observed experimentally for a wide range of semiconductors. This is a much studied problem, whose resolution certainly lies with going beyond the LDA (Sham 1991). However, somewhat curiously, this gross failure to predict the most significant optical property quantitatively appears not to translate into inaccuracies in the total energy calculations discussed above.

4.3.3 Metals

With jellium being a metal, one might think that the LDA would be a better approximation for systems with a metallic ground state. To some extent, in spite of the peculiarities of the metallic bond, this is indeed the case. From the point of view of density functional theory there are no fundamental difficulties, and calculations based on the LDA are making as dramatic an impact on current research as in the case of semiconductors. Indeed, as far as bonding is concerned, they cover roughly the same ground, dealing with bulk, point defects (Dederichs *et al.* 1989) and extended defect problems in turn. Technically, these calculations are different from those for semiconductors, mainly in the methods used for solving the Kohn–Sham equations. For the most interesting metals, like those with d- and f-bands, the pseudopotential method becomes, to say the least, inconvenient. The most commonly used methods are the linear muffin-tin orbital (LMTO; Moruzzi *et al.* 1978; Skriver 1983), the linearised augmented plane wave (LAPW; Loucks 1967) and the Korringa, Kohn, Rostoker (KKR) methods (Gonis 1992). But such technicalities are beyond the scope of the present discussion. Of course, there are qualitatively new features, namely superconductivity and itinerant magnetism, but these lie outside our present discussion.

I shall close this chapter with a timeless example which will prompt my concluding remark. It concerns the pioneering LDA calculations of the IBM group: Moruzzi, Janak and Williams (Moruzzi *et al.* 1978). Their principal result is displayed in Fig. 4.9 in self-explanatory fashion. The main point, of course, is that the LDA reproduces the experimentally observed trends of the cohesive energies, Wigner–Seitz radii (equilibrium volumes) and bulk moduli for 26 metals remarkably well. A secondary, but nevertheless striking, feature

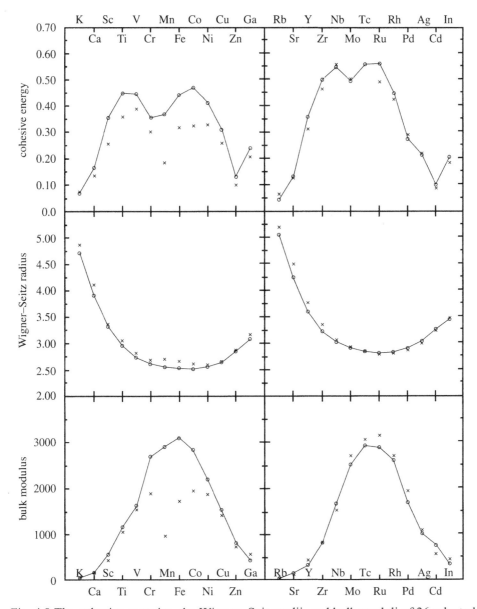

Fig. 4.9 The cohesive energies, the Wigner–Seitz radii, and bulk moduli of 26 selected metals. The points connected by a line are the results of first-principles, LDA calculations by Moruzzi *et al.* (1978) and the crosses are the experimental data.

of these results is the significant deviation between theory and experiments for metals with magnetic ground states. Clearly, this 'magnetic complication' is due to the fact that the calculations are for a paramagnetic ground state. Indeed, as Moruzzi *et al.* show elsewhere in their book, by using spin-polarised calculations not only does one obtain a satisfactory description of such magnetic properties as the magnetic moment per site, but the good general agreements in Fig. 4.9 are recovered for the magnets as well.

As is well known, the trends of Fig. 4.9 are easy to understand. Starting with neutral, open shell, atoms infinitely far apart, one builds up a solid, in a thought experiment, by bringing them together. As the atoms approach one other, their highest occupied state of energy, ε_{atom}, is broadened into a band centred on ε_{atom}. Evidently, the states of the solid in the lower half of the band have lower energy than the atomic state from which they originate. These are called bonding states. Naturally, the states with energy above ε_{atom} are referred to as antibonding. Clearly, as one moves across the first transition metal row from Sc to Ni, one progressively fills the d-band, which holds ten electrons. Up to Mn, where the d-band is half filled, the cohesive energy per atom increases for each additional electron because they occupy bonding states with energy less than ε_{atom}. Obviously, beyond Mn the cohesive energy per atom will decline as the antibonding states are filled, and this explains the trend (Friedel 1971).

Thus, a question one may wish to pose in connection with the above calculations, or many others discussed in this chapter, is: Does the above argument, or its slightly more sophisticated versions based on simple model Hamiltonians for the electrons, make first-principles calculations unnecessary? Of course, Wigner's remarks notwithstanding, the answer is: not at all. The point of ever larger and more accurate calculations is not to replace simple 'vivid pictures' or semi-phenomenological models, but rather to replace those experiments which use nature as a clumsy analogue computer. To put it another way: computations change not only the way we do theory, but also how we select experiments. Nowadays, no one should measure the critical fluctuations near the Curie point of iron in order to determine the critical exponents of the Heisenberg model. This is a job for simulations. The proper aim of such experiments should be to see where such models break down, as they surely must. Of course, by contrast, any experiment which can be interpreted as throwing light on the nature of the ground state of the periodic Anderson model is welcome because, as yet, computations are of no avail. Clearly, if a realistic model, like quantum electrodynamics, is known to describe the physics of interest, calculating its consequences is preferable to checking it with experiments. In addition to the fact that it is likely to be cheaper, in calculations complications can be eliminated or studied in isolation, and the emergence of a dominant mechanism, behind a physical phenomenon, can be reliably documented. Thus the desired end result, a 'vivid picture' which deserves the name of 'understanding', is more readily and reliably established by calculations, where practicable, than by experiments. The trouble is not 'too many boring calculations' but rather too few which are capable of doing justice to the complex behaviour of electrons in solids.

5

The electron fluid

P. COLEMAN
Rutgers University

5.1 Introduction

What is electricity? Though 100 years have passed since J. J. Thomson discovered the electron, this innocent question continues to fascinate us. Electrons inside conductors form a new type of fluid: a 'quantum fluid', where the particles move in a highly co-operative fashion. It possesses the ability to transform itself into new forms with unexpected properties, such as superconductivity. We have only just begun to understand the diversity this can give rise to.

This chapter will describe the evolution of our understanding of the electron fluid. We will talk about the wave-like nature of electrons, and see how this leads to the concept of the 'Fermi liquid'. We will discuss the idea of a 'Fermi surface' and speak of how it develops instabilities that promote quite new metallic states, such as the superconductor. I will end by giving a flavor of the great ferment of new ideas that surrounds the discovery of new classes of metallic behavior: 'non-Fermi liquid' metals and high-temperature superconductors.

Thomson's discovery of the electron in 1897 has been likened to the opening of a scientific Pandora's box (Pais 1986; March 1992). For the very first time, it became possible to envisage the internal structure of atoms and to address the very nature of electricity. Thomson inferred that the electron carried the basic unit of charge $e = -1.6 \times 10^{-19}$ C, but was thousands of times lighter than an atom. It could not be an atom or an ion, and must therefore, he concluded, be a new fundamental particle or 'atom' of electricity. This led very naturally to the idea that metals contain a fluid of such particles, and that the flow of the electron fluid gave rise to electricity. The earliest applications of this idea were made by Drude and Lorentz at the turn of the century. They supposed that one could think of the electron fluid as an almost completely free gas of electrons, whose coarse properties could be understood without worrying about the details of the ions surrounding them. The Drude 'free electron picture' of electricity is the basic elementary picture we present to students when first discussing electricity. Drude's picture accounts crudely for the resistivity of metals in terms of their scattering off vibrating atoms. Moreover, it predicts a fixed ratio between the amount of heat and the amount of charge carried by each electron: this ratio, the so-called 'Weidemann Franz' ratio, is in rough accord with that observed in most metals.

5.2 Electron waves

By the 1920s, physicists were in a deep quandary. It was increasingly clear that this picture of metals was inadequate. Electrons in metals carry far less heat than expected in a classical, charged gas. Moreover, it was clear from the work of Bohr, some ten years earlier, that electrons in atoms did not move classically, but rather moved in 'quantized orbits', where their energy and angular momentum assumed definite values. Something truly fundamental was missing.

The great conceptual breakthrough came in 1924, when a young French physicist, Louis de Broglie, boldly suggested in his celebrated Ph.D. thesis that the motion of electrons should be described as waves, with a wavelength λ related to their momentum $p = mv$ by the relation

$$\lambda = \frac{h}{p},\tag{1}$$

where $h = 6.626 \times 10^{-34}$ J s is Planck's constant. Twenty years earlier, Einstein had argued that the properties of light could only be understood if light waves were made up of particles, or 'quanta', of light, each carrying an energy $E = h\nu$, where ν is the frequency. De Broglie was young and bold enough to turn this argument around, suggesting that particulate electrons should be thought of as waves. With a beautiful and profound simplicity, he argued that if an electron wave followed an orbit of radius r, the orbit would only survive if the electron wave constructively interfered with itself. By setting the circumference of the allowed orbits equal to an integral number of wavelengths, $2\pi R = n\lambda$, de Broglie's basic relation (1) leads to the conclusion that the angular momentum $l = Rp$ of the electron is 'quantized' in units of $\hbar$, where $\hbar = h/2\pi$:

$$l = Rp = n\hbar, \quad (n = 0, 1, \ldots).\tag{2}$$

At a stroke, de Broglie had furnished a rationale for the mysterious quantization condition that Bohr had introduced a decade earlier to explain the quantized states of the hydrogen atom. Within a year, Davisson and Germer at

Bell telephone laboratories produced startling confirmation of de Broglie's hypothesis, demonstrating that electrons reflected from a crystal are Bragg diffracted, like X-rays. The angles at which the electrons came out were in accordance with the wavelength de Broglie had predicted.

Of course, every wave, be it a vibration on a guitar string or a ripple on a pond, is characterized by its amplitude $\psi(x)$. In each of these familiar cases, ψ represents the displacement of a medium from equilibrium, such as the height of the ripple above the level of the pond. But an electron is an indivisible particle, an object with a definite mass, a definite energy, so how can it possibly be thought of as a wave, and what is the meaning of its amplitude?

As the radical answer to this question unfolded in the late 1920s, it proved deeply unsettling. It meant giving up the most precious aspect of classical physics: the ability to predict the precise location and time evolution of an electron's motion. Heisenberg and Schrödinger took the next steps in 1925: Heisenberg introduced a new matrix mechanics and Schrödinger gave the physics a conceptual form by formulating the wave equation for the electron. This paved the way for Max Born to introduce the modern interpretation of the ψ field: he proposed that the probability of finding an electron at position x is determined by the squared amplitude

$$p(x) = |\psi(x)|^2. \tag{3}$$

Even from the perspective of 1997, Born's interpretation of the wavefunction stands out as one of the most radical and daring ideas in recent scientific history.

One of the inescapable consequences of an electron's wave-like nature is 'interference'. Classical particles, such as bullets, baseballs and other macroscopic projectiles, do not interfere: if a particle can arrive at a particular point via two paths '1' and '2', the probability of its arrival is merely the sum of the probabilities for each path:

$$p_{12} = p_1 + p_2, \quad \text{classical particle.} \tag{4}$$

By contrast, an electron's movement is governed by a *probability amplitude*. If an electron can travel to a point in space by two routes, the probability amplitude is the sum of two amplitudes $\psi = \psi_1 + \psi_2$. If path 2 is closed off, then the probability that the electron reaches x via path 1 is simply

$$p_1 = |\psi_1|^2. \tag{5}$$

Likewise, if path 1 is closed, the probability of arriving via path 2 alone is

$$p_2 = |\psi_2|^2. \tag{6}$$

But if both paths are open, the arrival probability

$$p_{12} = |\psi_1 + \psi_2|^2 \tag{7}$$

is not the sum of p_1 and p_2. By writing $\psi_1 = (p_1)^{1/2} e^{i\theta_1}$ and $\psi_2 = (p_2)^{1/2} e^{i\theta_2}$, we see that

$$p_{12} = p_1 + p_2 + 2(p_1 p_2)^{1/2} \cos(\theta_1 - \theta_2) \tag{8}$$

contains an additional 'interference' term. If $\psi_1 = \psi_2$, $p = p_1 = p_2$, constructive interference occurs and $p_{12} = 4p$ $(p = |\psi_1|)$ is twice that expected classically. But if $\psi_1 = -\psi_2$, destructive interference occurs, and then the effect of two open paths is to eradicate completely any chance of arrival. This is deeply counter-intuitive: we are used to planning our lives on the basis that independent probabilities are additive. When quantum effects are important, independent probabilities combine in a much more interesting way, canceling or enhancing one another via quantum interference. This is how nature works in the quantum world, and it has a profound influence on the properties of the electron fluid.

5.3 The quantum traffic jam

Our basic picture of metals, the 'Sommerfeld model', is named after Arnold Sommerfeld, one of the great teachers of the quantum revolution. His model gives a very simplified picture of the electron fluid, but it includes the most important effects of quantum mechanics (Kittel 1986). A conductor contains a vast number of electrons moving inside a crystal structure. Fortunately, we can make a lot of headway in our understanding by radically simplifying our model; by ignoring all the atoms and the details of the crystal structure, and instead just thinking of a lump of crystal as a featureless cube of side length L into which we can add electrons, one by one.

To an electron, the crystal in which it moves is like a huge 'squash court' of dimensions far greater than its own wavelength. Electron interference between the electron wave bouncing off the wall severely modifies electron motion. Like a vibration on a guitar string, the electron must form a standing wave, and its wavelength in each direction must be such that the wave fits precisely into the box. The wavefunction has the form

$$\Psi(x, y, z) = \left(\frac{2}{L}\right)^{3/2} \sin(p_x x/\hbar) \sin(p_y y/\hbar) \sin(p_z z/\hbar). \tag{9}$$

In order that the electron wave 'fits' into the box, the momentum in each direction is quantized according to

$$(p_x, p_y, p_z) = \frac{h}{2L}(l, m, n) \quad (l, m, n = 1, 2, \ldots) \tag{10}$$

(Fig. 5.1). It follows that the only allowed electron energies inside the box are

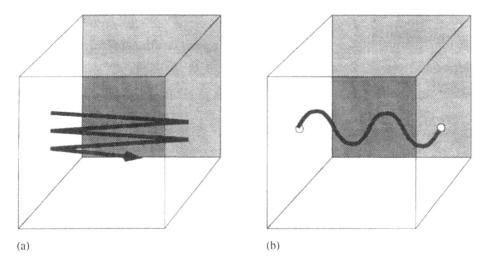

(a)   (b)

Fig. 5.1 'Electron in a squash court'. (a) Classical electron bouncing between the walls. (b) Wavelength selection for a quantum electron. Constructive interference between de Broglie electron waves can only occur when the width of the box, L, is an integral number of half wavelengths, which in part (b) corresponds to $L = 4(\lambda/2)$.

$$E = (p_x^2 + p_y^2 + p_z^2)/2m$$
$$= \frac{h^2}{8mL^2}(n^2 + k^2 + l^2), \tag{11}$$

where k, l and n are positive integers. These are the 'quantized energy levels' of an electron in the box that represents our vastly over-simplified metal.

What happens if we now add two or more electrons? Do they all go into the lowest energy state? No! This issue puzzled the first quantum physicists studying electrons in atoms. The diversity of chemistry and spectroscopy arises because electrons do not all go into the lowest energy state, but instead fill each electron state individually. Wolfgang Pauli made this observation in 1924 and postulated that *not more than one electron can ever go into the same quantum state*.* Surprisingly, Pauli initially resisted the idea that his exclusion principle would reach out beyond the atomic realm, to the much larger scales that are important for electrons in metals. In a letter written to Schrödinger in December 1926 he admitted his reluctant conversion (Pais 1986):

With a heavy heart, I have been converted to the idea that Fermi–Dirac, not Einstein–Bose is the correct statistics. I wish to write a short note on its application to paramagnetism.

---

* Of course, electrons carrying their own internal 'spin' angular momentum $(1/2)(h/2\pi)$, which can only have two values 'up' or 'down'. This means that for each momentum there are two spin orientations, which doubles the number of available states into which electrons can be added: each momentum state can actually hold two electrons.

It was actually Fermi who first worked out the consequences of the exclusion principle on a gas of identical particles, and who recognized that at a temperature where the average separation between the electrons is comparable with their typical wavelength,

$$a \sim \lambda_T \sim \left(\frac{\hbar^2}{m k_B T} \right)^{1/2}, \qquad (12)$$

the gas of particles changes radically. Nowadays, particles that obey the exclusion principle are called 'fermions', with the understanding that two identical fermions in the same quantum state destructively interfere to prevent this state from forming. The physics of the electron fluid is deeply influenced by this effect.

To continue our discussion, imagine 'pouring' electrons into our crystal to make up a metal: each electron can only go into a discrete momentum state. Visualize each electron state as a 'dot' in momentum space, positioned as shown in Fig. 5.2. The first electrons added go into low-momentum states, but, as more and more are added, they go into higher and higher momentum states. Gradually, as more and more electrons are added, we begin to build a 'sphere' of filled states in momentum space. The very last electrons to go into the crystal must go in at the highest energy at the surface of the sphere. The volume of the Fermi sphere, V_F, grows in direct proportion to the electron density,

$$n = 2 \frac{V_F}{(2\pi)^3}, \qquad (13)$$

where the factor of two derives from the spin 'up' and spin 'down' electrons. The surface of the Fermi sphere plays a central role in metal physics. In real metals, the Fermi surface is not spherical; nevertheless, the above equation proves to be the exact relation that we now call 'Luttinger's theorem'. All electrons at the Fermi surface have the same 'Fermi energy'. This energy is far higher than an electron would have if it were a classical particle. In most metals the Fermi energy is equivalent to the energy a gas molecule would have at roughly $10\,000$ K (1 K $\equiv 1$ kelvin). All electron states inside the Fermi surface are filled; all states outside the Fermi surface are empty. Fig. 5.2(b) shows the occupancy of each state, below and above the Fermi energy. This is the 'degenerate electron gas'.

With this simple picture, we can understand the basic physics of metals. With all the electron states below the Fermi energy filled, and all the states

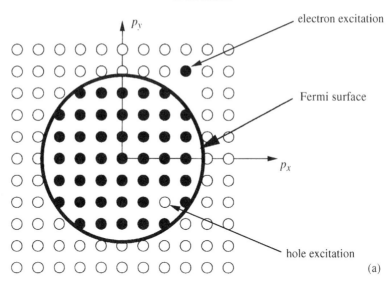

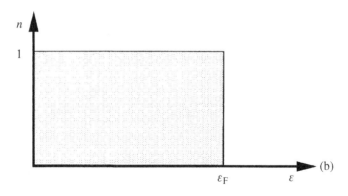

Fig. 5.2 (a) Electrons added to the metal must go into states with discrete momenta. Those states inside the Fermi surface are occupied by electrons; those outside are empty. (b) The occupancy $n(\varepsilon)$ as a function of energy, showing the sharp drop to zero at the Fermi energy.

above it empty, the metal is in its lowest energy state, or 'ground-state'. Pauli's exclusion principle means that there are strong restrictions on the way electrons can be added to or subtracted from the metal. To add electrons, we must put them outside the Fermi surface. To remove them, we must take them from inside the Fermi surface, creating a 'hole' in momentum space. By removing a negative charge, the total charge of the metal increases by one unit, so the 'hole' is a positive excitation. A helium gas balloon inside a car provides a crude example of the 'hole' concept. When the car decelerates, the balloon accelerates, because of the absence of air. A hole in a metal is a positive charge, the 'anti-particle' of the electron. Should it meet up with an electron, the two annihilate with the emission of light.

The exclusion principle lies at the core of electron physics. Its effect can be likened to a 'quantum traffic jam': like a car in a traffic jam, when one electron accelerates to a higher momentum, it can only do so if another electron moves on to vacate the new momentum state. Electrons cannot pile up at the same momentum, and this means that they must move in a correlated fashion. Furthermore, if the metal is heated, it is not possible to add thermal energy to all electrons, since most of them cannot be excited: only electrons near the Fermi energy can be excited. In a classical gas of atoms, each particle carries $(3/2)k_B T$ of thermal energy at a temperature T, and the specific heat capacity is $C_V = (3/2)R$ per mole ($R = 8.31$ J/(mol K) is the gas constant). In a metal only a tiny proportion of electrons and holes, $k_B T/\varepsilon_F$, are excited at temperature T, where ε_F is the Fermi energy. Since $\varepsilon_F \approx 10\,000$ K, it follows that only $300/10\,000 \approx 3\%$ of the electrons are excited at room temperature. This has the effect of blurring the Fermi surface, slightly smoothing out the sharp precipice in occupation at the Fermi surface over an energy range of order $k_B T$. The electronic specific heat capacity of a metal is usually very small and linear in temperature. In the Sommerfeld model

$$C_V = \frac{3}{2} R \left(\frac{\pi^2 k_B T}{3\varepsilon_F} \right), \tag{14}$$

which is far smaller than for a classical gas. The Pauli principle also affects the magnetic properties of a Fermi–Dirac gas. The application of a magnetic field aligns the magnetic moments of each atom, producing a small magnetization proportional to the applied field, $M = \chi B$. We call the constant of proportionality the magnetic susceptibility. In a conventional gas, the magnetic susceptibility grows, at low temperatures, proportionally to the inverse temperature $\chi(T) \propto \mu_B^2/k_B T$. In a Fermi liquid, fermions deep inside the Fermi surface cannot change their spin direction, because of the Pauli principle. Only those fermions within an energy $k_B T$ are polarized by the magnetic field: this gives rise to a susceptibility

$$\chi(T) \propto \frac{n\mu_B^2}{k_B T} \times \left(\frac{3k_B T}{4\varepsilon_F} \right) \sim \text{constant}, \tag{15}$$

where n is the electron density. This is the 'Pauli susceptibility' of the electron fluid.

5.4 The quantum precipice and the Fermi liquid

Sommerfeld's model leads to the remarkable conclusion that electrons fill up a metal to a maximum energy and momentum. The occupancy of states is unity below the Fermi surface, and then, like a precipice, shoots to zero at the Fermi energy. Unfortunately, this lovely picture ignores a crucial feature of electrons. Electrons repel one another due to the Coulomb repulsive forces between them. The potential energies associated with this repulsion can be as large as the Fermi energy itself. If temperature slightly blurs out the Fermi surface over an energy $k_B T$, then surely the Coulomb interaction will completely wipe out the Fermi surface? (See Fig. 5.3(a).)

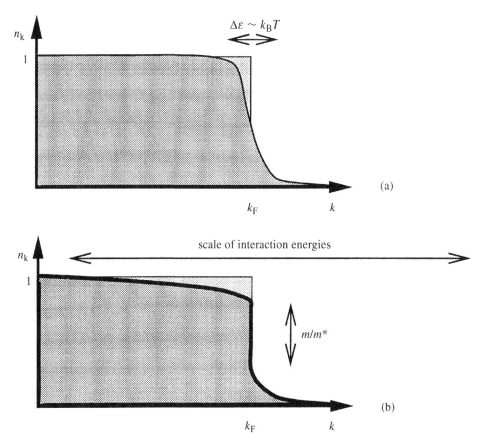

Fig. 5.3 (a) A temperature T that is smaller than the Fermi energy, slightly 'blurs' the Fermi surface; (b) even though the interaction energy is greater than the temperature, often greater than the Fermi energy, in a Fermi liquid, the exclusion principle stabilizes the jump in occupancy at the Fermi surface.

This proves to be a very tough question. The short and conservative answer is that the Pauli exclusion principle actually prevents electrons from scattering off one another at the Fermi surface. Electrons near the Fermi surface cannot find any available states to scatter to, and since they cannot scatter to new momentum states, the Fermi surface remains sharp and the quantum precipice survives. This is the basis of the standard modern picture of metals. Alas, like all simple answers, we now know it does not always work and the Fermi liquid does break down. When it does so, wonderful things can happen.

Let us now discuss this remarkable robustness of the Fermi surface. First imagine two electrons, called '1' and '2', in our quantum squash court. The Coulomb force between them causes them to scatter off one another. After they scatter, their individual momenta and kinetic energies change. However, the total momentum and energy do not change, so that

$$\text{initial momentum} = p_1 + p_2 = p_1' + p_2' = \text{final momentum},$$

$$\text{initial kinetic energy} = \varepsilon_1 + \varepsilon_2 = \varepsilon_1' + \varepsilon_2' = \text{final kinetic energy}, \quad (16)$$

where 1 and 1' refer to electron 1, before and after the collision. Now consider what happens when two electrons scatter in the presence of the entire Fermi sea of other electrons. This is a different story, for when they scatter, they must scatter to empty states that are outside the Fermi surface. If our two electrons are on the Fermi surface, the only states that are vacant have a larger energy, so they cannot scatter: to scatter, electrons must be away from the Fermi surface. The Indian physicist Homi Bhaba was one of the first to appreciate this point. He calculated the 'lifetime' τ that an electron stays, on average, in a given momentum state, and found that it grows as the inverse square of the electron's distance from the Fermi energy:

$$\tau^{-1} \propto (p - p_F)^2. \quad (17)$$

In the 1950s, the Russian physicist Lev Landau crystallized and matured these ideas into a new theoretical construct: the 'Landau Fermi liquid theory'. This model has become the standard model of metallic behavior. Advances in cryogenics and isotope separation made it possible, for the first time, to liquefy large amounts of the rare isotope of helium, helium-3. Atoms of helium-3 have a spin one-half, just like electrons, and they therefore behave as fermions. Since the late 1940s, physicists had speculated that helium-3 might form a degenerate Fermi gas. Experiments conducted by Fairbank, Ard and Walters at Duke University in the USA showed that, like a metal, at temperatures below 0.3 K, liquid helium-3 has a constant magnetic susceptibility. Surprisingly, the value of this susceptibility was some ten times larger than had been suspected. Landau realized that this meant that helium-3 must be a Fermi–Dirac fluid: he became fascinated as to why, despite very strong repulsive forces between the helium atoms, the helium-3 Fermi surface remained intact, and he set out to characterize how these interactions could modify the Fermi liquid that results.

Landau's approach to this problem was elegantly simple. He imagined starting with a hypothetical gas of helium atoms with no forces of repulsion between them: this case he could understand using Sommerfeld's model. Landau imagined gradually turning back on the interactions between the atoms until they reached their full value. What would happen if one could do this extremely slowly, or 'adiabatically'? Nothing! Landau argued that since the fermions near the Fermi surface had nowhere to scatter to, then, as interactions were smoothly turned on, the quantum states of the metal would evolve smoothly, without any of the energies of the quantum states crossing one another to become equal. With this reasoning, Landau concluded that each quantum state of the fully interacting liquid helium-3 would be in *precise one-to-one correspondence with a state of the idealized 'non-interacting' Fermi liquid* (Pines and Nozières 1966).

Each one-particle of the original non-interacting Fermi liquid has a momentum p and a spin $\sigma = \pm 1/2$. The number of fermions which occupy it, $n_{p\sigma}$, is either one or zero. The complete quantum state of the non-interacting system is labeled by these occupancies. We write

$$\Psi = |n_{p_1\sigma_1}, n_{p_2\sigma_2}, \ldots \rangle. \tag{18}$$

In the ground-state Ψ_0, all states with momentum p less than the Fermi momentum are occupied; all states above the Fermi surface are empty:

$$\text{ground-state } \Psi_0: \quad n_{p\sigma} = \begin{cases} 1 & (p < p_F) \\ 0 & (\text{otherwise}) \end{cases}. \tag{19}$$

Landau argued that if one turned on the interactions slowly, then this state would evolve smoothly into the ground-state of the interacting Fermi liquid. The energy of this state is unknown, but we can call it E_0. Suppose we now add one fermion to the original state with momentum $p_0 > p_F$. Once again we can slowly turn on the interactions. This does not change the momentum of the state, and, providing nothing catastrophic takes place, our original state will now evolve smoothly into the final quasi-particle state that we can label as follows:

$$\text{quasi-particle } \Psi_{p_0}: \quad n_p = \begin{cases} 1 & (p < p_F \text{ and } p = p_0) \\ 0 & (\text{otherwise}) \end{cases}. \tag{20}$$

This state has total momentum p_0 and an energy $E(p_0) > E_0$ larger than the ground-state. It is called a quasi-particle state because it behaves in every respect like a single particle. This concept is a triumph of Landau's theory, for it enables us to continue using the idea of an independent particle, even in the presence of strong interactions; it also provides a framework for understanding the robustness of the Fermi surface whilst accounting for the effects of interactions. The 'excitation energy' required to create a single quasi-particle is

$$\varepsilon_{p_0} = E(p_0) - E_0. \tag{21}$$

The quasi-particle concept would be of limited value if it was restricted to individual excitations. At a finite temperature, a dilute gas of these particles is excited around the Fermi surface and these particles interact. At first sight, this is a hopeless situation: how can the particle concept survive once one has a finite density of excitations?

Landau's appreciation of a very subtle point enabled him to go much further: he realized that near the Fermi surface electron–electron scattering is severely limited by the phase space constraints of the exclusion principle. In particular, the amount of momentum that two particles can exchange in a collision goes to zero for particles that are on the Fermi surface:

$$(p_1, p_2) \rightarrow (p_1 - q, p_2 + q) \quad (q = 0 \text{ on Fermi surface}). \tag{22}$$

In the low-density Fermi gas formed around the Fermi surface, particles only scatter in the *forward direction*, and in the asymptotic low-energy limit, the number of particles at a given momentum is unchanged by scattering, becoming a constant of the motion.

These physical considerations led Landau to conclude that the energy of a gas of quasi-particles could be expressed as a functional of the number of quasi-particles, n_p, at momentum p. Since the density of quasi-particles is low, it is sufficient to expand the energy in the small deviations in particle number $\delta n_{p\sigma} = n_{p\sigma} - n_{p\sigma}^{(0)}$ from equilibrium. This leads to the Landau energy functional

$$E(\{n_{p\sigma}\}) = E_0 + \sum_{p\sigma} \varepsilon_p \delta n_{p\sigma} + \frac{1}{2} \sum_{p,p',\sigma,\sigma'} f_{p\sigma,p'\sigma'} \delta n_{p\sigma} \delta n_{p'\sigma'} + \cdots, \tag{23}$$

where

$$\varepsilon_p = \frac{\delta E}{\delta n_{p\sigma}} \tag{24a}$$

and

$$f_{p\sigma,p'\sigma'} = \frac{\delta^2 E}{\delta n_{p\sigma} \delta n_{p'\sigma'}} \tag{24b}$$

are the first and second derivatives of the functional, evaluated in the ground-state ($\delta n_p = 0$). This functional proves to be very useful in understanding the properties of interacting Fermi gases, for it turns out that many physical quantities can be directly related to the first two terms in this expansion, and higher order terms are generally not needed. The first term in the expansion, ε_p, is just the single quasi-particle energy mentioned above. It is easily seen that the change in total energy when $n_p \to n_p + \delta n_p$ is

$$\frac{\delta E}{\delta n_{p\sigma}} = \varepsilon_p + \sum_{p'\sigma'} f_{p\sigma,p'\sigma'} \delta n_{p'\sigma'}. \tag{25}$$

The second term is naturally interpreted as the change in the quasi-particle energy due to the low density of quasi-particles $\delta n_{p'\sigma'}$ at momentum p'. In helium-3, which is an isotropic Fermi liquid, the interaction is invariant under both spin rotations, which means that

$$f_{p\sigma,p'\sigma'} = f^s_{p,p'} + f^a_{p\sigma,p'\sigma'} \tag{26}$$

Furthermore, since the interaction is invariant under spatial rotations on the Fermi surface, these terms only depend on the relative angle between p and p', which permits them to be expanded in terms of Legendre polynomials. By convention one writes

$$N(0) f^{s,a}_{p,p'} = \sum_{l=0}^{\infty} (2l+1) F^{s,a}_l P_l(\cos\theta_{p,p'}), \tag{27}$$

where $N(0)$ is the density of one-particle quasi-particle states at the Fermi energy. The parameters F^s_l and F^a_l are called the Landau parameters.

Using this formulation of the interacting Fermi gas, Landau was able to deduce three major consequences. The first consequence of the Landau quasi-particle interactions is the renormalization of the quasi-particle mass. As the fermion moves through the medium, it produces a 'wake' of other quasi-particle excitations which moves along with it. This changes its inertia, modifying its energy, as follows:

$$m \to m^*$$

$$\varepsilon(p) = \frac{p^2}{2m^*} - \varepsilon_F. \tag{28}$$

Landau was able to relate the renormalized mass m^* to the interactions by considering the equivalent problem where the quasi-particle is stationary but the entire Fermi sea is moving backwards at a velocity v. Using the 'Galilean' equivalence of the two situations, he argued that the momentum carried by the quasi-particle could be written as

$$p = mv(1 + F^s_1), \tag{29}$$

where the additional term is derived from the back-flow of particles displaced by the quasi-particle. Since the velocity at which the particle moves is given by $v = p/m^*$, it follows that $p = (m/m^*)(1 + F^s_1)p$ or

$$\frac{m^*}{m} = 1 + F_1^s. \tag{30}$$

'Mass renormalization' has the effect of compressing the spacing between the fermion energy levels, which increases the number of quasi-particles that are excited at a given temperature by a factor m^*/m: this enhances the linear specific heat to the renormalized value

$$C_V^* = \frac{m^*}{m} C_V, \tag{31}$$

where C_V is the Sommerfeld value for the specific heat capacity. Experimentally, the specific heat of helium-3 *is* greater than that expected from the Sommerfeld model, and from the observed enhancement we know that the quasi-particle mass in helium-3 is $m^* \approx 3m$.

Lastly, when a Fermi gas is spin polarized by a magnetic field, interactions polarize the cloud surrounding each quasi-particle. This leads to an additional renormalization of the susceptibility relative to the specific heat capacity,

$$\chi^* = \frac{m^*}{m(1 - F_0^a)} \chi. \tag{32}$$

This result enabled Landau to account for the observation that the Pauli susceptibility of helium-3 is enhanced four times more than the linear specific heat capacity.

Helium-3 is a much simpler Fermi liquid than real metals; nevertheless, many aspects of the Fermi-liquid picture generalize to the electron fluid. Amazingly, even when the Coulomb interactions are comparable with or bigger than the electron kinetic energies, the effects of the Pauli exclusion principle are seen to be strong enough to sustain the Fermi liquid in simple metals. In the early 1960s, Luttinger and Ward provided a detailed mathematical foundation for Landau's picture, and showed that, provided one can adiabatically turn on the interactions, the 'precipice', or jump in occupation, survives (Migdal 1957), but that it is reduced by the factor m/m^* (see Fig. 5.3(b)). In the simplest metals, the effect of interactions is rather weak. In a typical metal of this type, the mass enhancement m^*/m is close to unity and the Fermi energy is several electron-volts in size. A good example of such a metal is sodium. In more complex metals, such as those containing transition metals, or rare-earth metals, the interactions between the electrons become much stronger, and this causes the effective mass of the electrons to grow substantially. The most extreme example of this effect is found in the so-called 'heavy fermion metals'. In these metals, the electrons are very tightly bound close to rare-earth or actinide atoms, with effective masses up to 1000 times larger than a free electron. The inertial mass of quasi-particles in these metals is almost as big as a proton (see chapter 8)!

5.5 Beyond the Fermi liquid

I would like to end this chapter by giving a flavor of the fascinating physics of metals that lies beyond the Fermi liquid. Since the 1950s, physicists have become increasingly aware of the limitations of the Fermi-liquid description. Fermi-liquid theory works because the Pauli exclusion effect builds in strong correlations into the motions of the electrons that are not easily modified by interactions. But, even in a Fermi liquid, interactions lead to low-temperature *instabilities* of the Fermi surface. These result in phase transitions into a wide range of low-temperature 'broken symmetry' phases, with new types of properties, such as superconductivity or magnetism. The most strongly interacting metals, such as the heavy fermion metals, are particularly prone to these instabilities.

Sometimes interactions between electrons can give rise to a new type of metallic fluid even without a phase transition. This happens in one-dimensional conductors, which form a metal we call a 'Luttinger liquid'. The possible existence of these new types of metal outside one dimension is a matter of great current interest, for the metallic state that gives rise to high-temperature superconductivity in the newly discovered cuprate perovskites appears to be a new type of 'non-Fermi liquid'. Fig. 5.4 summarizes these points.

Fermi-liquid instabilities lead to new ordered electronic states. Such instabilities give rise to 'spontaneously broken symmetry', so called because an ordered state, like a crystal, a magnet or a superconductor, has a lower symmetry than the featureless fluid out of which it forms. Broken symmetry is a concept that enables us to understand how the macroscopic world we live in emerges from the microscopic quantum world. A good example of spontaneously broken symmetry is crystallization. Whereas a liquid is smooth and translationally invariant, the crystal has a definite location: it is only invariant under discrete, rather than continuous, translations. Crystallization breaks translational invariance, and it is this loss of symmetry which causes the solid

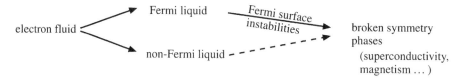

Fig. 5.4 Routes to broken symmetry ground-states.

to develop its mechanical rigidity. Just think about how our world depends on the concept of mechanical rigidity, and how different it would be if there were only fluids and liquids.

Broken symmetry enables us to contemplate other forms of macroscopic rigidity which are far from our everyday experience, for emergence of new macroscopic properties is a quite general property of spontaneous symmetry breaking. We may formulate this idea as a general principle:

symmetry breaking → new forms of rigidity → new macroscopic properties.

A solid's rigidity permits it to transmit a continuous shear-stress: stress is actually a persistent current of momentum that develops whenever the crystal is deformed. The same general principles also apply to the great diversity of broken symmetry phases of an electron fluid (Anderson 1984). In a magnet, the electron spins align, and magnetism results from their orientational rigidity. A superconductor has a more subtle kind of rigidity: here, the quantum-mechanical phase of the underlying electron waves develops a rigidity. It turns out that we can twist the electron phase by applying a magnetic field, and when we do this we induce a persistent current or 'supercurrent'. Superconductors carry persistent charge currents. Crystals carry persistent momentum currents. The two phenomena are directly analogous:

$$\text{stress} \propto \text{gradient of atomic displacements,}$$

$$\text{supercurrent} \propto \text{gradient of electron phase.} \qquad (33)$$

Superconductivity was discovered in Kamerlingh Onnes' laboratory in Leiden in 1911. The link between electron phase rigidity and superconductivity was not advanced until 30 years later, by the brothers Fritz and Heinz London (London 1950). This was already a great conceptual leap that required a strong familiarity with the principles of quantum mechanics. It took another 20 years before it was discovered *how* this rigidity develops.

One of the most dramatic Fermi-liquid instabilities is the 'Cooper instability', which leads to superconductivity. Leon Cooper showed in 1956 that, in the presence of a weak attractive interaction, two electrons on opposite sides of the Fermi surface will bind into what we now call 'Cooper pairs' (chapter 7). In conventional metals, the displacement of the lattice made by one electron attracts other electrons into its wake. Cooper pair formation correlates the electron motions to take advantage of this attraction. In 1957, John Bardeen, Leon Cooper and Robert Schrieffer (referred to as 'BCS' in the following), at the University of Illinois, showed how pair formation leads to superconductivity. BCS had the great insight to simplify the problem of the interacting electron gas down to a bare minimum required to capture the essence of the superconducting instability. They formulated the basic physics of the Cooper instability; one of the first models of strongly correlated electrons. A Cooper pair is a 'boson'. Whereas fermions in the same quantum state destructively interfere with each other to cancel one another out, pairs of fermions behave in the opposite way: bosons in the same state *constructively interfere* with each other, enhancing the probability that they go into the same state. For this reason, an arbitrarily large number of electron pairs *can* be placed into one quantum state, and the more that are present, the higher the probability that others will join!

In typical superconductors, the superconducting transition temperature, T_c, is of order 10 K, and, until 1986, superconductivity had never been observed above 25 K. Of course, many other types of instability can occur in a Fermi liquid. For example, it is possible for bound states to form between electrons and holes. These can give rise to magnetic instabilities, or to the formation of a charge-density wave. Typically, the stronger the interactions amongst the electrons, the more prone the metal to such Fermi surface instabilities. Amongst the heavy fermion metals, for instance, only a tiny minority fail to develop into some kind of magnet or superconductor at low temperatures (Cox and Maple 1995).

I want to end by touching on one of the most exciting aspects of interacting electron systems: the potential for the formation of new kinds of metal: 'non-Fermi liquids'. Since the 1960s, we have come to realize that a Fermi liquid, despite its extraordinary stability, is probably only one of many different types of metallic state that are possible. For example, the Fermi liquid is always unstable in a perfectly one-dimensional conductor. It turns out that, in one dimension, the physics of interacting metals can be solved exactly. The physics community has coined the name 'Luttinger liquid' to describe the general class of metallic behavior that develops in one dimension, naming the fluid after Joaquin Luttinger who developed one of these early solutions. Duncan Haldane, one of the key investigators in this field, has emphasized that Landau quasi-particles do not exist in one dimension: instead, charge and spin are carried individually by two separate types of quasi-particle that we call 'holons' and 'spinons' that move at different velocities. Although material science has not yet advanced to the point where such conductors can be made without a lot of disorder, experimentalists are optimistic that these new types of particle will be observed within the next few years.

One of the most active areas of current research on the electron fluid has been stimulated by the discovery due to Georg Bednorz and Alex Müller, in 1986, of a new class of ceramic material that exhibits superconductivity at temperatures as high as 135 K. This unexpected discovery has created a 'Wild-

West' quality in condensed matter physics, spawning an explosion of experiments and generating a fertile environment for new theoretical ideas (Anderson and Schrieffer 1991).

The new materials contain metallic layers of copper oxide sandwiched between insulating layers that donate the charged carriers into the conducting planes (Fig. 5.5). What is fascinating is that, although they are superb superconductors, at temperatures above their transition temperature they are very poor metals. Indeed, their metallic properties are quite different from *any* known metals. For example, their resistances are proportional to the temperature up to their melting point; they are exceptionally anisotropic conductors,

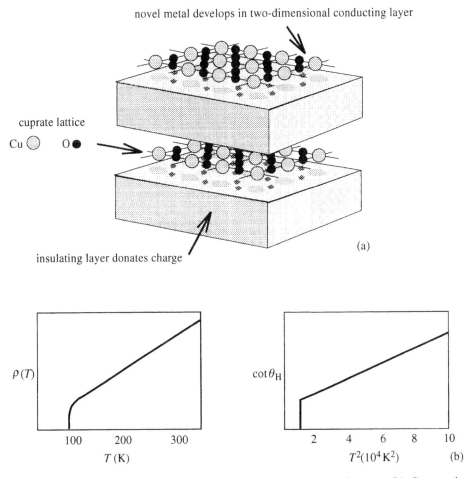

Fig. 5.5 (a) Generic layered structure of the cuprate superconductors. (b) Contrasting the linear temperature dependence of the resistivity with the *quadratic* temperature dependence of the Hall angle. These results indicate that electric and Hall currents relax at qualitatively different rates.

and a small reduction in the density of carriers in the copper oxide layers converts them to magnetically ordered insulators.

The coincidence of high-temperature superconductivity with anomalous metallic behavior has led many to conclude that the cuprate metals are not a type of Fermi liquid, but a new form of highly correlated metallic state (Anderson 1995; Coleman 1995b). One indication is provided by their resistivity. In a Fermi liquid, Bhaba scattering of electrons off one another leads to a transport scattering rate τ_{tr}^{-1} and resistivity $\rho(T)$ that grow quadratically with temperature. The linear resistivity of the cuprates implies that a much stronger scattering mechanism is at work, with a scattering rate that grows linearly with temperature, $\tau_{tr}^{-1} \propto T$ (Varma *et al.* 1989).

Yet hidden within this violently incoherent electron fluid there does seem to be a quasi-particle type which propagates coherently with a T^2 scattering rate. This is suggested by the temperature dependence of the Hall angle. The Hall angle, given by the ratio of the Hall and electrical conductivities,

$$\tan \theta_H = \frac{\sigma_{xy}}{\sigma_{xx}}$$

provides a measure of the relaxation time of a Hall current $\tau_H \propto \tan \theta_H$. In a conventional metal, Hall and transport relaxation times have the same qualitative temperature dependence, but measurements by Phuan Ong's group at Princeton University (since confirmed by many other groups) show that the Hall relaxation rate grows quadratically with temperature $\tau_{tr}^{-1} \propto \cot \theta_H \propto T^2$ (see Fig. 5.5(b)).

In a Fermi liquid, the Hall conductivity is given by an average of the squared relaxation time around the Fermi surface

$$\sigma_{xy} \propto \langle \tau_{tr}(\boldsymbol{k})^2 \rangle_{FS}. \tag{34}$$

The above discussion leads us to believe that the Hall conductivity in the cuprate metals has a fundamentally different structure: it is not an *additive* combination of relaxation times, but rather a *product* of two independent relaxation times:

$$\sigma_{xy} \propto (\tau_{tr} \tau_H). \tag{35}$$

One view (Stojkovic and Pines 1996), is that the two relaxation times reflect different parts of the Fermi surface, a 'hot' region and a 'cold' region with linear and quadratic scattering rates. In this theory, the product form in (35) appears as a cross-over between a low-temperature region, where the 'cold' quasi-particles short-circuit the Hall conductivity, and a high-temperature regime, where 'hot' quasi-particles dominate. An alternative view, due to Anderson (1991), proposes that the cuprate metal is a two-dimensional 'Luttinger liquid' and that the two relaxation rates appear because of spin-charge decoupling. In a one-dimensional Luttinger liquid, spin and charge decouple to form two new kinds of excitation: the holon, carrying just charge, and the spinon, carrying just spin. Were this physics to extend into two dimensions, then an electron and a hole moving with the same velocity would decay by holon emission into a common spinon state as follows:

$$e^- \rightleftharpoons \text{spinon} \rightleftharpoons h^+. \tag{36}$$

This kind of phenomenon occurs in the decay of K-mesons, leading to 'kaon oscillations' (Gell-Mann and Pais 1955). The analogous phenomenon (Coleman *et al.* 1996a,b) in the cuprates would mean that an electron splits into a 'short'-lived component (S) and a 'long'-lived component (L):

$$e = \frac{1}{\sqrt{2}}(e^{-t/\tau_{tr}}e_S + e^{-t/\tau_H}e_L), \tag{37}$$

where

$$e_{S,L} = \frac{1}{\sqrt{2}}(e^- \pm h^+). \tag{38}$$

The short-lived part decays rapidly, leaving behind a long-lived component that is orthogonal to spinon decay. Although a current of long-lived electrons contains electrons and holes moving in the same direction, and is thus neutral, a magnetic field beam-splits the current to produce a Hall current with a long relaxation time τ_H.

In the last three or four years, a series of experiments have made it increasingly clear that Cooper pairs in high-T_c superconductors appear to possess 'd-wave symmetry' (chapter 7). Unlike conventional superconductors, the pair wavefunction contains nodes where it changes sign. Electrons with momenta that line up with the nodes do not condense into the pair condensate, leading to many more quasi-particles in the superconducting state than would be expected in a conventional superconductor. We know almost nothing about the 'glue' that binds these pairs. One school maintains they are bound by the exchange of magnetic quanta, or 'spin-fluctuations' within a Fermi liquid. Others seek a new type of pairing mechanism which draws on the non-Fermi-liquid properties of the normal state for its development. Actual realization of either scenario would be quite remarkable.

It is appropriate to end on this note of uncertainty and promise: the controversy surrounding the new cuprate metals seems so typical of the state in which the physics of the electron has found itself during the past 100 years. Each time 'complete' understanding is around the corner, the paradigm shifts a little, and new surprises appear in experiments, the interpretation of which is often painful to those attached to the earlier view.

So, 100 years after the discovery of the electron, we are still deeply engaged in the pursuit of greater insight into electricity. We no longer think of the electron fluid as a simple gas of electrons, but rather as a highly correlated entity, which, like a chameleon, has the ability to transform radically and alter its properties to suit the environment in which it flows. What makes it so exciting is that the chameleon shows absolutely no signs of being tamed.

6

The magnetic electron

G. G. LONZARICH
University of Cambridge

6.1 Introduction

The aim of this chapter is to provide an introduction to the role of spin in systems of interacting electrons in conducting materials. To make this discussion manageable, we limit ourselves mainly to states with no long-range magnetic order and in vanishing external magnetic fields. Even under these simplifying conditions, the electron spin, in conjunction with the electron–electron interaction, can lead to subtle and remarkable co-operative phenomena. These include new kinds of 'normal' states and forms of superconductivity in which binding of Cooper pairs arises not from phonons, as in simple metals, but via the effects of spin fluctuations in the electron system itself.

6.1.1 Boundary between magnetic and non-magnetic order

The above phenomena may be found in conductors in which the Coulomb interaction leads to strong local, but not static, spin alignment. In this spin 'liquid' state, the electrons themselves may move rapidly, but the spin density they produce will appear to be undergoing slow and large-amplitude spontaneous fluctuations.

In a magnetic state of a conventional kind, e.g. in the case of ferromagnetism or anti-ferromagnetism, fluctuations of a particular wave vector have a static component of finite amplitude. We may imagine arriving at our spin liquid state by 'melting' this frozen-in magnetic order in some appropriate manner.

This crossover from the frozen to the liquid state is quite different from that of normal phase transitions. In the latter, an ordered state melts with increasing temperature into a disordered state of higher entropy. But the melting process we are considering is at a *fixed* temperature as a function of some other control parameter, such as the strength of the screened electron–electron interaction.

In practice, this may be achieved by varying the lattice spacing (via hydrostatic pressure) or the composition of the underlying atomic structure.

In the zero-temperature limit, such transitions involve no change in entropy. Hence, they cannot be viewed as transformations from order to disorder, but rather from conventional order to some more subtle 'hidden' order. In the latter, the conventional order parameter undergoes spontaneous quantum fluctuations with zero mean.

An example of this kind of quantum melting is the transition of solid helium-4 (^{4}He) to a liquid as a function of pressure at a low temperature below the λ point. In the low-temperature limit, the latter is not less ordered than the former, i.e. both have zero entropy and are in well defined quantum states. But the nature of the underlying order in the quantum liquid state is subtle and more difficult to visualise than that of the solid.

Our spin problem is partly, but not wholly, analogous to that of ^{4}He. The solid would correspond to conventional magnetic alignment and the quantum liquid to a state characterised in part by quantum fluctuations in the spin density. But the spin is only one of the variables which can play a role in determining the nature of the new ordered state. Since the electrons are itinerant, we must consider not only the spin, but also the charge and transverse current densities. The latter are clearly important in determining thermal and electrical conductivities as well as instabilities to superconducting phases. Since all of these variables are in general coupled to each other, the number of possible ordered states may be considerable, and the results could depend sensitively on fine details of the electron orbitals and of the underlying lattice.

In section 6.1.2 we introduce a model which provides us with a tractable way of thinking about electron systems near the boundary separating magnetic and quantum spin liquid phases. The model has helped to shed light on a wide range of unusual phenomena observed in nearly magnetic d-metals. It also provides a qualitative background for the study of the heavy fermion compounds and of the copper oxide conductors. Systems such as these, which find themselves near the boundary between magnetic and non-magnetic order, are remarkably widespread. A reason for their pervasiveness arises quite naturally out of the model we shall develop.

6.1.2 Fermions and interaction fields

It is conventional to begin a treatment of the properties of conducting materials with a model of electrons interacting with each other and an underlying lattice of ions via a static Coulomb interaction. For reasons which will become clear below, it is useful to begin at the still more fundamental level of quantum electrodynamics, namely the theory of matter and light (Feynman 1985; Feynman and Hibbs 1965). In this framework, the ground state is imagined to be a vacuum and the elementary excitations above this vacuum are *Fermi* and *Bose particles*, i.e. electrons and photons.*

The bosons are the well known quanta of Maxwell's electromagnetic *interaction field*. The fermions, though usually treated as particles from the start, may also be imagined to be quanta of some fundamental field. The latter satisfies the Dirac equation (chapter 3) and its quantisation proceeds in a way analogous to that of Maxwell's field, but the operator used to replace the classical field is constructed in such a way as to guarantee that the resulting quanta will obey the Pauli principle.

This leads to a picture of two fields forming initially separate linear subsystems which are then coupled. The quanta of the matter field do not interact with each other directly, but only with the quanta of the interaction field. If the bosons are in a formal sense 'filtered out', we would then 'observe' fermions that appear to interact with each other via a retarded potential. The latter is the indirect effect of the exchange between fermions of the now 'hidden' bosons. This procedure, known as integrating out the boson field, can be made mathematically precise.[†] For our purposes, however, it will be sufficient to rely on an analogy to conventional filtering.

6.2 Low-frequency and long-wavelength phenomena

With this brief outline of key ideas from quantum electrodynamics, we now return to our original problem. For simplicity, we confine our attention to phenomena involving spontaneous fluctuations of low frequency ($\omega < \omega_c$) and small wave vectors ($q < q_c$). For now, we assume only that the cut-off wave vector q_c and frequency ω_c are small compared with the Brillouin zone dimension and the Fermi energy, respectively. It will turn out that some properties predicted by the model presented below depend mainly on the behaviour of the system in the limit of low q, ω and hence are relatively insensitive to the precise choice of q_c and ω_c.[‡]

---

* Anti-particles are formally included by allowing particles to propagate not only forward but also backward in time. For an overall neutral system the description would also include a compensating ionic background.

† At the heart of the procedure, which we shall employ qualitatively below, is the simple idea that the unconditional statistical behaviour of, say, two variables may be found from a joint probability density function conditional on one or more new variables, by simply summing over all possible values of (i.e. integrating out) these additional variables.

‡ We note that field theories, including quantum electrodynamics, quite generally involve implicit cut-offs in space and time. Results which are independent of such cut-offs are obtained by the method of renormalisation discussed briefly in section 6.3.

To focus attention on processes within $q < q_c$ and $\omega < \omega_c$, we may imagine formally filtering out all Fourier components of fluctuating variables outside of this window. In technical terms, this means integrating out the phonon and photon field components with $q > q_c$ and $\omega > \omega_c$, and the electron field components with wave vectors outside of a window of width of the order of q_c about the Fermi surface (Shankar 1994).

This procedure might be expected to lead to a picture somewhat analogous to that of the previous section. But the fermions in the new model will be defined only on a shell near the Fermi surface and will have a modified energy spectrum. The bosons in the filtered description can also be radically different from the original phonons and photons. As in quantum electrodynamics, the fermions emit and absorb these bosons and hence experience a new kind of retarded interaction, which may be viewed as a screened version of the original potential.

Within this picture, the key problem is to identify the dominant part of the boson field, or the induced interaction it produces, and attempt to model its main properties in a tractable fashion. At this stage, it might be supposed that a reasonable starting point for this effective interaction is simply the conventional Thomas–Fermi potential or its generalisation in terms of standard dielectric theory. But for a system close to a magnetic phase transition, a still more important interaction field emerges out of the complex filtering process which we have carried out.

6.2.1 The exchange field

To understand the nature of this new dynamical interaction field, we return to the original concept of an exchange molecular field introduced in phenomenological descriptions of magnetism (White 1983). In this treatment, the effects of the electron–electron interaction acting through constraints imposed by the Pauli principle may be described in terms of an effective magnetic field which couples to the electron spin (and not to the current) and which varies in proportion to the uniform magnetisation. This implies that in an applied magnetic field H, the magnetisation M will satisfy $M = \chi_0(H + \lambda M)$, where λM is the exchange field, λ is a coupling constant, and χ_0 is the magnetic susceptibility when $\lambda \to 0$. Solving for M, we obtain a final susceptibility of the well known form $\chi = \chi_0/(1 - \lambda\chi_0)$, which is enhanced over χ_0 in the normal case where λ is positive. For $\lambda\chi_0 \geqslant 1$, the feedback produced by the exchange field is critical or over-critical, and magnetic polarisation can occur spontaneously in the absence of the applied field. As in the analogous problem of a self-oscillating feedback amplifier, the amplitude of the spontaneous polarisation in the over-critical state is governed by anharmonicities. In the model developed in section 6.3, the non-linear dependence of M against H may be described in the form $H = aM + bM^3$, where a is the inverse of χ given above and b is the anharmonicity parameter (Edwards and Wohlfarth 1968). In the over-critical regime, a is negative and the magnetisation in zero external field grows spontaneously to a magnitude $(-a/b)^{1/2}$ limited by the coefficient of the non-linear term in H against M.

This conventional description is normally applied to the uniform average magnetisation alone. It is now proposed to extend the concept of an *exchange field* to deal with space- and time-varying processes with q, ω components in our specified window. This interaction field is then dynamical and forms a sub-system in analogy to the electromagnetic field in the original problem.

In this way, we arrive at a low-frequency and long-wavelength model consisting of fermions in a narrow range close to the Fermi surface plus the exchange field. As in the starting theory of section 6.1, the fermions do not interact directly with each other, but only with the exchange field itself. And if we filter out the latter, we are led to a picture of fermions which appear to be coupled via a retarded and spin-dependent interaction that originates from the now 'invisible' exchange field.

It is helpful to view this unfamiliar interaction field as analogous to the more primitive counterpart in quantum electrodynamics. It fluctuates spontaneously, and on average will 'follow' the time- and space-varying magnetisation set up by the spins of the fermions qualitatively in the way that the electromagnetic field, through Maxwell's equations, may respond on average to the charge and current densities of the electrons.

The model of fermions and an interaction field related to the magnetisation incorporates the twin features raised in section 6.1. Namely, it describes both the itineracy of the electrons and the importance of slow magnetisation fluctuations in the spin liquid state. The fermions in the model describe the former, while the interaction field will allow us to incorporate the latter feature.

6.2.2 Fermi liquid model

The basic scattering processes which may occur in this low-frequency and long-wavelength world can be summarised as follows. First, the fermions may undergo single transitions (within their restricted shell near the Fermi level) via the emission or absorption of the exchange field. The strong constraint imposed by the Pauli principle on these transitions at low temperatures normally allows us to extract useful results in leading order in perturbation theory, i.e. in the Born approximation which yields the fermion scattering rate or lifetime.

Secondly, fermions may first emit and then recapture the same exchange field. This process may be interpreted as a self-interaction, which leads, as in the corresponding classical theory of the electron, to a renormalisation of the effective mass, or more generally of the apparent energy spectrum, of the fermions. This spectrum for a chosen state near the Fermi level may be determined via a final filtering process in which the interaction field and all states, except the one in question, are formally integrated out. The fermions characterised by this fully renormalised spectrum will be called *quasi-particles*.

We note that the self-interaction leads us not only to the quasi-particle mass but also to the lifetime which we discussed above in terms of single emission and absorption processes. The mass and lifetime of a quasi-particle, or more generally the real and imaginary parts of the quasi-particle energy spectrum, normally determine the heat capacity and (under suitable conditions) the resistivity of the interacting electrons.

In a paramagnetic state well away from a magnetic instability, the above model yields linear and quadratic temperature dependencies, respectively, for the heat capacity C and the resistivity ρ at a low temperature T. We note that C/T is proportional to the average of the quasi-particle mass on the Fermi surface at low temperatures. In a similar way, it is useful to view ρ/T^2 as a measure of an effective scattering cross-section. This cross-section is that for a quasi-particle colliding with another quasi-particle via the full induced interaction which we have described, and before account is taken of the T^2 factor arising from the constraint imposed by the Pauli principle on allowed scattering processes.

This behaviour of C and ρ is that expected for the Standard Model, known as the Landau *Fermi liquid theory*, which may be viewed as an extension and generalisation of the traditional single-particle description of conduction electron systems (Baym and Pethick 1991). The Fermi liquid model, in both its early and final modern forms, has played a central role in the physics of condensed matter since the development of quantum mechanics (chapter 5).

6.2.3 Breakdown of the Fermi liquid description

As the magnetic instability is approached, and spontaneous magnetic fluctuations slow down and grow in amplitude, the normal Fermi liquid description for the temperature dependence of thermal and transport properties may break down.

For simplicity, we consider a homogeneous isotropic model of an electron system near a ferromagnetic instability in three spatial dimensions. It will be argued in section 6.4 that the temperature dependence of the heat capacity and resistivity should in this case have the anomalous forms $C \propto T \ln(T^*/T)$ and $\rho \propto T^{5/3}$, in place of the usual linear and quadratic variations, respectively, of the normal Fermi liquid model. This anomalous behaviour, which is found to be consistent with recent high-pressure studies of certain cubic d-metal ferromagnets, is associated with a state known as the *marginal Fermi liquid* (Baym and Pethick 1991).

If we interpret, as in section 6.2.2, $C/T \propto \ln(T^*/T)$ and $\rho/T^2 \propto T^{-1/3}$ as the quasi-particle mass and cross-section, respectively, we are led to the conclusion that, in the above state, the quasi-particle properties are singular and hence ill defined in the $T \to 0$ limit.

This apparent divergence of the quasi-particle 'size', as defined via the mass and cross-section, may be attributed to the fact that the range of the induced interaction tends to grow without limit at the critical point, i.e. where the magnetic transition temperature $T_c \to 0$. This induced interaction arises essentially as follows. A fermion introduced into the system on the verge of a magnetic instability will nucleate local spin order which radiates outward. The magnetisation induced by this first fermion then produces an exchange field that can act on the spin of a second fermion some distance away. The range of this indirect coupling should grow as the tendency for long-range order increases, i.e. the range should vary as the magnetic correlation length which diverges at a continuous phase transition. It is plausible that the cross-section for quasi-particle–quasi-particle scattering and the quasi-particle mass would then also exhibit anomalous behaviour.[*]

Let us recall that this spin–spin interaction originates from the spin-independent Coulomb potential itself in conjunction with the spin-dependent effects of the Pauli principle. The true magnetic interaction arising from Maxwell's fields, i.e. the current–current and dipole–dipole couplings, can also remain long range at the end of the screening process. But these are very weak compared with the Coulomb potential and (at least in three dimensions) are expected to lead to a breakdown of the Fermi liquid description only at extremely low temperatures and in metals of extraordinarily high conductivity.

We may then summarise our results as follows. In the paramagnetic state, well away from the critical point, the range of the residual spin–spin interaction is short, and the normal Fermi liquid model describes the low-temperature properties with essentially constant quasi-particle mass and cross-section.

---

[*] The normal Fermi liquid model is based on the assumption that screening leads to an effective interaction of *short* range. This is normally the case for the *charge* interaction. But the complex process of screening also leads to new effective (e.g. spin–spin or current–current) interactions, which can become long range and, hence, destabilise the usual Fermi liquid description.

As the critical point is approached and the interaction grows in intensity and range, the mass and cross-section increase in magnitude and, though still finite as $T \rightarrow 0$, become strongly temperature dependent. This is the *heavy fermion* behaviour found frequently among nearly ferromagnetic or anti-ferromagnetic f-electron compounds as well as certain d-metals and low-dimensional conductors (see Coleman 1995a, b and chapter 5 in this volume). Finally, at the critical point of a continuous phase transition, the effective interaction and, hence, the quasi-particle properties become strongly temperature dependent and singular as $T \rightarrow 0$. In this limit, the concept of a fermion quasi-particle becomes ill defined. A singular mass would imply, for example, that a quasi-particle state of a given wave vector has a vanishing probability of being connected with a normal fermion state of the same wave vector. Hence, the conventional association of quasi-particles with single electrons breaks down.

Finally, we comment that the range of applicability of non-Fermi liquid models as developed in section 6.3 is not restricted to the critical point alone. Anomalous temperature dependences of C and ρ are both expected and observed over wide ranges in both temperature and the control parameter (e.g. pressure) used to cross the boundary between magnetic and non-magnetic states.

6.2.4 Anisotropic superconductivity

In section 6.2.3 we considered the breakdown of the Fermi liquid description due to the long-range nature of the induced interaction. But this is not the only way in which the effective interaction can destabilise the normal state. In particular, a second condition for the stability of the conventional Fermi liquid, namely the requirement that the effective interaction must be repulsive, can also be violated for our indirect spin–spin coupling. In an isotropic model, this coupling is proportional to the inner product of the spins of interacting fermion pairs and hence can be attractive or repulsive depending on the relative spin orientation.

In nearly ferromagnetic systems, the effective spin coupling tends to be attractive for fermions of the same spin. This favours the formation of Cooper pairs of parallel spins and, under suitable conditions at sufficiently low temperatures, can lead to an unusual kind of superconductivity.

It differs from conventional superconductivity in that the pairing interaction is due to the exchange field itself rather than phonons and, since the spins are in a triplet state, the Cooper pairs form in p-wave rather than s-wave orbital states. The internal spin and orbital degrees of freedom in the former lead to superconducting order parameters with multiple components (chapter 7) and anisotropies not found in the conventional singlet s-wave counterparts (Leggett 1975; Sigrist and Ueda 1991).

Due to their anisotropic nature, p-wave, d-wave and higher-order super-conducting states can be destabilised by scattering processes which are normally harmless to the isotropic s-wave systems. In particular, the anisotropic states are expected to be observed only when the carrier mean free path l exceeds the superconducting coherence length ξ (i.e. the characteristic spatial extent of the Cooper pair orbitals).

Anisotropic superconductivity is therefore most stable in systems in which ξ is small. Since the latter varies as the Fermi velocity and inversely as the superconducting energy gap, the heavy fermion systems (Steglich *et al.* 1979; Ott *et al.* 1983; Stewart *et al.* 1984 and others) and the high-temperature (e.g. copper oxide) superconductors (Bednorz and Müller 1986; Wu *et al.* 1987 and others) are naturally favoured. Evidence for the existence of anisotropic super-conductivity in some of these systems is now considerable. In most cases, however, the anisotropic states appear to be spin singlets with essentially d-wave character. The latter would arise in the presence of interaction fields which have dominant components at high rather than low q, e.g. as in incipient anti-ferromagnets rather than ferromagnets.

The simplest p-wave superconductivity consistent with the present model would be expected to exist in nearly ferromagnetic d-metals with weak spin–orbit interaction. In the known candidates in this class, however, ξ is expected to be one or two orders of magnitude greater than in the heavy fermions or high-temperature superconductors, respectively. This imposes conditions on carrier mean free paths beyond that achieved thus far, with the exception of the binary compound MnSi* and the elemental metals. Among the latter, perhaps the most promising are the ferromagnets Co and Ni at pressures high enough to suppress long-range magnetic order.[†] A p-wave superconducting state might be found in one of these systems provided the magnetic to non-magnetic transition is not too abrupt and the inhomogeneities in the applied pressure do not lead to a breakdown of the requirement $l \gg \xi$.

The high-pressure non-magnetic state of Ni might be similar to that of the isostructural and isoelectronic relatives Pd and Pt. Estimates based on the present model suggest, however, that the latter systems are already too far

---

* Lack of inversion symmetry of the cubic (B20) lattice of MnSi implies that the usual degeneracy of parallel spin states of opposite momentum will be lifted by the spin–orbit interaction. This in turn suggests that pairing, if it occurs, would be more exotic than envisioned here.
[†] Numerical calculations suggest that the required critical pressures should, in all cases, be within the range of current technology.

removed from the ferromagnetic instability to be viable candidates for the kind of superconductivity we are considering.

The simplest form of p-wave superconductivity remains elusive. A closely related but more complex state may exist in some of the uranium-based heavy fermion compounds. Also, a neutral analogue has been identified in the *super-fluid* phases of liquid ^{3}He in the low-millikelvin temperature range. The normal and superfluid properties of this neutral Fermi liquid are consistent with the model introduced in sections 6.3 and 6.4 for nearly ferromagnetic itinerant electron systems.

6.2.5 Additional interaction fields

Our focus on the exchange field associated with long-wavelength spin fluctuations limits our discussion to systems near ferromagnetic instabilities. The model presented in the next section can, however, be extended to describe the simplest examples of incipient anti-ferromagnetism in conducting materials.

Other fluctuating variables, such as the charge density (for both the electrons and the ions) and the transverse current density, will also have their own corresponding molecular fields. The present treatment is restricted to those phenomena in which the exchange field, as defined more fully in the next section, plays the leading role.

6.3 Phenomenological model

6.3.1 Outline

Let us summarise the strategy intended to lead us to a tractable model for the phenomena reviewed in section 6.2. The fundamental starting point, quantum electrodynamics of a dense system of electrons and ions interacting with photons, is of unimaginable complexity. Since we are concerned only with low-temperature phenomena, it is natural to reduce the apparent complexity by filtering out all but the low-frequency fluctuations in the relevant fields. Furthermore, except where otherwise stated, we also filter out all but the relatively low-wave-vector fluctuations expected to play the central role in the nearly ferromagnetic systems.

We assume that the system which we observe through our filter can still be described in terms of a model analogous in some respects to the starting theory. Thus, in the window defined by the cut-off parameters q_c and ω_c introduced in section 6.2, we have, once again, fermions and interaction fields which are analogous to the electrons and the electromagnetic field, respectively, of the starting theory.

The analogy is, however, imperfect in several ways. In particular, the new model is not explicitly relativistic and the fermions are now defined only in a narrow shell (with thickness of the order q_c) near the Fermi level. The latter implies that we cannot explicitly describe in the new model processes which involve transitions outside of this shell.

Except when prohibited by strict conservation principles, the physical properties of these fermions will also be different in general from those of the bare electrons. Thus, the fermion mass will be renormalised by the effects of all fields which have been integrated out. The quasi-particle mass, as we have noted, arises from the final level of renormalisation, when all influences, including those arising within our window, are taken into account. At least near the critical point ($T_c \to 0$), it is the interaction field which we retain explicitly in our model that is expected to produce the dominant renormalisation of the quasi-particle properties. As shown below, the singularities in the latter in fact arise from the infrared components of the field (i.e. those in the low q, ω limit) and hence are weakly dependent on our somewhat arbitrary cut-offs q_c and ω_c.

The central task, as we have said, is to identify the dominant part of the interaction field. Our key assumption is that this can be described as an exchange field $\boldsymbol{h}_m$ which couples to the magnetic moment μ (but not the velocity) of a fermion via a potential energy of the form $-\mu \cdot \boldsymbol{h}_m$. In the simplest case, $\boldsymbol{h}_m$ is proportional to the local magnetisation $\boldsymbol{m}(\boldsymbol{r}, t)$, so that we may write $\boldsymbol{h}_m = \lambda \boldsymbol{m}$, where λ is the exchange field parameter (a phenomenological constant).

In a more general treatment, $\boldsymbol{h}_m$ would be a *functional* of $\boldsymbol{m}$, involving in lowest order a convolution of an exchange field function $\lambda(\boldsymbol{r}, t)$ with $\boldsymbol{m}(\boldsymbol{r}, t)$. In next order we would retain an appropriate term cubic in $\boldsymbol{m}$. For a statistical treatment, the full interaction field must also include, in addition to the component which depends explicitly on $\boldsymbol{m}$, a randomly fluctuating part with zero mean. Near the critical point, however, the fluctuations in the overall exchange field are expected to follow closely those of $\boldsymbol{m}$ itself.

We stress that the Fourier components of both $\boldsymbol{h}_m$ and $\boldsymbol{m}$ are finite in our model only for $q < q_c$ and $\omega < \omega_c$. We also restrict ourselves to cubic lattices and to the normal state (i.e. a state in the absence of symmetry breaking transitions). Under these conditions, the filtered system will appear essentially homogeneous and isotropic.

Alternative descriptions

The model outlined above, of fermions and the exchange field coupled to each other, will be referred to as description (I). If we filter out (i.e. integrate out) the exchange field, we arrive at a description (II) in which we have fermions alone coupled to each other via a retarded spin–spin interaction. The fermion scattering produced by this interaction leads us directly to the quasi-particle spectrum, thermal and transport properties, and conditions for pair instabilities.

It is clear that there exists also a description (III) in which we imagine filtering out the fermions (i.e. integrating out the fermion field). This leads us to a picture of a fluctuating anharmonic field alone (Moriya 1985). In this way, we arrive at a framework analogous to that normally used in the study of critical phenomena which focuses on the statistical properties of fluctuations in an order parameter field. The difference is that for our problem it is not sufficient to treat the relevant field in a classical fashion. For the limit $T_c \to 0$, description (III) can thus be viewed as a *quantum* theory of *critical phenomena* (Hertz 1976; Millis 1993).

The advantages of the latter approach are that (i) it highlights one of the two main features we have noted, the importance of spin fluctuations, and (ii) it leads us most readily, with only elementary techniques, to the thermodynamic properties.* The limitation is that the itinerant character of the system, though implicitly included via the behaviour of the field, is not explicitly present in the model. For a description of the resistivity and pair instabilities, we must then generally return to descriptions (I) or (II). But even for calculations of these properties, description (III) is fruitful in providing a self-consistent model for the (final) way in which the field fluctuates or propagates. This model may then be used directly to determine the resistivity via an approach analogous to that used for electron–phonon scattering (Ziman 1960), and the pair instability via a generalisation of the BCS theory (Schrieffer 1964).

Our technical discussion will initially focus on description (III) and hence on the statistical mechanics of the field. It is important to stress that the field does not form a closed system. It is coupled strongly to the fermions which have been filtered from view. Thus, the modes of the field are necessarily heavily damped and, indeed, over-damped in the non-magnetic state we shall consider. To deal with this problem, the normal theory of statistical mechanics must therefore be generalised. The key ingredients which will lead to a simplification of our treatment of the temperature dependences of the magnetic susceptibility and heat capacity will be the *fluctuation dissipation (Nyquist's) theorem* and the related concept of the *entropy* of a *dissipative mode* of a system.

From the coupling of the field to the fermion moment $(-\mu \cdot h_m)$, we may then develop models for the resistivity and pair instability. The key results of

---

* Note that the heat capacity of the exchange field should be identical to the *shift* in the heat capacity of the quasi-particles arising from the fermion–field or the fermion–fermion couplings in descriptions (I) or (II), respectively.

this second stage of the programme will be summarised but not derived in full detail here.

As in the quantum theory of electromagnetism, the starting point in our discussion is the space and time behaviour of the interaction field. Since in the limit we shall consider the relevant interaction field is essentially always proportional to the local magnetisation $m(r, t)$, it will be sufficient to focus on the dynamics of the latter more physical variable.

6.3.2 Field equation

In principle, the dynamics of the interaction field may be obtained by the formal method of integrating out essentially all other variables in the problem. This is, of course, impractical for real systems without the use of auxiliary assumptions, the validity of which is difficult to assess.

An alternative approach is to postulate from the start, on the basis of empirical evidence and other constraints, the form of the field equation. From this intermediate starting point we may go 'forward' to determine the implications of the model for various observable properties, or else 'backward' towards a deeper understanding of the origin of the field equation and of its limitations in terms of a more microscopic model. The latter link between the microscopic and macroscopic levels clearly becomes more tenuous as we explore systems of increasing levels of complexity.

We will begin with an intermediate starting point and consider the implications for experimentally observable properties. Where possible, the known links to the more microscopic description will be indicated, but not developed explicitly. In this way, the phenomena discussed in sections 6.1 and 6.2 may be seen to arise in simple terms from a plausible common origin, namely the field equation for $m(r, t)$, which will be developed below.

Static scalar field

For simplicity, we consider in this and the following sub-section, a scalar model of the field. Our goal is to obtain a plausible dynamical equation for the space and time dependence of the *average* magnetisation $M(r, t)$ in the presence of an applied magnetic field $H(r, t)$. The latter is employed as a device for developing a model of the non-local magnetic susceptibility, which plays a central role in our discussions.

Consider first the simplest case of a uniform static field H which stabilises a uniform magnetisation M given by some relation $H = H(M)$. This function must be odd in M for our isotropic system. Assuming that a Taylor expansion is possible, we then have, in the absence of the feedback effect of the exchange field, $H = a_0 M + b_0 M^3$, where a_0 is the inverse of the Pauli susceptibility and b_0 is the anharmonicity parameter. We now include the effect of the exchange field by adding to H the effective field λM. This yields $H(M) = aM + bM^3$, where $a = a_0 - \lambda$ is the inverse of the enhanced susceptibility χ if $a > 0$, and b is b_0 if the exchange field is strictly linear in M. We assume $b > 0$, so that when $a < 0$ the spontaneous magnetisation is finite and vanishes continuously as $a \to 0$. For systems in which $b < 0$, higher-order terms must be retained to describe an ordered state. The extension of our treatment to the latter case, which can exhibit transitions of the first kind, will be briefly discussed in the final sub-sections.

With this background for orientation, we now turn to the case of a spatially varying but still static field $H(r)$ which stabilises $M(r)$. We drop explicit reference to r and write $H = H[M]$, where the square brackets now imply a mapping of one function of r to another. Now H depends not only on M but also in general on the spatial gradient of M. For an analytic function in an isotropic and homogeneous system, only even gradients enter, so that in a minimal extension of the uniform model (Landau and Lifshitz 1980)

$$H = H[M] = aM + bM^3 - c\nabla^2 M, \tag{1}$$

where c is a new parameter which measures the resistance of the system against spatial modulations in M. For stability of a ferromagnetic (rather than an anti-ferromagnetic) phase, we require $c > 0$. Note that c reduces to c_0, the stiffness for the non-interacting fermions, only if the exchange field itself is independent of $\nabla^2 M$.

It will be useful in the following to introduce an effective field

$$H_{\text{eff}} = H - H[M], \tag{2}$$

where H is the applied field and $H[M]$ is the right-hand side of eq. (1). H_{eff} vanishes in the equilibrium state when eq. (1) is satisfied.

Scalar dynamical field

We next turn to the time evolution of the magnetisation. In the paramagnetic state, we expect M to relax towards equilibrium given by $H_{\text{eff}} = 0$. Thus, we assume that for small H_{eff} and slow variations

$$\dot{M} = \gamma * H_{\text{eff}}, \tag{3}$$

where $\dot{}$ and $*$ denote time derivative and spatial convolution, respectively, and $\gamma(r)$ is a relaxation function to be described below. Note eqs. (1)–(3) imply that, for $H = 0$ and in leading order in M, a Fourier component of M relaxes to zero exponentially in time.

To simplify our treatment we decompose the field in a Fourier series as

$$\chi(\boldsymbol{\rho}) = \frac{1}{\Omega^{1/2}} \sum_k \chi_k \, e^{i k \cdot \rho}, \tag{4}$$

where $\boldsymbol{\rho} = (\boldsymbol{r}, t)$, $k = (\boldsymbol{q}, \omega)$, $k \cdot \boldsymbol{\rho} = \boldsymbol{q} \cdot \boldsymbol{r} - \omega t$, and Ω is the product of the volume and a long time interval in which the expansion is performed. In the limit of large Ω, the sum goes over to an integral of the usual form $\Omega \int d^{d+1}k/(2\pi)^{d+1}$, where d is the spatial dimension. In terms of Fourier components, the content of eqs. (1)–(3) in leading order in M can be expressed as

$$H_k = \chi_k^{-1} M_k, \tag{5a}$$

where

$$\chi_k^{-1} = \chi_q^{-1} \left(1 - i \frac{\omega}{\Gamma_q} \right), \tag{5b}$$

$$\chi_q^{-1} = \chi^{-1} + c q^2 \tag{5c}$$

and

$$\Gamma_q = \gamma_q \chi_k^{-1}. \tag{5d}$$

Note that the anharmonic term in eq. (1), which does not simplify in terms of Fourier components, is not explicitly given above. The quantity χ_k is the *generalised linear susceptibility*, defined as the ratio of M_k to H_k in analogy to the uniform case. It will be of central importance throughout. When $\omega \to 0$, χ_k reduces to χ_q, the static wave-vector-dependent susceptibility, which goes to χ as $\boldsymbol{q} \to 0$. In the more complete analysis given below, the form of eqs. (5a)–(5d) will survive, but χ^{-1} will no longer simply be the difference between the inverse Pauli susceptibility and the bare exchange constant.

The spectrum Γ_q, which plays the key role in our discussions, may be interpreted as the rate at which a component $M_q(t)$ will relax to equilibrium for $H = 0$. It depends on the inverse susceptibility and on γ_q, the Fourier component of $\gamma(\boldsymbol{r})$ defined as in eq. (4) but with the factor $1/\Omega^{1/2}$ omitted. If the exchange field is independent of $\dot{M}$, then γ_q depends only on the properties of the non-interacting fermions and has low $\boldsymbol{q}$ form

$$\gamma_q = \gamma q^n, \tag{5e}$$

where γ is a constant and n, as we argue below, is normally unity.

To understand the latter result, let us consider the way in which a disturbance of wavelength $2\pi/q$ would tend to relax in the absence of an external field. In a homogeneous non-interacting Fermi system, relaxation proceeds via the *ballistic* motion of fermions moving typically at the Fermi velocity. Hence, decay takes place in a time proportional to wavelength, which implies $\Gamma_q \propto q$. Thus, from eq. (5d) we find $n = 1$.

This linear relaxation at low $\boldsymbol{q}$, known as *Landau damping* (Baym and

Pethick 1991), survives in a Fermi liquid even in the presence of the interaction field which is assumed to alter χ_q, but not γ_q itself. In this case Γ_q is quite generally proportional to $q\chi_q^{-1}$. The last factor correctly reflects the fact that as we approach the critical point the fluctuations at the critical wave vector tend to freeze. This slowing down of fluctuations is a general property of phase transitions of the second kind.

The assumption of ballistic motion of the fermions, and, in the Landau theory, of the quasi-particles themselves at the Fermi level, breaks down at low q in the presence of frozen-in disorder or of thermal excitations at finite T. For wavelengths $2\pi/q$ much greater than the carrier mean free path, we expect diffusive rather than ballistic motion and hence a relaxation time proportional to the wavelength squared. In this case, $\Gamma_q \propto q^2$ at small q, which implies $n = 2$.

The above argument assumes that scattering does not flip the spin locally but merely slows down the transport of spin by the fermions. In the presence of mechanisms which lead to local decay of the magnetisation, we may then have a non-vanishing Γ_q even as $q \to 0$. In this case, n is then essentially zero.

The experimental evidence suggests that for d-metals with weak spin–orbit interaction, a model with a single relaxation process at low q with $n = 1$ proves adequate (Ishikawa *et al.* 1985; Bernhoeft *et al.* 1989; Lonzarich *et al.* 1989). However, for the f-metals with strong spin–orbit forces, two or more processes may be required for a minimal description of low-temperature properties (Bernhoeft and Lonzarich 1995).

Vector dynamic field

Thus far the magnetisation has been treated as a scalar. This is adequate for an isotropic system in the paramagnetic state in the linear regime, i.e. for small M as $H \to 0$. But to investigate effects of the random field in the statistical treatment of the next section, we need to go beyond the harmonic model for the dynamics of a vector $\boldsymbol{M}$. This leads us to consider two types of non-linear terms. The first, already present in the scalar model, is the cubic term in eq. (1) which we now write in a slightly more general vector form $\boldsymbol{M}(\boldsymbol{M} \cdot \boldsymbol{M})$ in the vector extension $\boldsymbol{H}[\boldsymbol{M}]$ of eq. (1).

The second form of anharmonicity (present in the vector model alone) arises from the fact that $\boldsymbol{M}$ will precess about some effective field $\boldsymbol{H}'_{\mathrm{eff}}$, a part of which depends on $\boldsymbol{M}$ itself. $\dot{\boldsymbol{M}}$ will vanish as required in equilibrium if we take $\boldsymbol{H}'_{\mathrm{eff}} = \boldsymbol{H}_{\mathrm{eff}}$ given by the vector extension of eq. (2). We are thus led to postulate a vector generalisation of eq. (3)

$$M = \zeta(M \times H_{\text{eff}}) + \gamma * H_{\text{eff}}, \tag{6}$$

where $H_{\text{eff}} = H - H[M]$ and ζ is an appropriate gyroscopic ratio. Since $M \times M = 0$, we may replace H_{eff} in the cross-product in eq. (6) by $H + c\nabla^2 M$, an effective field which also appears in the *Landau–Lifshitz model* for a ferromagnet with a local magnetisation of fixed amplitude and in the absence of damping (Herring 1966; Lifshitz and Pitaevskii 1980).

For a paramagnet, it follows from the fact that M vanishes as $H \to 0$ that the precession term in eq. (6) is strictly non-linear in H. Thus, to leading order each component of M is correctly described, as expected, by our original scalar model (3). Non-linear precession, however, will be important in the presence of some additional field (e.g. a uniform component on top of a modulation) and, as we have said, in the analysis of the effect of the random field (see section 6.3.4).

We arrive then at a dynamical non-local model which describes precession, decay and two types of anharmonicity. The parameters of the model a, b, c, γ and ζ are phenomenological constants. If the exchange field is strictly proportional to m, then only a differs from that of the non-interacting fermions. The parameters apart from a may then be read off by comparing the prediction of eq. (6) for the generalised susceptibility with the known *Lindhard function*, i.e. the generalised susceptibility for non-interacting fermions[*] (Moriya 1985). This comparison also provides us with a first-principle justification of eq. (6), which we have introduced on physical grounds.

The microscopic evaluation of the remaining parameter a $(= a_0 - \lambda)$ is made difficult by the fact that, near the critical point, it is the difference of two almost equal numbers. Thus, in contrast to the other quantities, a is very sensitive to small corrections. Of particular interest are corrections due to thermal excitations, which we attempt to describe in terms of temperature-independent parameters and a universal thermal function. The dominant effect comes not from the Fermi function defining the parameters of the non-interacting fermions (e.g. the Pauli susceptibility), but rather the Bose function which describes the amplitude of spontaneous fluctuations in the local magnetisation. The way in which these fluctuations affect the thermal properties in general will be the subject of sections 6.3.3 and 6.3.4.

---

[*] The anharmonicity parameters appear in the Lindhard function for a given q, ω, but in the presence of a uniform applied field. For completeness, we point out that the limiting form of the dynamical equation at low q, ω depends on how the origin in q, ω is approached, i.e. on whether $\omega/q \to 0$ or $q/\omega \to 0$; here we shall be concerned with the limit $\omega/q \to 0$.

6.3.3 Statistical description of open systems

Quantisation

The field equation (6) may be taken as a classical starting point for a quantum description. The quantisation of the field proceeds along one of two routes. The better known is the canonical, or operator, method, in which quantisation emerges from the discreteness of the eigensolutions. The second is the path integral technique, which involves an extension of stochastic theory to deal essentially with complex weights for configurations. In this approach, quantisation arises out of interference between alternative configurations.

The results given in the remainder of this chapter may be developed via the latter method by introducing an action whose stationary solution yields the field equation (6). Here we attempt to arrive at the same conclusions via more elementary arguments based on analogies to known results for the statistical mechanics of a quantum oscillator with dissipation.

Fluctuation dissipation theorem

We begin with a brief review of the statistical properties of an oscillator with a linear response function defined as $\alpha_\omega = x_\omega / f_\omega$, where x is the amplitude and f is the external force. From the standard equation of motion of a simple damped oscillator, we find

$$\alpha_\omega^{-1} = \alpha^{-1}\left(1 - \frac{i\omega}{\Gamma} - \frac{\omega^2}{\Omega^2}\right), \tag{7}$$

where α is the static response function, Ω is the free resonance frequency and Γ describes damping. In the zero damping limit, $\Gamma/\Omega \to \infty$, the imaginary part $\mathrm{Im}\,\alpha_\omega$ reduces to two delta functions centred on $\pm\Omega$ with weights $\pm\pi\alpha\Omega/2$. From the well known mean energy of an oscillator, we may then express the variance of x as an integral over frequency in the form

$$\overline{x^2} = \frac{2}{\pi}\int_0^\infty d\omega(\tfrac{1}{2} + n_\omega)\,\mathrm{Im}\,\alpha_\omega, \tag{8}$$

where $n_\omega = (e^{\omega/T} - 1)^{-1}$ is the Bose function. We adopt a system of units in which $\hbar = k_B = 1$ throughout. This simply means that energy, frequency and temperature are measured in the same way (e.g. in milli-electron-volts). It is clear from $\mathrm{Im}\,\alpha_\omega$ with zero damping that eq. (8) does indeed yield the expected results $\alpha^{-1}\overline{x^2} = ((1/2) + n_\Omega)\Omega$ for the thermal energy of a quantum oscillator at temperature T. The first term is the zero point contribution. The second is the thermal part, which is a consequence of quantisation followed by averaging over an ensemble at finite T.

The last result for the variance of x (in terms of n_Ω) holds only for zero damping. But eq. (8) is completely general. It holds not only for eq. (7) but also for an arbitrary linear response function and in the presence of anharmonicities. This universal result is known as the *fluctuation dissipation (Nyquist's) theorem* (Landau and Lifshitz 1980), which is well known in connection with the related problem of Johnson noise. It provides us with a direct way of quantising the field at a level sufficient for our discussion of the variance of m, and hence of the temperature dependence of the susceptibility.

Thermal population for over-damped modes

We are interested in a susceptibility of the form of eq. (5b) which corresponds to eq. (7) in the extreme over-damping limit $\Omega/\Gamma \to \infty$. In this case, it proves useful to introduce a dimensionless thermal population defined as $n(\Gamma/T) = \overline{x_T^2}/\Gamma\alpha$, where the subscript T denotes the thermal part (i.e. the second term on the right-hand side of eq. (8)). From eq. (7) in this limit together with eq. (8) we find[*]

$$n(\Gamma/T) = \frac{2}{\pi} \int_0^\infty d\omega\, n_\omega \frac{\omega}{\omega^2 + \Gamma^2} \tag{9a}$$

$$\cong \frac{T}{\Gamma(1 + (3\Gamma/\pi T))}. \tag{9b}$$

Note that the low- and high-T limits of $n\,(\Gamma/T)$ are, respectively, $(\pi/3)T^2/\Gamma^2$ and T/Γ. It is interesting to compare this with the corresponding limits of the thermal population for an undamped oscillator of frequency Ω, given by the Bose function as $\exp(-\Omega/T)$ and T/Ω, respectively. We see that the high-T limits are of the same qualitative form. But at low T, in place of an exponential freezing out of the population, we find, for the over-damped limit, a more gradual quadratic temperature dependence. The form of $n(\Gamma/T)$ will play a crucial role in our discussion of the Fermi liquid and non-Fermi liquid exponents in section 6.4.

Entropy of dissipative modes

For a discussion of the heat capacity, we shall also need to generalise the concept of the entropy of an oscillator to include the effects of damping. We begin with a closed system consisting of an oscillator coupled to a reservoir. For sufficiently strong coupling, the standard treatment of the oscillator as a well defined normal mode breaks down. Nevertheless, the shift of the entropy

---

[*] Equation (9b) is an approximation for the exact integral on the right-hand side of eq. (9a) given by $n(x) = [\ln(x/2\pi) - (\pi/x) - \psi(x/2\pi)]/\pi$, where $x = \Gamma/T$ and ψ is the digamma function. The approximate and exact forms for $xn(x)$ differ by at most a small percentage for all x, and are identical to leading order in x and $1/x$. The approximate form (9b) will prove adequate for all of our purposes.

of the closed system as a whole due to this coupling may be calculated in principle using conventional statistical mechanics. We now define the entropy of the *damped oscillator* to include this shift as if it were entirely the property of the oscillator itself, and attempt to express it in terms of the final oscillator response function α_ω (Lonzarich 1986; Pippard 1987; Edwards and Lonzarich 1992). The required generalisation for the entropy is found to be

$$S = \int_0^\infty d\omega S_\omega \left(\mathrm{Im}\, \frac{\partial \ln \alpha_\omega}{\pi \partial \omega} \right), \tag{10}$$

where S_ω is the usual entropy of an undamped oscillator of frequency ω. The latter may be obtained from n_ω via the relation $T \partial S_\omega / \partial T = \omega \partial n_\omega / \partial T$.

To gain some understanding of this result, let us consider a model in which the average potential and kinetic energies are $\overline{x^2}/2\alpha$ and $\overline{\dot{x}^2}/2\alpha\Omega^2$, respectively. From Nyquist's theorem applied to both x and $\dot{x}$, we arrive at a thermal energy given by $(1/2\alpha)$ times the right-hand side of eq. (8), but with a factor $(1 + (\omega^2/\Omega^2))$ in the integral. It is now straightforward to show that this thermal energy yields an entropy of the form (10) for a (temperature-independent) α_ω given by eq. (7). The result for S is more general than this analysis would suggest. In particular, eq. (10) holds for an arbitrary α_ω. But, in contrast to Nyquist's theorem, it may be limited to low temperatures except, as discussed in section 6.3.4, when the mean field approximation is appropriate.

The last factor in eq. (10) reduces to $\delta(\omega - \Omega)$ for no damping ($\Gamma/\Omega \to \infty$) and to $(\Gamma/\pi)/(\omega^2 + \Gamma^2)$ in the opposite limit of over-damping, when only the lowest imaginary pole of α_ω in eq. (7) is retained. The Lorentzian form is quite natural and reflects the spectral composition of an amplitude which does not oscillate, but decays exponentially with a rate Γ. Since only the low-frequency pole is retained, the integrated weight is here $1/2$ and not unity. Also, we note that the contribution of the ω^2/Ω^2 term discussed above vanishes in this limit, so that the energy is now essentially purely potential.

Heat capacity of over-damped modes

For the problem of interest here, in which the relevant oscillator modes are strongly damped, the heat capacity $C = T \partial S / \partial T$ takes a particularly simple form:

$$C = \frac{\Gamma}{2} \frac{\partial}{\partial T} n \left(\frac{\Gamma}{T} \right). \tag{11}$$

This follows directly from eq. (7) for α_ω in the limit $\Omega/\Gamma \to \infty$, eq. (10) for S and eq. (9a) for the thermal population. We note that eq. (11) holds under these conditions even if $\partial \Gamma / \partial T$ is non-zero.

It is interesting to compare eq. (11) with the corresponding heat capacity for an undamped oscillator of frequency Ω, namely $C = \Omega \partial n_\Omega / \partial T$. The latter gives $C = 1$ and $C = (\Omega^2 / T^2) \exp(-\Omega/T)$ in the high- and low-temperature limits, respectively; while eqs. (11) and (9) yield, for the same limits, $1/2$ and $(\pi/3)T/\Gamma$, respectively. The difference at high T arises from the fact that an over-damped oscillator, as we have said, has only potential energy. At low temperatures, instead of an exponential collapse of C we find, in the over-damped limit, a *linear* heat capacity. The latter will be related in section 6.4 to the renormalisation of the quasi-particle mass in the Fermi liquid state.

6.3.4 *Mean field approximation for the mode–mode coupling*

With the above background for the statistical mechanics of a dissipative mode, we are now ready to deal with the exchange field, which, as we have stressed, corresponds to an open system strongly coupled to a reservoir, i.e. the fermions are now filtered from view. We think of each Fourier component of this field as an oscillator variable, or mode. Each mode, labelled by a cartesian component (x, y or z) and the wave vector q, is coupled not only to the reservoir, but also to other modes of the field. The latter, or the mode–mode coupling, arises in the model through the anharmonic terms in the field equation (6).

Our basic strategy is to deal with the coupling to the bath via the formalism developed in the preceding section for dissipative modes, and the coupling between modes themselves by the mean field approximation. The latter assumes that (i) the modes are statistically independent and that (ii) the characteristic energy (in our case Γ_q, eq. (5d)) for any one mode depends on the thermal population $n_q = n(\Gamma_q/T)$ of all other modes q in a self-consistent fashion. Since each mode includes corrections due to interaction with the others, the thermal energy of the field is not additive. But because of assumed statistical independence, the entropy itself *is* additive and is thus more convenient to use than the thermal energy. This is the chief reason for our focus on the entropy in the preceding section.

Thus, we arrive at a picture of a collection of independent dissipative modes described by a relaxation spectrum which depends, in the mean field approximation for the effect of anharmonicities, on the thermal population. Since the latter also depends on the spectrum, we are led to a self-consistent equation for the temperature variation of Γ_q, and hence of the thermal properties, in terms of the $T = 0$ parameters alone.

We begin with a discussion of the heat capacity, assuming Γ_q is known at each T, and only later consider how the temperature dependence of Γ_q may be determined from the anharmonicities built into the field equation. Finally, we shall consider the validity of the mean field approximation itself.

Heat capacity

If the modes of the field can be treated as statistically independent in the above sense, then the entropy of the field reduces to the sum of contributions of the form (10). In the limit of strong dissipation, we then obtain from eq. (11) a heat capacity

$$\Delta C = \tfrac{3}{2} \sum_q \Gamma_q \frac{\partial n_q}{\partial T}, \tag{12}$$

where Γ_q and n_q are the spectrum and population given, respectively, by eqs. (5) and (9). The factor of 3 in eq. (12) represents the degeneracy for each q, i.e. the three cartesian components for the field, assumed to be equivalent throughout.[*]

Since $n(\Gamma_q/T)$ varies as T^2/Γ_q^2 in the low-T limit, eq. (12) seems to lead to a *shift in the linear* heat capacity by an amount proportional to the q-space average of $1/\Gamma_q$. In this case, interactions do not alter the qualitative temperature dependence of the system in lowest order in T, but lead only to a renormalisation of the linear coefficient.

This analysis is clearly incomplete. We have ignored the temperature dependence of Γ_q and the fact that, for any finite T, there may exist a range in q at small q for which the assumption $T \ll \Gamma_q$ breaks down. As we shall see in section 6.4, the latter in particular can lead to anomalous corrections to the linear heat capacity which have no counterpart in the non-interacting electron system, and in extreme cases can also give rise to a *leading* temperature dependence which is qualitatively different from that of the conventional Fermi liquid. Since these are associated with the low-q behaviour of the sum in eq. (12), they may be called *infrared anomalies*.

---

[*] For over-damped modes, eq. (12) follows identically from eq. (10) even when Γ_q is temperature dependent. The total heat capacity of the system includes a contribution C_0 due to the non-interacting fermions of the model. Thus, ΔC, which we are viewing in our present description (III) as the heat capacity of the field, is equivalent to the correction due to interactions, which we might have obtained via a different analysis starting with descriptions (I) or (II) defined in section 6.3.1. In arriving at eq. (12), we have assumed, as is appropriate near the critical point, that the exchange field and the magnetisation follow each other and hence are governed by the same dynamical equations. More generally, we must subtract from eq. (12) a contribution of essentially the same form (and in the same range in $q < q_c$), but now for non-interacting fermions, so that in the absence of interactions ΔC correctly vanishes.

Non-linear effects of the exchange field

We now turn to the more difficult task of evaluating the temperature dependence of the spectrum Γ_q or, in particular, of the most sensitive variable in Γ_q, namely the susceptibility χ.

In our treatment in section 6.3.2, we have taken into account that part of the exchange field which follows the average M in the presence of H. There also exists a spontaneous stochastic component with zero mean which drives random fluctuations in the local magnetisation. In the absence of anharmonicities, this additional field would have no consequences for the ensemble average of the field equation, and the results of section 6.3.2 would remain unchanged. But this is not the case in the presence of terms non-linear in the magnetisation, which appear in eq. (6) both via precession and damping.

To see what can happen, let us look first at the static field equation (1). In the presence of a random exchange field h, we may make the substitution $H \rightarrow H + h$ and $M \rightarrow M + m$, where m is the additional component induced by h. Taking an ensemble average with $\bar{h} = \bar{m} = 0$, we see that only the cubic term $(M + m)^3$ gives a non-vanishing contribution through the averages $3bM\overline{m^2}$ and $b\overline{m^3}$. Note that $\overline{m^2}$ is finite even as $M \rightarrow 0$, while $\overline{m^3}$ must vanish in this limit.

In general, these moments lead to corrections to all terms in $H[M]$. In the mean field approximation, only the second moment as $M \rightarrow 0$ is retained. This then leads to a shift of the sensitive linear term in $H[M]$ from a to $a + 3b\overline{m^2}$.*

The same statistical analysis may now be carried out for the vector model, which includes a second independent type of anharmonicity, i.e. non-linear precession. The averaging over the random field is facilitated by first inverting eq. (6) to express H as a function of M, $\nabla^2 M$ and $\dot{M}$. This is achieved by taking the cross-product of M with each side of eq. (6), rearranging and dropping terms beyond second order in M in the cross-product. Making the substitutions $H \rightarrow H + h$ and $M \rightarrow M + m$, and averaging with $\bar{h} = \bar{m} = 0$, we then find that the coefficients of the linear and cubic terms in $H[M]$ are now replaced by $a + \Delta a$ and $b + \Delta b$, where

$$\Delta a = \frac{5}{3}\left(b + \frac{\Delta b}{2}\right)\overline{|m|^2} - \zeta\gamma^{-2} * \nabla_M\overline{m \times \dot{m}} \tag{13a}$$

and

$$\Delta b = \tfrac{14}{3}g\overline{|m|^2}. \tag{13b}$$

---

* Note that $\overline{m^2}$ is the variance due to the effects of the random exchange field. In the absence of interactions, $\overline{m^2}$ must therefore be zero. The *full* variance of m does not, of course, strictly vanish in the non-interacting system. Near the critical point, where magnetic fluctuations are strongly enhanced, the latter contribution becomes negligibly small in our low-q, ω model. A discussion of the validity of the mean field approximation, i.e. of our neglect of the M dependences of $\overline{m^2}$ and $\overline{m^3}$, will be postponed until the end of section 6.3.4.

As in the scalar model, the M dependences of the moments of m are not retained. The shift in b arises qualitatively from the same origin as the earlier shift in a, but now from a fifth-order term in $H[M]$ of the form $gM(M \cdot M)^2$. The latter must be retained when b is negative or, as in the case of a itself, nearly critical. The only qualitatively new term is the last one in eq. (13a) due to non-linear precession. Here $\gamma^{-1}*$ is the inverse of the operation $\gamma*$ and ∇_M denotes a gradient in M in the limit $M \to 0$. We retain the first gradient because it leads to corrections of the same order in T as that of the mean field approximation.

The precession correction (the last term in Δa) is strictly negative and grows in magnitude with increasing T. This is in contrast with the first term, whose sign depends on b, which can be positive or negative depending, for example, on the form of the density of states of non-interacting fermions at the Fermi level. It is interesting that thermal fluctuations in the presence of non-linear precession alone tend to promote rather than to destroy magnetic order, i.e. χ in the paramagnetic state would in this case grow rather than fall with increasing T. Eventually, the effects of more conventional anharmonicities in Δa (e.g. the $\overline{|m|^2}$ term in eq. (13a)) will dominate and lead to a downturn in χ at high T.

The ubiquitous peaks in χ against T observed in nearly magnetic metals normally arise essentially from a negative b with positive higher-order coefficients such as g in eq. (13b). But, as the above argument suggests, a peak can also arise if b is positive but sufficiently small for the precession correction in Δa to dominate at low T. We note, however, that for fermions with a simple parabolic spectrum, we expect the first term in eq. (13a) to be positive and entirely dominant over the precession correction (Ramakrishnan 1974; McMullan 1989).

Susceptibility

For simplicity, we restrict ourselves to the $\overline{|m|^2}$ term in eq. (13a), which defines the mean field approximation for the susceptibility. Our goal is to express this in terms of the population n_q and the $T = 0$ parameters alone. We note that $\overline{|m|^2}$ is the q-space sum per unit volume of $\overline{|m_q|^2}$ and that the latter is given via Nyquist's theorem (8) in terms of the generalised susceptibility in eq. (5). The thermal component of $\overline{|m|^2}$ is now seen to be explicitly a function of n_q, eq. (9), while the zero point part is implicitly a function of this quantity through the correction (13) to χ^{-1} which defines Γ_q. Our procedure is to expand the zero point part to first order in χ^{-1} and solve for χ^{-1} self-consistently.

In leading order in n_q the shift of χ^{-1} from its $T = 0$ value a is then found to be

$$\Delta a = 5b\widehat{\sum_q}\gamma_q n_q = 5b\widehat{\sum_q}\gamma_q n\left(\frac{\gamma_q}{T}(a + \Delta a + cq^2)\right), \qquad (14)$$

where $\widehat{}$ denotes a sum per unit volume. The $T = 0$ parameters a and b are now as renormalised by the effects of zero point fluctuations (see also, e.g., Solontsov and Wagner 1995). From the explicit form of n_q (eq. (9)) and Γ_q (eqs. (5d, e)), we see that eq. (14) is an *implicit* equation for Δa, or $\Delta\chi^{-1}$, in terms of T and the basic parameters of the model a, b, c and γ.[*]

We note that, since n_q is proportional to T^2/Γ_q^2 at low T, $\Delta\chi^{-1}$ might be expected to vary initially as T^2. This corresponds to the normal Fermi liquid regime. Since $\Delta\chi^{-1}/T^2$ depends on q-space averages of γ_q/Γ_q^2, it tends to be even more strongly enhanced as the critical point is approached than the linear heat capacity $\Delta C/T$, which depends on an average of $1/\Gamma_q$. But as we shall see, sufficiently close to the critical point, where infrared fluctuations are important, the above qualitative temperature dependence will, in general, break down.

In summary, we have arrived at a simple model for $\Delta\chi^{-1}$ which depends on the population n_q and the $T = 0$ parameters. Since n_q is a function of $\Delta\chi^{-1}$ itself, we obtain a self-consistent equation, the solution of which yields the temperature dependence of the susceptibility and hence of the spectrum Γ_q. From the latter we may then obtain ΔC from eq. (12) and also, as shown in the next section, the electrical resistivity $\Delta\rho$ of the fermions in leading order in perturbation theory.

Resistivity

For an analysis of the electrical resistivity we now return to description (I), in which fermions interact with the exchange field via a potential $-\mu \cdot \lambda m$. The rate of scattering of fermions with the field may be determined in perturbation theory, and the resistivity is then obtained by means of the Boltzmann transport equation, as in the case of phonon scattering (Moriya 1985).

A standard analysis in the Born approximation then leads to a resistivity, which may be expressed as

$$\Delta\rho = \eta \sum_q q^k \left(\frac{T\partial n_q}{\partial T}\right)_\Gamma, \qquad (15)$$

where normally $k = 2$, the suffix Γ denotes a derivative at constant Γ_q, and η

---

[*] The original cut-off ω_c appears only implicitly in the renormalised $T = 0$ parameters, which are to be determined by experiment. The thermal population is essentially independent of ω_c because the ω integral is naturally cut off by the Bose function in our limit $T \ll \omega_c$. The cut-off q_c similarly does not enter the description of infrared anomalies nor of the qualitative forms of the temperature dependences.

depends on the strength of the coupling between the fermions and the exchange field, which is essentially temperature independent at the level of approximation employed.* This result has a plausible form. The factor in the brackets in eq. (15) is simply proportional to n_q in leading order in T and reflects the fact that $\Delta\rho$ grows with increasing thermal disorder. On the other hand, the q^2 factor describes the fact that fluctuations of high q, which produce scattering at large angles, should be more effective in reducing the current than those at low q that scatter fermions only at low angles. We note that eq. (15) is of the same form as for phonon scattering if by n_q we mean the Bose function evaluated at the phonon propagation frequency Ω_q.

In leading order in T we expect $\Delta\rho$ to vary as T^2 times a q-space average of q^2/Γ_q^2. We see that, in the Fermi liquid regime, the quadratic coefficient is even more strongly enhanced than the linear coefficient of the heat capacity, which depends on an average of $1/\Gamma_q$.

The above analysis is, however, incomplete for the reasons given in our discussion of ΔC. As we shall show in section 6.4, the Fermi liquid form of $\Delta\rho$ and that of ΔC can break down under certain conditions, which lead to the infrared anomalies.

Extensions of the model

For completeness, we also note that, when the anharmonicity parameter b is small or negative, it may be necessary to retain the additional correction in eqs. (13a) and (13b) even at low temperatures. An analysis similar to that outlined above yields an expression for $\Delta\chi^{-1}$ equivalent to eq. (14), but with the parameter b in the right-hand side replaced by $b + (\Delta b/2)$, where Δb is now the shift of the cubic coefficient in $H(M)$ from its $T = 0$ value,

$$\Delta b = 14g\widehat{\sum_q \gamma_q n_q}. \tag{16}$$

In the limit of small b, we must also take account of the non-linear precession term in eq. (13a), the qualitative effects of which have already been discussed.

Throughout, we have assumed that the number of equivalent field components for each q is 3. It is straightforward to generalise our analysis for an arbitrary number N. The result is that the numerical factors 3, 5 and 14 in eqs. (12), (14) and (16) are replaced respectively, by N, $2 + N$ and $8 + 2N$.

It is also useful to extend the results, where possible, to describe fluctuations

---

* The standard analysis, which yields eq. (15), is based on an implicit assumption that the momentum of the combined system of fermions and the exchange field can be efficiently transferred to the underlying lattice.

which are nearly critical about some large wave vector Q near the Brillouin zone boundary rather than, as we have assumed thus far, about the origin. In this case, we may shift origin in q-space, $q \rightarrow Q + q$, and continue to use the above model, but with the following *minimal* modifications (Moriya 1985). Since in lowest order in q, $\gamma_{Q+q} \cong \gamma_Q$ is a finite constant, we may take n to be zero in eq. (5e). Furthermore, since $|Q + q|^2$ is also a constant in leading order in q, the exponent k in $\Delta\rho$, eq. (15), similarly reduces to zero. Finally, the susceptibility and anharmonicity parameters, eqs. (14), (16), must now be considered to correspond to the anti-ferromagnetic magnetisation at wave vector Q. This extension to the case of incipient anti-ferromagnetism has not been tested experimentally as thoroughly as for the ferromagnetic counterpart. In particular, the range of validity of the model of $\Delta\rho$ is now more restricted than for the latter, and the assumptions in section 6.3.2 on the analytical properties of the relevant magnetic equation of state may apply only to the simplest examples of conductors near anti-ferromagnetic instabilities (Hlubina and Rice 1995; Anderson 1997a).

Range of validity of the mean field approximation – a general Ginzburg criterion

We conclude our technical discussion with a brief examination of corrections to the mean field approximation for the mode–mode coupling. The key result is that the latter should break down qualitatively in a finite interval ΔT_G near T_c when $d \leqslant 4$. But this interval vanishes more rapidly than T_c as $T_c \rightarrow 0$, if the *effective dimension $d + z$* exceeds 4, where z is the *dynamical exponent*. The latter gives the leading q dependence of $\Gamma_q \propto q^z$ near the critical point and for our model, eqs. (5d, e), reduces to $2 + n$. The conclusion is that when $d + z > 4$, the mean field model is expected to be qualitatively correct for describing quantum critical phenomena (Hertz 1976; Millis 1993; Sachdev 1996).

The effective dimension is now greater than the spatial dimension which enters the classical theory, because a quantum description must be based not on a free energy for a spatially varying variable, but rather on an action for an order parameter which depends on both space and an effective time (Tsvelik 1995). The contribution of the latter is, in general, not equivalent to that of a space dimension unless the dynamical exponent z which relates space and time in the spectrum is unity.

The first correction we wish to consider is the last term in eq. (13), which arises from non-linear precession. A calculation analogous to that given above shows that this correction has the same initial temperature dependence as that of the first term, and that, for a parabolic band model for the fermions, its magnitude is small and ignorable (McMullan 1989). As we have already discussed, however, non-linear precession can play an important role when the anharmonicity parameter b is small.

The next correction arises from the dependence of the variance of $\boldsymbol{m}$ on the average magnetisation $\boldsymbol{M}$. If we expand $\overline{m^2}$ in a power series in M in the scalar model discussed earlier, we obtain a correction to b of the form $\Delta b = 3b\,\partial\overline{m^2}/\partial(M^2)$, where $\overline{m^2}$ includes both zero point and thermal contributions, and the derivative is evaluated for $M \to 0$. The M dependence appears in $\overline{m^2}$ through the inverse susceptibility $\partial H/\partial M = a + 3bM^2$. From Nyquist's theorem and our model for χ_k (eq. (5)), we readily find

$$\Delta b = -9b^2 \widehat{\sum_q} \gamma_q \chi_q [1 - (x\partial n(x)/\partial x)]_q, \tag{17}$$

where $x = \Gamma_q/T$ and $n(x)$ is the population function (9).

Consider first the immediate vicinity of a non-vanishing T_c. The low-q modes satisfying $\Gamma_q \ll T_c$ may be treated in the classical limit ($n(x) \to 1/x$) and hence they lead to a contribution to the right-hand side of eq. (17) proportional to a low-q sum of $T\chi_q^2$. At T_c, where χ_q^{-1} reduces to cq^2, it is clear that this sum is finite only when $d > 4$. For $d \leqslant 4$, the correction to b is singular and, hence, the mean field model breaks down. A range in temperature ΔT_G about T_c, where a qualitative breakdown occurs, may be defined by the condition $|\Delta b/b| > 1$. Where $\Delta T_G/T_C \ll 1$, we find from eq. (17) and for $d < 4$

$$\Delta T_G \sim (b^2 T_c^2/c^d u^{4-d})^{\frac{1}{v(4-d)}}, \tag{18}$$

where u and v are defined by $\chi^{-1} = u(T - T_c)^v$ in the breakdown region.

As $T_c \to 0$, the classical treatment above is no longer applicable. Returning to the more general result (17), we see that, in this limit ($n(x) \to 0$), Δb is non-singular if $d + z > 4$. Thus, the mean field approximation is more stable in the quantum than in the classical limit. The improved stability arises from the contribution of the dynamics of the fluctuations, which effectively leads to the replacement of T in the classical limit of eq. (17) by Γ_q. Thus, the sum of $T\chi_q^2$, which is singular for $d \leqslant 4$, is replaced by a sum over $\Gamma_q\chi_q^2 = \gamma_q\chi_q$, which is singular only if $d + z \leqslant 4$.

In conclusion, the mean field approximation for the mode–mode coupling is more stable for quantum than for classical phase transitions, and, in the absence of new instabilities, the predictions of the model discussed in the preceding section may be expected to be qualitatively correct in essentially all cases considered.

6.4 Predictions and comparisons with experiment

In this final section, the predictions of the model developed in section 6.3 will be summarised and briefly compared with observations. The Fermi liquid limit of the model has already been mentioned, and our focus will be mainly on its breakdown, which, in the language of description (II), may be attributed either to the long range or to the attractive nature of the induced fermion–fermion interaction.

It is useful to summarise the key elements of the model which we shall need to carry with us in this section. The central quantity is the spectrum Γ_q (eqs. (5d, e)), which defines the population $n_q = n(\Gamma_q/T)$ (eq. (9)). From this and γ_q, we arrive at a self-consistent equation for the susceptibility χ, eq. (14). The latter yields the temperature dependence of Γ_q and hence of ΔC (eq. (12)) and $\Delta\rho$ (eq. (15)). The effective induced interaction between fermions, which we shall need explicitly to discuss the pair instability, can be shown to be given by the real part of the generalised susceptibility, equation (5), which is also fully determined by the above procedure.

The predictions of the model depend primarily on Γ_q, which has been investigated by inelastic neutron scattering in a number of nearly magnetic metals (see references given in section 6.3).

6.4.1 General form of low-temperature exponents

We consider first in a general way the prediction of the model for the temperature dependences in the low-temperature limit. In all cases, this is given by a q-space average of the population n_q weighted by a q-dependent factor of the form q^m at low q. Thus, for $\Delta\chi^{-1}$ and $\Delta\rho$, m reduces to n and k, respectively (see eqs. (5e), (14) and (15)). For ΔC, the low-q form of Γ_q leads to a value of m equal to n away from the critical point, but to $z = 2 + n$ near the critical point when χ^{-1} in Γ_q can be essentially ignored.

From our expression for the population (9), we find that the required sum in d dimensions reduces to

$$\sum_q q^m n_q \propto T^s \int_0^{x_c} \frac{x^{s-2}\,\mathrm{d}x}{1+x}, \tag{19}$$

where s is $(d+m)/n$ or $(d+m)/z$, respectively, for the above two limits of Γ_q, and x_c is a cut-off inversely proportional to T. This simple result describes in a compact way all of the predictions of the model for the asymptotic low-temperature exponents for the susceptibility, heat capacity and resistivity in the paramagnetic state.

6.4.2 The Fermi liquid

We consider first the most common limit $s > 2$, which is normally appropriate for $d > 2$ well away from the critical point. In this case, the integral in eq. (19) is determined by the upper cut-off $x_c \propto 1/T$. In leading order in T, we see that s drops out and eq. (19) reduces to the universal T^2 form which is independent of d, m or n. This is the behaviour of $T\Delta C$, $\Delta \chi^{-1}$ and $\Delta \rho$ in the *normal Fermi liquid* description, which, as we have stressed, characterises the vast majority of conducting systems at low temperature, but above the temperature at which pairs are formed.

These temperature dependences are qualitatively the same as that for non-interacting or weakly interacting fermions. But the corrections to the linear heat capacity are in general *not* qualitatively the same as for non-interacting fermions. Thus, for Γ_q of the Landau damping form away from the critical point in three dimensions and with three components of m, eqs. (5d, e), (9) and (12) yield

$$\Delta C = \alpha T + \beta T^3 \ln(T/T^*) + \cdots , \qquad (20a)$$

where

$$\alpha = \pi \sum_q (1/\Gamma_q). \qquad (20b)$$

The second term in eq. (20a), which has no counterpart in a non-interacting system, arises from the behaviour of Γ_q at low q and is an example of an infrared anomaly discussed earlier. The existence of the $T^3 \ln(T/T^*)$ term is well established for liquid ^{3}He in the normal state (Baym and Pethick 1991). But in metals the range of validity of this term can be small, and a comparison with experiment beyond leading order in T normally requires a numerical integration of eq. (12).

We note that, for our model, $\alpha = \ln(1 + c\chi q_c^2)/4\pi\gamma c$, $\beta = 2\pi\chi^3/5\gamma^3$, and T^* is of the order of $\gamma/(c\chi^3)^{1/2}$. If we substitute for γ and c the parameters appropriate for non-interacting fermions with a parabolic spectrum, we arrive, via a new route, precisely at the results of the *paramagnon theory* originally developed for the normal state of liquid ^{3}He (Izuyama *et al.* 1963; Berk and Schrieffer 1966; Doniach and Engelsberg 1966; Béal-Monod *et al.* 1968; Brinkman and Engelsberg 1968 and others).

The coefficient of the first term in eq. (20a) is a measure of the enhancement of the quasi-particle mass due to the induced interaction associated with the exchange field in this model. We see that a *heavy fermion state* is favoured where Γ_q is small over a wide range in q. For our form (5d) for Γ_q, this is expected near the instability where χ^{-1} is small and in systems where the stiffness c is small. These conditions are met in some d-transition-metals which are both near the middle of the series and on the border of magnetic instabilities, and especially in many f-metals.

The above model for the linear coefficient has provided a quantitative description of heavy quasi-particle masses observed in the nearly magnetic d-metals (Lonzarich 1986). It may also offer a useful starting point for describing the more strongly renormalised masses in many f systems.

We have arrived at a model of ΔC via an unconventional route, namely description (III) of dissipative modes of the interaction field. It may seem from this viewpoint that our attribution of the excess linear term, eqs. (20a, b), to fermion quasi-particles has only a formal significance. However, it can be shown via descriptions (I) or (II) that the elementary excitations of the combined system of fermions and exchange field are indeed fermions again, but with a renormalised mass given by the starting mass plus a correction consistent with eqs. (20a, b).

Direct experimental evidence in support of the existence of such quasi-particles with extraordinarily high effective masses has been collected in the last decade in a number of d-metals (Taillefer *et al.* 1986) and f-metal heavy fermion compounds (see chapter 8; also Reinders *et al.* 1986; Taillefer and Lonzarich 1988; Aoki *et al.* 1993 and others).

In a few cases, these investigations, via quantum oscillatory phenomena in intense magnetic fields and very low temperatures, have established that the contribution to the linear heat capacity expected to arise from the quasi-particle system as a whole (Fulde *et al.* 1988) is indeed comparable to that actually observed in bulk heat capacity experiments (Julian *et al.* 1997a). These findings, together with theoretical analyses of the heat capacity via eq. (20) directly, have helped to confirm the consistency of results obtained from the different starting descriptions (I) to (III).

6.4.3 Non-Fermi liquids

Let us now return to the original q-space sum (19) and consider the opposite case where the exponent $s \leq 2$. In contrast to the results of the previous subsection, we now find that the upper cut-off x_c does not play the dominant role. The case $s = 2$ is marginal and eq. (19) yields again a universal function, independent of d, n and m, but now of the non-analytic form $T^2 \ln (T^*/T)$, where T^* is a constant. For $s < 2$, the relevant exponent is non-universal and falls below that of the usual Fermi liquid model. For $1 < s < 2$, in particular, eq. (19) is well defined and varies as T^s at low T.

The limit $s < 2$ may be satisfied in reduced dimensions in general or in three dimensions near a critical point where $s = (d + m)/z$, $z = 2 + n$. Let us consider first the latter in the presence of Landau damping so that $d = z = 3$. From eq. (19), we may immediately infer that ΔC, $\Delta \chi^{-1}$ and $\Delta \rho$ have leading temperature dependences of the form $T \ln (T^*/T)$, $T^{4/3}$, and $T^{5/3}$, respectively. This defines a state known as the *marginal Fermi liquid*. (The $T^{5/3}$ form for $\Delta \rho$ was first obtained by Mathon (1968).) Note that s has precisely the marginal value of 2 for $T\Delta C$, which may be obtained from eq. (19) when we set $m = z$.

The latter limit has provided a useful framework for describing certain d-metals near ferromagnetic instabilities at low temperature (Pfleiderer *et al.* 1994, 1995; Grosche *et al.* 1995; Thessieu *et al.* 1995). The marginal Fermi liquid model has also appeared in theoretical treatments of the very weak, but long-range, Lorentz interaction of moving charges in an ideal conductor (Holstein *et al.* 1973; Reizer 1989). Furthermore, a state with the same $T \ln (T^*/T)$ form for the heat capacity has been introduced in the study of the normal state of the high-temperature superconductors (Varma *et al.* 1989).

The departure from the Fermi liquid form is even more extreme near a critical point in lower dimensions. Thus, for $z = 3$ and $d = 2$, eq. (19) yields $T^{2/3}$, T and $T^{4/3}$, respectively, for ΔC, $\Delta \chi^{-1}$ and $\Delta \rho$. These results for ΔC and $\Delta \rho$ have also been obtained by a different technique for a two-dimensional system in which the critical fluctuations are not of the magnetisation, but of a higher-order object, known as a *chiral field* (Lee and Nagaosa 1992).[*]

We also consider briefly the implications of this model for the nearly anti-ferromagnetic metals discussed at the end of section 6.3 for which, effectively, $n = k = 0$ (Moriya 1985). At the critical point, therefore, the dynamical exponent is $z = 2$. In three dimensions, we find from eq. (19) that the heat capacity is linear in first order, as expected for a Fermi liquid. However, the linear coefficient can be very large, particularly when the stiffness c is small, as in the heavy fermion f-metals. Furthermore, the correction to the leading term has the anomalous form $T^{3/2}$. Thus, the ratio of $\Delta C/T$ can show a marked upturn at low temperature, which, as numerical analyses suggest, can resemble, over a wide range in temperature, the divergence associated with a non-Fermi liquid state. We also note that both $\Delta \chi_Q^{-1}$ and $\Delta \rho$ are (from eq. (19)) now expected to have a $T^{3/2}$ temperature dependence, which differs from that of the usual Fermi liquid.[†]

We note that in all of the examples considered above the requirements

---

[*] A chiral field may be constructed from inner- and cross-products of three magnetic moments on different points on the unit cell.

[†] The range of validity of the present model for $\Delta \rho$ in an incipient anti-ferromagnet, however, remains unclear (Hlubina and Rice 1995).

$d + z > 4$ and $s > 1$ are indeed satisfied.[*] In lower dimensions, one or both of these may be violated and the treatment given here then breaks down. In a one-dimensional conductor, this leads us to the *Luttinger liquid* state which lies outside of the present framework (Anderson 1997b).

6.4.4 Magnetic instability

The great abundance of systems which appear to be on the border of long-range magnetic order, and for which therefore the models of the last section are relevant, is, at first sight, surprising. Part of the explanation lies in the role of the anharmonicities in the field equation, which, in the presence of zero point and thermal fluctuations, generally tend to suppress T_c below the value T_c^0 in the absence of either anharmonicities or fluctuations. In our model, T_c^0 is essentially given by the condition $a_0 - \lambda = 0$, where a_0 is the temperature-dependent inverse Pauli susceptibility of the non-interacting fermions. The final T_c is then given by $a_0 - \lambda + \Delta a = 0$, where Δa is the correction, eq. (13), which arises from the combined effect of anharmonicities and of spontaneous fluctuations. For sufficiently large fluctuations, Δa is, in general, positive, and hence T_c is driven down below T_c^0.

This suppression of T_c increases with decreasing dimension and with decreasing value of the stiffness c. Among the low-dimensional systems, or in the heavy fermion f-metals where c is small, we therefore expect, and find, numerous examples of systems with low or vanishing magnetic critical points.

Even for three-dimensional d-metals, the reduction of the magnetic transition by fluctuations is very pronounced, and typically T_c is one or more orders of magnitude below T_c^0. In such cases, where the temperature dependence of a_0 can be ignored, T_c may be obtained in our model by the condition $a + \Delta a = 0$, where a is a constant and Δa is given by eq. (14). From Γ_q (eqs. (5d, e)) and n_q (eq. (9)), we then find for $d = 3$ and three equivalent components of $\boldsymbol{m}$,

$$T_c = v c \gamma^{1/(1+z)} \left| \frac{a}{b} \right|^{z/(1+z)}, \tag{21}$$

where a, b, c and γ are the basic parameters of the model and v is a pure number approximately equal to 2.39 for $z = 3$ and 0.256 for $z = 2$, corresponding to the Curie and Néel transitions, respectively. This model has provided the starting point for our current understanding of magnetic transitions in

---

[*] We have assumed that the temperature dependence of Γ_q (i.e. of χ^{-1}) could be ignored in evaluating eq. (19) in leading order in T. This is valid if, for small T, the correlation wave vector $\kappa = (1/c\chi)^{1/2}$ is vanishingly small compared with the typical thermal wave vector q_T, for which the relaxation rate is of order T. From $T \propto q_T^z$ and $\chi^{-1} \propto T^{(d+n)/z}$, we find that the above condition is satisfied if $d + z > 4$ (see also section 6.3.4).

conducting materials. For the low-temperature ferromagnets, it leads to predictions of T_c in close agreement with observation (Lonzarich 1984, 1986;
Lonzarich and Taillefer 1985; Takahashi and Moriya 1985).

The above treatment also allows us to determine the way in which T_c may
depend on a control parameter such as pressure p about the critical p_c, where
a and T_c vanish. A leading order expansion of a is expected to be linear in
$(p - p_c)$, so that, from eq. (19) with $m = n$, we find

$$T_c^{(d+n)/z} \propto (p_c - p). \tag{22}$$

For $n = 1$, $d = z = 3$, the exponent of $4/3$ in eq. (22) is consistent with observation in a d-metal ferromagnet with a continuous phase transition near p_c
(Grosche *et al.* 1995).

It is important to note that the above treatments assume that b is positive and
not very small. When this is not the case, the analysis must be extended to
take account of the effects of higher-order anharmonicities (eqs. (13), (16)).
The problem is then considerably more complicated, and near p_c the phase
transition may cross over from second to first order (Yamada 1993; Pfleiderer
et al. 1995).

Finally, we note that eq. (14) for $\Delta\chi^{-1}$ also allows us to understand the
ubiquitous linear temperature dependence of χ^{-1} at high T in d-metals. For
the normal case, $n = 1$, $d = z = 3$, χ^{-1} varies as $T^{4/3}$ just above T_c. With
increasing χ^{-1}, and hence Γ_q, the curve bends below $T^{4/3}$ and, as confirmed
by self-consistent numerical analysis, tends to stabilise at a linear temperature
dependence over a wide range (Murata and Doniach 1972; Moriya and
Kawabata 1973a, b).

6.4.5 Spin fluctuation mediated interaction

The magnetic instability discussed above arises when a is negative. From a
somewhat unconventional point of view, this instability is due to a tendency of
particles and holes of opposite spins to attract and pair. When a is positive, the
latter repel, but now an attractive induced interaction can arise between
particles or between holes of the appropriate relative spins. This pairing now
drives a pure system at sufficiently low temperature towards a superconducting
instability. Thus, the normal states described in earlier sections are not expected to be the true ground states of the model we are considering.

The way the induced interaction arises via the exchange field has already
been discussed qualitatively. Let us now consider the form of this interaction
more quantitatively. Our goal is to determine the potential energy $v(r)$ of a
fermion at r due to another at the origin. This potential, however, arises from

the exchange of fields which propagate slowly. Thus, we are led to consider a potential energy per unit time $v(r, t)$ of a first fermion at r and t due to a second at the origin in both space *and* time. The first experiences an interaction with the exchange field of the form $v(r, t) = -\mu \cdot \lambda m(r, t)$ where m arises from an effective field h produced by the second via a convolution with the non-local susceptibility $\chi(r, t)$. Taking $h = \lambda\mu'\delta(r)\delta(t)$, where μ' is the moment of the second fermion, we then find $v(r, t) = -\mu \cdot \mu'\lambda^2\chi(r, t)$.

The final potential is the average of the latter plus the inverse process in which fermion 1 transmits an equivalent exchange field to fermion 2.* Thus we are led to a Fourier transform of $v(r, t)$ of the form

$$v_{q\omega} = -\mu \cdot \mu'\lambda^2 \operatorname{Re} \chi_{q\omega}, \tag{23}$$

where Re denotes the real part of $\chi_{q\omega}$, which is given by eq. (5).

We note that $v_{q\omega}$ is negative for particles of the same spin, and hence we might expect eq. (23) to lead only to pairing in the *spin triplet state* of necessarily *odd parity*. This conclusion is valid for our model (5) of a nearly ferromagnetic system ($c > 0$), for which the static non-local susceptibility $\chi(r)$ is everywhere positive.

On the other hand, for an incipient anti-ferromagnet ($c < 0$, see below), even where χ_q is everywhere positive, $\chi(r)$ oscillates and has a negative regime at small r, which can favour the *spin singlet state of even parity*. It is interesting that, in this case, pairing can arise even when the transform (23) is everywhere greater than zero. A positive binding energy is achieved via the construction of an anisotropic pair wavefunction, which allows the particles to sample mainly the attractive regions of the space-time potential.

Also note that, for an adequate description of the latter, we now need to begin with a model for fermions and the exchange field which takes explicit account of the lattice. The potential now depends not only on the separation r, but also on the coordinate r' of the second fermion. We may average over r' to obtain essentially $v(r)$ and χ_q as discussed above. In an extension of eq. (5), we replace expansions in q by leading sinusoidal functions periodic in the crystalline lattice. For a simple cubic unit cell with side a, χ_q^{-1} in particular would then take the form $\chi^{-1} + (2c/a^2)(3 - \cos q_x a - \cos q_y a - \cos q_z a)$. Incipient ferromagnetism and anti-ferromagnetism can now be described in essentially the same model with positive and negative values of c, respectively.

It is interesting to compare the above potential (eqs. (5), (23)) with the more familiar case arising from the exchange of massive propagating bosons. For

---

* Note that m and h are now defined per unit time. Since fermions in our low-frequency model cannot be strictly localised in space-time, h must be replaced by a more gradual function of r and t than used here. The above argument should, however, correctly give the low-q and ω behaviour in the linear response approximation.

the latter, $v_{q\omega}$ is inversely proportional to $(q^2 + \kappa^2)(\omega^2 - s^2q^2)$, where s is the speed of propagation, κ^2 is a measure of mass and $1/\kappa$ is the interaction range. We see from eqs. (5) and (23) that, in our problem, κ is $(1/c\chi)^{1/2}$ and hence that the range, as already argued qualitatively, diverges at the critical point. It also follows from eqs. (5) and (23) that the frequency of propagation of the interaction is now not real but purely imaginary. Thus, by analogy, we may describe our interaction as arising from the exchange of massive bosons of imaginary energy, which become essentially massless (i.e. $\kappa \to 0$) as the critical point is approached. It is the latter, and hence long-range, character of the induced interaction which leads in the present language (II) to the infrared anomalies and non-Fermi liquid exponents discussed above.

We also note that analogous phenomena arise in particle physics under similar circumstances when a gauge boson becomes massless. Thus, even Thomson's electron in free space interacting with the massless photons is *not* expected to propagate as a free particle at very long times and long distances (Karanikas *et al.* 1992). This infrared anomaly, that would appear only over cosmic dimensions, has been found to have related (though not entirely equivalent) counterparts in dense electron systems which may be studied under normal laboratory conditions.

6.4.6 Anisotropic superconductivity

The pairing potential (23) forms the starting point for an analysis analogous to that of BCS for a superconducting instability. It is instructive to begin with the Cooper problem in which we initially drop the ω dependence of the interaction and hence solve the Schrödinger equation of two particles in a static potential. The problem is simplified by working with the Fourier components of the potential v_q and of the pair wavefunction ψ_q, which is expressed as a product of a 'radial' part R_q and an angular part $\eta_{\hat{q}}$ where, for the isotropic limit, $\hat{q}$ is a unit vector parallel to q. Note that the spins are in a triplet or singlet state depending on whether $\eta_{\hat{q}}$ has odd or even parity, respectively. The effect of the finite range of $v_{q\omega}$ in ω is then put in approximately by limiting contributions of the transition matrix element $v_{q-q'}$ to values of q and q' within an appropriate range about the Fermi level. We also include the effect of the Pauli principle by restricting transitions mainly to states above the Fermi surface. This leads us to a pair wavefunction and a binding energy which is always finite if the appropriate matrix element of the interaction is negative.

An extension of this Cooper problem to take account of the co-operative effect of all pairs ultimately leads us to the BCS transition temperature T_s. For a representation defined by $\eta_{\hat{q}}$ for which binding occurs, we then find in the weak coupling limit

$$T_s \cong \theta \exp(-m/g\Delta m), \tag{24}$$

where θ is a scale set by the finite frequency range of the pair interaction, m is the total mass of the quasi-particles, Δm is that part of the mass which arises from the self-interaction for the potential (23), and g measures the effectiveness of the pair wavefunction in sampling the attractive part of the pair potential (Millis *et al.* 1988).

In terms of the angular function $\eta_{\hat{q}}$ given above this critical parameter g is defined, in the isotropic limit, as the integral of $v_{q-q'}\eta_{\hat{q}}\eta_{\hat{q}'}$ over the solid angles in both variables divided by the same without the η functions. Also, we note that $\Delta m/m$ is $\Delta C/C$ in leading order in T at T_s, where ΔC is given by eq. (20) and C is the total heat capacity of the interacting fermion system. Finally, θ is some appropriate average over the relaxation spectrum Γ_q.

We consider first an incipient ferromagnet defined as in the preceding section for $c > 0$. Then pairing is expected to occur for the spin triplet state, the simplest representation ($\eta_{\hat{q}}$) for which corresponds to the *p-wave state*. When the correlation wave vector $\kappa = (1/c\chi)^{1/2}$ is well below the characteristic Fermi wave vector k_F, we then find that g reduces to $1/3$ and that θ is of the order of $\gamma\chi^{-1}k_F$ (Fey and Appel 1980; Levin and Valls 1983). As κ increases, g tends to fall, and vanishes as the potential becomes essentially local in space where the p-wave function has negligible weight.[*]

It is interesting to consider the orders of magnitude for T_s predicted by the above model. For p-wave pairing in liquid ^{3}He, $m/g\Delta m$ is $3m/(m - m_0)$, the empirical value of which is approximately 1.5. Also $\theta \sim k_F^2\chi_0/m_0\chi$ is of the order of 0.5 K. Thus, eq. (24) predicts a value of a few millikelvin, in order of magnitude agreement with a transition to a triplet p-wave superfluid state observed at approximately 2 mK.[†]

Let us now turn to the analogue of this superfluid state in metals, in which the carriers move in a lattice and are charged rather than neutral. If the expressions for ^{3}He carry over, then for the same χ/χ_0 we might expect T_s to be greater by a factor of the order of the ratio of the nuclear to the relevant

[*] We note that for liquid ^{3}He, galilean invariance leads us to identify $g\Delta m$ as approximately $1/3$ of the mass correction $m - m_0$, where m_0 is the bare mass of ^{3}He atoms. This is not valid, in general, for a metal in which galilean invariance is broken by the underlying lattice.

[†] Let us also examine the prediction for $m/(m - m_0)$ of our expression (20b) for ΔC. For a parabolic band, it yields $\Delta m/m_0 = (9/2)\ln(1 + \chi q_c^2/12\chi_0 k_F^2)$, which, in contrast to the infrared anomalies discussed earlier, *is* dependent on the cut-off q_c (albeit weakly in this case) and hence cannot be evaluated in the model except in order of magnitude. To obtain the expected value $\Delta m/m_0 \cong 2$ for the known susceptibility enhancement $\chi/\chi_0 \cong 9$, we need a cut-off q_c of the order of k_F.

electron mass, i.e. by perhaps three orders of magnitude. Examples of p-wave superconductivity of this *simple* kind have not, however, been unambiguously identified, even in the low-millikelvin regime.

It is worth listing some of the effects which arise in the lattice, but not in liquid ^{3}He, which may account for the elusiveness of p-wave superconductivity. The list would include (i) the non-parabolic and multi-band spectrum of real metals, (ii) lack of galilean invariance, (iii) the pair breaking effects of frozen-in disorder, (iv) the spin–orbit interaction, (v) additional low-frequency branches of the spin fluctuation spectrum, (vi) corrections to the linear response approximation and (vii) further components of the true induced interaction which we have discussed only briefly. Some of these factors could have a deleterious effect on the stability of the form of superconductivity we are considering.

As an example, consider the case of Pd, for which χ/χ_0 is indeed comparable to ^{3}He and for which the starting mass m_0 may be taken to be the conventional band calculated value, which is nearly three orders of magnitude below that of ^{3}He atoms. The model used for ^{3}He would then lead to T_s of the order of 1 K. But a parabolic band model provides very poor estimates of the basic parameters, in particular of c, which define Δm and θ in the more general expressions above. From the known χ and estimates of γ and c from neutron scattering data, as well as from first-principle calculations of the generalised susceptibility (McMullan 1989; Dungate 1990), we find that $\Delta m/m$ is in fact much smaller in Pd than in ^{3}He, *despite* the fact that χ/χ_0 is virtually the same in both cases. The revised estimate of T_s is now in the low- or sub-millikelvin range and, because of the effects of frozen-in disorder already discussed in section 6.2, p-wave superconductivity may be suppressed altogether. Most candidate materials which have been considered theoretically have met with a similar fate from one or another of the effects (i)–(v) listed above. Promising cases not yet investigated experimentally are summarised in section 6.2 (see also Foulkes and Gyorffy (1977) for a discussion of p-wave pairing in a more conventional BCS model). Also, we do not rule out the possibility that more exotic forms of odd-parity super-conductivity already exist in some cases among heavy fermion f-electron systems and the low-dimensional conductors (Sigrist and Ueda 1991).

Finally, we consider briefly singlet pairing for an even-parity anisotropic state of an incipient anti-ferromagnet. For the simple cubic lattice discussed in the preceding sub-section, a plausible representation describes an anisotropy of a 'clover leaf' form with orthogonal positive and negative lobes along the cube axes. For this choice, the parameter g increases to a maximum of $1/6$ (for $d = 3$), rather than $1/3$ for the p-wave model, as the critical point is approached (Millis *et al.* 1988). We may think of the above configuration with even parity as a *d-wave state*.

The exponential factor $m/g\Delta m$ is minimised by approaching the critical point, i.e. the critical pressure p_c, where v_q is typically large and strongly q dependent. Thus we might expect the peak in T_s to be found close to p_c if the magnetic transition goes continuously to zero at the critical pressure. This behaviour has recently been observed in a heavy fermion anti-ferromagnet, in which the Néel temperature can be suppressed continously with hydrostatic pressure (Grosche *et al.* 1996; Julian *et al.* 1997b).

The possible applicability of anisotropic even-parity models in other systems has already been discussed in section 6.2. The development of a quantitative theory, which accounts both for the appearance of anisotropic superconductivity in some cases as well as for its absence in other materials of a similar kind, remains an outstanding unsolved problem.

Acknowledgements

I am indebted to my co-workers, in particular, N. R. Bernhoeft, C. Pfleiderer, F. M. Grosche, S. R. Julian, A. P. Mackenzie, G. J. McMullan, L. Taillefer and I. R. Walker, some of whose contributions relevant to this review are cited in the text.

It is also a pleasure to thank Y. Chen, P. Coleman, D. M. Edwards, J. Flouquet, A. J. Hertz, D. Khmelnitskii, A. J. Millis, T. Moriya, A. B. Pippard, S. Sachdev, A. Z. Solontsov and A. M. Tsvelik for a number of stimulating discussions on these problems over the years.

This chapter could not have been completed without the vision, interest and considerable patience of the editor of this volume, Professor Mike Springford. Most of all, I am deeply grateful to Gerie Lonzarich for her support, inspiration and encouragement throughout and for her contributions to the completion of the final text.

The research programme on which this chapter is partly based has been supported by the EPSRC of the UK, the EC and the Isaac Newton Trust of Trinity College in Cambridge.

7

The paired electron

A. J. LEGGETT
University of Illinois

In one sense, the idea that the properties of two electrons paired up are qualitatively different from those of a collection of single unpaired electrons is as old as the theory of the chemical bond. However, that is not what this chapter is about; in that case, the effects of the pairing extend only over a single molecule, whereas I shall be concerned with the much more spectacular effects which occur, as a result of pairing of electrons, in the phenomenon of super-conductivity and extend over the whole of a macroscopically large sample.

7.1 The phenomenon of superconductivity

From a modern point of view, 'superconductivity' is actually not a single phenomenon but a complex of phenomena which occur, usually together, in some but not all metals below a certain 'transition temperature' T_c. The most obvious indicator of superconductivity, and the one which gives it its name, is the property of *zero resistance*. Suppose a metal which is superconducting at low temperature, let us say, for example, Al, is incorporated in a circuit, the rest of which is composed of a non-superconducting metal (say Cu), and a voltage is applied across the circuit. Then at room temperature the voltage will be distributed across both the Cu and the Al, in proportion to their resistance; but if the circuit is cooled below the 'transition temperature' of Al (about 1.2 K), all the voltage will appear across the Cu, i.e. it appears that the resistance of the Al is zero. Similarly, if a current is set up in a ring of Al which is maintained below T_c, it will continue to flow indefinitely. This phenomenon, which seems to set in discontinuously at T_c, was first discovered by the Dutch experimental physicist Kamerlingh Onnes in Hg in 1911, and for two decades thereafter was the only property by which the superconducting state was thought to differ qualitatively from the 'normal' (non-superconducting) one. With hindsight, this is somewhat ironic, since nowadays many theorists hold

148

the view that zero resistance is *not* a necessary characteristic of the 'super-conducting' state under all circumstances! In the early years of research, a great deal of phenomenological information was amassed about the occurrence of superconductivity, one of the most surprising results to emerge being that it appeared at least as common in alloys as in pure metals; but all attempts to explain the phenomenon failed dismally.

The picture changed overnight in 1933 with the discovery by Meissner and Ochsenfeld that if a simply connected sample (e.g. a solid sphere) of a superconducting material is placed, while above its transition temperature T_c, in a weak magnetic field (which then penetrates, it almost completely) and is then cooled through T_c, the field is suddenly expelled and the sample in its superconducting state behaves as a perfect diamagnet. Like the phenomenon of zero resistance, this 'Meissner effect' seems to represent a qualitative difference between the normal and superconducting states which sets in discontinuously at T_c; however, unlike zero resistance, it is quite unambiguously a manifesta-tion of a *thermodynamic equilibrium* property and cannot be explained, e.g., by a sudden disappearance of the scattering mechanisms which operate in the normal phase.

More recently, other phenomena beside zero resistance and the Meissner effect have been discovered which appear to be rather generally characteristic of the superconducting state; since these were mainly predicted on the basis of the theory of pairing, I postpone discussion of them until later.

7.2 Phenomenology

Within two years of the announcement of the Meissner effect, the brothers Fritz and Heinz London had formulated a phenomenological theory which is still believed to capture the essentials of what is going on in the superconducting state. The simplest way to convey a qualitative understanding of the Londons' ideas may be to use the analogy with the phenomenon of atomic diamagnetism. Consider an atom such as neutral He, with a closed electronic shell and hence no spin angular momentum, which initially is in zero magnetic field. On average, the angular velocity of the atomic electrons relative to the nucleus is zero, so that there are no circulating currents. If now a magnetic field is applied, the effect is to induce an average circulation of the electrons, in such a sense that the magnetic field produced by the resulting current tends to oppose the external one ('diamagnetism'). The Londons suppose that something similar happens at the surface of a superconductor. As a result, as we go further into the body of the sample, the magnetic field of the induced currents more and more nearly cancels the external field, and beyond a certain characteristic distance, the so-called London penetration depth, both the total field and the induced current essentially vanish. Since the London penetration depth is typically around 10^{-5} cm and hence too small to be accessible in most experiments, the 'observed' effect is that the magnetic field appears to be completely expelled from the sample – the Meissner effect. (By contrast, on the scale of a single He atom ($\ll 10^{-5}$ cm) the attenuation of the magnetic field, though finite and measurable, e.g. by precise nuclear magnetic resonance techniques, is very small compared with unity and the effects are not at all spectacular.)

What is the origin of the diamagnetism of the He atom? The first point to be made is that it is an essentially quantum-mechanical effect; indeed, a famous theorem, that of Bohr and Van Leeuwen, tells us that in equilibrium *classical* statistical mechanics no diamagnetism is possible. So consider the quantum-mechanical description of an atomic electron in terms of a de Broglie wave. In the absence of a magnetic field, this leads to the quantization of angular momentum, by the following well-known argument (see also chapter 5): if the radius of the electron orbit is r, then, since an integral number of wavelengths must fit into the circumference, we must have $n\lambda = 2\pi r$, where n is any integer including zero. On the other hand, the de Broglie relation tells us that $\lambda = h/p$, where p, the momentum, is simply the mass m times velocity v. Putting these two results together, we find that the angular momentum mvr is just $nh/2\pi \equiv n\hbar$ as posited in the Bohr model of the atom.

If now a magnetic field is applied, the only difference is that the 'momentum' p which enters the de Broglie relation $\lambda = h/p$ is no longer given simply by mv but is the so-called 'canonical' momentum $mv + eA$, where e is the electronic charge and A is the tangential component of the electromagnetic vector potential. As a result, the quantization condition becomes

$$mvr = n\hbar - eAr. \tag{1}$$

In particular, if we take the integer n to be zero, then we simply have $mv = -eA$, and correspondingly the electric current $j = ev$ carried by a single electron is $j = -e^2A/m$. This gives quantitatively the experimentally observed atomic diamagnetism.

The Londons apply essentially the same argument to a superconductor, with the result that if the density of 'superconducting' electrons is $n_s(T)$ (which may be less than the total density), then the induced electric current density $J(r)$ is proportional to the electromagnetic vector potential $A(r)$:

$$J(r) = \frac{-n_s(T)}{m} e^2 A(r). \tag{2}$$

Notice that this equation is quite different from the usual expression (Ohm's law) for the electric current in a normal metal, which relates $J(r)$ not to the electromagnetic vector potential A but to its time derivative, the electric *field* $E(r)$; in particular, eq. (2) allows finite electric currents to flow in thermo-dynamic equilibrium. On combining eq. (2) with the standard Maxwell equations of electromagnetism, we find that both the magnetic field $B \equiv \nabla \times A$ and the current J fall off exponentially with distance into the superconductor, with a characteristic length $\lambda_L(T)$ which is given by

$$\lambda_L(T) = (n_s(T)\mu_0 e^2/m)^{-1/2}. \tag{3}$$

If we assume that at zero temperature n_s is equal to the total electron density, we find $\lambda_L(0) \sim 10^{-5}$ cm as stated above.* Thus the fundamental London equation (2) indeed explains the experimentally observed expulsion of the magnetic field (Meissner effect).

7.3 Problems of the London phenomenology

The most obvious question raised by the Londons' theory of superconductivity is: why are not all metals superconducting, at least at zero temperature? A difficulty which is closely related to this is the following: the superficially attractive analogy with atomic diamagnetism is actually much less convincing than it looks, for the following reason. Suppose that in eq. (1) we start with $A = 0$ (i.e. zero magnetic field) and ask what value of n we should assign to the electrons. Since the kinetic energy of circulation, $(1/2)mv^2$, is given according to eq. (1) by $n^2\hbar^2/2mr^2$, it is clear that it is a minimum for $n = 0$, and hence, at zero temperature, this assignment is the correct one. However, for finite temperature the crucial question is the ratio of the quantity $\hbar^2/mr^2$ to the thermal energy $k_B T$, and here there is a qualitative difference between the He atom and the superconductor. In the former case, $\hbar^2/mr^2$ is typically of the order of the rydberg (13.5 eV) and hence is enormous compared with $k_B T$ at any temperature likely to be relevant; as a result, the only relevant value of n is zero, and this is maintained when a magnetic field of any reasonable magnitude is applied, as we implicitly assumed in the argument for atomic diamagnetism given above. By contrast, for bulk superconductors the characteristic value of r is typically of the order of 1 cm, and thus the quantity $\hbar^2/mr^2$ is of order 10^{-16} rydberg – much less than the thermal energy at any temperature likely to be attainable in the foreseeable future! Under these circumstances, we should expect the electrons to be thermally distributed, in the absence of magnetic

---

* In the limit $T \to T_c$ we have $n_s(T) \to 0$ and hence $\lambda_L(T) \to \infty$; thus for any sample of finite size there is a region just below T_c where expulsion of the field is not complete. However, this region is so tiny that in practice the expulsion looks discontinuous.

field, over a large number of different levels n; and a straightforward argument* then shows that, when the field is applied, the distribution shifts in just such a way as to produce zero net circulating current and thus no diamagnetism. While this naive analysis essentially ignores the indistinguishability of the electrons and thus the Fermi statistics obeyed by them (see below), it turns out that considering these complications only makes matters worse.

In their original paper, the Londons remark that this difficulty (which they stated in a different but related form) might be overcome, if it should turn out that, for some reason, there was a gap in the single-electron energy spectrum (i.e. in the language of the atomic analogy, that states with low n had an excitation energy considerably larger than $n^2\hbar^2/2mr^2$). As we shall see, this remark turned out to be highly prescient.

A second major difficulty with the London phenomenology relates to the stability of circulating supercurrents. Suppose for simplicity we consider a ring so thin ($\ll \lambda_L$) that the magnetic field induced by the supercurrent is negligible, and set any external field equal to zero so that in eq. (1) we can set $A = 0$. Then it is clear that any state carrying a circulating supercurrent must correspond to a non-zero value of n and cannot therefore be the ground state; indeed, it is easy to show that in some experimentally observed cases n is large enough that the corresponding single-electron excitation energy $n^2\hbar^2/mr^2$ is even large compared with $k_B T$. Now if we consider the atomic analogy, an electron in a high-n state will rapidly drop back to the ground state, losing energy, e.g., by emission of a photon; why does something similar not happen to the electrons carrying the supercurrent in a superconductor? That it does not seems to indicate rather strongly that in some sense we are not allowed to think of the electrons as behaving independently, as would be our picture of a normal metal.

Thus, the work of the Londons suggested two important features which would have to characterize any viable microscopic theory of superconductivity: first, contrary to many of the (now largely forgotten) theories of the pre-Meissner era, the origin of the phenomenon had to lie in the interaction between electrons; and, secondly, it had to result *inter alia* in a gap in the single-electron excitation spectrum. We shall shortly see that the eventually successful 'pairing' theory of Bardeen, Cooper and Schrieffer indeed includes both these features. Before doing so, however, it is necessary to make a brief digression.

---

* The argument in question is essentially the original Bohr–van Leeuwen one; whereas the former implicitly takes the classical limit $\hbar \to 0$, we take $r \to \infty$, with the result in each case that $\hbar^2/mr^2 \to 0$ (or at least is very much smaller than $k_B T$).

7.4 Digression: Bose condensation

Let us pretend for the moment that the electrons in metals were bosons rather than fermions (i.e. subject not to Fermi–Dirac but to Bose–Einstein statistics), and ask whether this would help to explain any of the phenomena of super-conductivity. One reason for conducting this apparently artificial exercise is that it has been recognized since the late 1930s that the superconducting phase of metals has much in common with the superfluid phase of liquid helium-4 (^{4}He), the so-called He-II phase which occurs below the 'lambda-point' at 2.17 K; indeed, from a modern point of view, superconductivity *is* nothing but superfluidity occurring in an electrically charged system. And the ^{4}He atom, having spin zero, really is a boson!

Whereas the distribution of independent particles obeying Fermi statistics is the well-known Fermi–Dirac distribution ($n_k = (\exp[(\varepsilon_k - \mu)/k_B T] + 1)^{-1}$), the behavior of non-interacting bosons whose total number is conserved is considerably more interesting. At sufficiently high temperature the distribution is the standard Bose–Einstein one:

$$n_k = (\exp[(\varepsilon_k - \mu)/k_B T] - 1)^{-1}, \tag{4}$$

where μ is the chemical potential. At high temperatures, μ is negative and very large (so that the Bose–Einstein distribution (4) approximates the classical Maxwell one); as T decreases, μ must increase so as to keep $N \equiv \sum_k n_k$ constant. Eventually, μ tends to zero from below, and at this point we get the phenomenon of *Bose–Einstein condensation* (BEC): a finite fraction of the total number of particles start to pile up in a *single* one-particle state, namely that of lowest energy (but see below). In the limit $T \to 0$, all the particles end up in this state. This phenomenon has long been assumed to occur in the interacting system of atoms formed by liquid ^{4}He, setting in at the experimen-tally observed lambda-point, and to be responsible for the superfluidity of the He-II phase; however, the evidence here is circumstantial, and it is actually only in the last few months that BEC has been *directly* observed in an atomic system (actually in various dilute gases of alkali atoms).

Why might BEC be relevant to the problem of superfluidity and/or super-conductivity? Let us go back to the analogy with atomic diamagnetism, but now imagine that we are dealing with a large assembly of non-interacting bosons below their condensation temperature. Now, in the absence of a field, despite the fact that $k_B T$ is (in the case of the bulk superconductor) large compared with the level spacing ($\sim \hbar^2/mr^2$), a finite fraction of all the particles will go into the single lowest state ('condense'), namely that with $n = 0$; the rest will continue to occupy the excited states according to eq. (4)

(with $\mu = 0$). Now, when we apply a magnetic field, the $n = 0$ state evolves according to eq. (1); and thus the contribution of the condensed particles will be given by eq. (2), with $n_s(T)$ representing the number density of those particles. The non-condensed particles, meanwhile, will shift their distribution in response to the field so as to give zero net current. Thus, this (artificial) model could, in principle, explain the Meissner effect.

However, quite apart from the fact that in real life electrons are fermions not bosons, the model of non-interacting bosons has another serious defect: it cannot explain the persistence of supercurrents, e.g. in a ring geometry. To be sure, we can form a state carrying a large current simply by requiring condensation to take place into a state with large n rather than the ground state ($n = 0$); but there is then nothing to prevent the condensed particles from dropping back, either one by one or cooperatively, into the ground state (it is easily verified that for a non-interacting system this process is 'downhill (in energy) all the way'). Thus, a minimum additional requirement is that interparticle interactions somehow stabilize the metastable (circulating-current) state against decay. It is believed that this is so in the case of superfluid ^{4}He, and we will see below that a similar result is achieved in the pairing theory of superconductivity.

7.5 The BCS (pairing) theory*

The first suggestion that perhaps pairs of electrons could combine to form effectively a boson, and that these bosons might then undergo BEC, was made by R. A. Ogg in 1946, in the context of a very specific problem in chemical physics; he also commented that such a mechanism might explain superconductivity in metals. This idea was taken up more explicitly by Schafroth in 1954; however, the scenario he proposed was only qualitative and afforded no basis for concrete calculations. What we now believe to be the correct microscopic theory of superconductivity, at least in the vast majority of superconductors discovered prior to 1986, was formulated by Bardeen, Cooper and Schrieffer (BCS) in their epoch-making paper of 1957 (Bardeen *et al.* 1957), which has been the basis for most work in the area ever since. The extent to which the BCS theory can be regarded, qualitatively, as a picture in which pairs of electrons form effective 'molecules' which behave as bosons and in particular undergo BEC is a matter of legitimate debate; in the early days of

---

* In the following I shall distinguish between the BCS *theory* and the BCS *model*, using the first to denote a generic theory in which Cooper pairs (see below) are formed, and the latter to include the specific assumptions made in the original BCS paper about the form of the attractive potential etc.

the theory there was an understandable tendency to downplay this aspect and to emphasize the important ways in which the 'Cooper pairs' of BCS theory (see below) differ from ordinary diatomic molecules. On the other hand, the 'molecular' point of view may be quite useful to one's intuition in certain contexts, particularly when dealing with possible 'exotic' forms of pairing where the 'internal' degrees of freedom of the Cooper pairs are important. In any case, this issue does not affect the quantitative formulation of the BCS theory to which I now turn.

A crucial insight, which lies at the heart of the theory, is the observation due to Cooper that while in three dimensions an attractive potential between two electrons has to be of a certain minimum strength (which depends in detail on its shape) in order to form a bound state, if the electrons are restricted to the states outside an already occupied Fermi sea then an arbitrarily weak attractive potential will form a bound state with zero center-of-mass (COM) momentum (i.e. the energy of a state in which the electrons are constrained to be close together in position space is less, by a finite amount, than that of a state in which they scatter to infinity). It is possible to understand this result by noticing that whether or not a bound state is formed depends, apart from the potential, only on the 'density of states', i.e. on the way in which the number of available plane-wave states varies with energy E. As is well known, for two particles in free three-dimensional space, the density of states is proportional to $E^{1/2}$ and hence tends to zero with E; by contrast, in two dimensions it is constant in this limit. If, on the other hand, the particles are constrained to occupy only states above the Fermi sea, the relevant density of states is that at or near the Fermi surface, which to a first approximation is constant. Thus the problem is essentially equivalent to that of a weak attractive potential in *two*-dimensional free space, a case in which it is known that a bound state is always formed.

Let us go through the argument a little more formally. If we write down the Schrödinger equation for two particles with COM momentum zero in free space and take the Fourier transform Ψ_k of the relative wave function $\Psi(r_1 - r_2)$, the equation reads

$$\Psi_k = \frac{1}{2\varepsilon_k - E} \sum_{k'} V_{k-k'} \Psi_{k'} \tag{5}$$

Suppose, for simplicity, that $V_{k-k'}$ is a constant V_0 up to some upper cutoff energy $\varepsilon_c,$* and sum both sides over k. The quantity $\sum_k \Psi_k = \sum_{k'} \Psi_{k'}$ can then be canceled, and one obtains

---

* The alert reader will notice that this is actually not compatible with V being a function only of the difference $k - k'$, i.e. a local function of relative position. This is one of the many serendipitous oversimplifications of the BCS model which have been shown by subsequent work not to affect the results qualitatively.

$$\frac{1}{V_0} = \sum_{\substack{k \\ (\varepsilon_k < \varepsilon_c)}} \frac{1}{2\varepsilon_k - E} \equiv \int_0^{\varepsilon_c} \frac{\rho(\varepsilon)\,\mathrm{d}\varepsilon}{2\varepsilon - E}, \tag{6}$$

where $\rho(\varepsilon)$ is the density of one-particle states per unit energy. If $\rho(\varepsilon)$ is proportional to $\varepsilon^{1/2}$ for small ε, as is the case for two free particles in three-dimensional space, then the right-hand side of eq. (6) is finite for all negative E, and for small enough V_0 the equation has no solution. If, on the other hand, $\rho(\varepsilon)$ is a constant ρ_0 for $\varepsilon \to 0$, then the integral is proportional to $\log(\varepsilon_c/|E|)$ for $E < 0$ (and $|E| \ll \varepsilon_c$), and hence diverges as $E \to 0$ from below. Thus the equation always has a solution, namely

$$E = -2\varepsilon_c \exp(-1/V_0\rho_0). \tag{7}$$

Moreover, it is easy to show that (as we should intuitively expect) the state is indeed 'bound' in position space, that is $\Psi(r_1 - r_2) \equiv \Psi(r)$ vanishes exponentially as $r \to \infty$. This is the case not only for two particles in free two-dimensional space, but also for the Cooper problem, provided that we interpret ε as the excitation energy relative to the Fermi energy and E similarly as the energy of the bound state relative to 2μ. Note the extreme sensitivity of the binding energy $|E|$ to the strength V_0 of the interaction in the limit $V_0 \to 0$.

So far, so good: Cooper's calculation at least strongly suggests that electrons in the presence of an attractive interaction, however weak, close to the Fermi surface would find it energetically advantageous to form a bound state (a 'Cooper pair'). However, it is artificial in that it treats the two electrons whose interaction is considered on a different footing from all the others, whose role is only to occupy the Fermi sea and thereby exclude the states within from occupation by the principal actors. Clearly, a satisfactory calculation has to treat all the electrons on an equal footing, and in particular respect the Fermi statistics, i.e. the requirement that the many-body wave function be antisymmetric under interchange of the coordinates and spin indices of any two electrons. This was the major technical problem that BCS had to overcome; in the present context, the details of the technique used are not of particular interest and I will proceed directly to quote results obtained for the many-body wave function in the so-called 'number conserving' form, actually giving it (with an eye to future generalizations) in a form rather more general than follows from the simple ansatz used in the original BCS paper.

Schematically, then, the generic form of the ground-state many-body wave function $\Psi(r_1\sigma_1 : r_2\sigma_2 \ldots r_N\sigma_N)$ of N electrons[*] in the pairing theory of BCS is as follows:

---

[*] I assume for simplicity of notation that N is even.

$$\Psi(r_1\sigma_1:r_2\sigma_2 \ldots r_N\sigma_N) = \mathcal{N}\mathcal{A}\varphi(r_1\sigma_1 r_2\sigma_2)\varphi(r_3\sigma_3 r_4\sigma_4) \ldots \varphi(r_{N-1}\sigma_{N-1}r_N\sigma_N).$$

$$(8)$$

Here $\mathcal{N}$ is a normalization factor and $\mathcal{A}$ is an operator which antisymmetrizes under exchanges of the coordinates and spins of *any* two electrons; thus, not only must $\varphi(r_1\sigma_1 r_2\sigma_2)$ be antisymmetric under the exchange $r_1 \leftrightarrow r_2$, $\sigma_1 \leftrightarrow \sigma_2$, but there is also, e.g., an explicit term of the form $-\varphi(r_1\sigma_1 r_3\sigma_3)\varphi(r_2\sigma_2 r_4\sigma_4)$ $\ldots \varphi(r_{N-1}\sigma_{N-1}r_N\sigma_N)$ and so on. It is also necessary to specify explicitly that the function $\varphi(r_1\sigma_1 r_2\sigma_2)$ falls off sufficiently fast (usually exponentially) for $|r_1 - r_2| \to \infty$ (if one does not specify this, then the state (8) actually describes, as a special case, the normal Fermi-sea ground state of non-interacting electrons). When I say that eq. (8) is 'schematically' the form of the many-body wave function in the pairing theory, what I mean is that one can add all sorts of complications to eq. (8), but, provided they preserve the basic 'topology', the resulting state will still have all the essential properties of a superconducting state.

To what extent can we regard the wave function (8) as describing a fully Bose-condensed set of $N/2$ diatomic (actually 'dielectronic') molecules? If we could neglect the antisymmetrization operator it would be just that, with $\varphi(r_1\sigma_1 r_2\sigma_2)$ simply the wave function of one of the molecules. Moreover, let us consider the artificial limit (probably not relevant to the problem of real superconductors, as we shall see) in which the 'range' of $\varphi(r_1\sigma_1 r_2\sigma_2)$ in the relative coordinate $|r_1 - r_2|$ is small compared with the mean spacing between electrons. In that case the antisymmetrization has very little effect,[*] i.e. all physical properties of the system are much the same without it as with it. In this case it is quite natural to interpret eq. (8) as describing a set of $N/2$ 'molecules' whose size is small compared with their average spacing, and which behave as free independent bosons and thus undergo BEC (note that the spin of the 'molecule' must be integral and hence we would indeed expect the relevant statistics to be of the Bose type).

In real-life superconductors, however, the range R of the function $\varphi(r_1\sigma_1 r_2\sigma_2)$ in $|r_1 - r_2|$ is sufficiently large that the pairs overlap heavily; in fact, typically, within the 'molecular' (?) volume R^3 there are billions of other electrons belonging to different Cooper pairs! Under these circumstances, the interpretation of eq. (8) as describing a set of $N/2$ independent molecules which undergo BEC is somewhat problematic. In particular, it turns out that in this limit the quantity $\varphi(r_1\sigma_1 r_2\sigma_2)$ which occurs formally in eq. (8) has very little direct physical significance; rather, all physical properties depend on a

---

[*] I refer here to its effect in interchanging, e.g., $r_2\sigma_2$ and $r_3\sigma_3$ in eq. (8). Antisymmetrization 'within' a pair (e.g. $r_1\sigma_1 \rightleftarrows r_2\sigma_2$) may still be important.

related but different quantity (sometimes called the 'anomalous average' of electron operators), which is usually written

$$F(r_1\sigma_1 r_2\sigma_2) \quad \text{or} \quad F(r\sigma: r'\sigma').$$

For example, just as the average potential energy of a two-particle system may be written (suppressing the spins) in the form

$$\langle V \rangle = \iint dr\, dr'\, V(r - r')|\Psi(rr')|^2, \tag{9a}$$

so in the case of a superconductor it can be written

$$\langle V \rangle = \iint dr\, dr'\, V(r - r')|F(rr')|^2. \tag{9b}$$

Thus, F plays, almost everywhere, the same role for two electrons in a superconducting system as the wave function $\Psi(r_1\sigma_1 : r_2\sigma_2)$ does for a two-particle system in free space, and it is therefore natural to regard it as the 'wave function of the Cooper pairs'. In the following we shall study its properties in detail.

Let us specialize, for simplicity of presentation, to the original BCS model, in which the pairs form with spin and orbital angular momentum both equal to zero, and moreover assume for the moment that their centers of mass are at rest. (We also assume $T = 0$.) In this case, the quantity F has the very simple form

$$F(r_1\sigma_1 : r_2\sigma_2) = \frac{1}{2^{1/2}}(\uparrow_1\downarrow_2 - \downarrow_1\uparrow_2)f(|r_1 - r_2|), \tag{10}$$

where the quantity in the first bracket is a symbolic representation of the spin singlet ($S = 0$) wave function of two spin-1/2 particles. It is important to realize that given a particular Hamiltonian the quantity $f(|r_1 - r_2|)$ is not arbitrary but is fixed by the energetics. Thus, in this model *the Cooper pairs have no internal degrees of freedom*.

The quantity $f(|r_1 - r_2|) \equiv f(r)$ clearly plays the role of the wave function of the relative coordinate of the pairs, and it is of interest to investigate the general nature of this dependence on r. At very short distances (of the order of a lattice spacing), this may be complicated, and may depend in detail on the nature of the interaction and the band structure, in a way specific to the individual metal in question. For rather larger distances, the quantity $f(r)$ looks very much like the wave function of two free electrons scattering in states close to the Fermi energy; in particular, it will have a characteristic oscillation with a period of approximately $(2k_f)^{-1}$. At this stage, it is impossible to tell, by inspection of $f(r)$, whether it corresponds to a bound state or not. However, if we go to large enough values of r, we find that the envelope of the oscillating function starts to fall off exponentially; typically, as $r \to \infty$ we have, apart from power-law prefactors,

$$f(r) \sim \exp(-r/\xi_0) \qquad (11)$$

so that the quantity ξ_0 is effectively the 'radius' of the pair. ξ_0 is usually known (up to a factor of order unity) as the (Pippard) coherence length[*] of the superconductor in question; in a clean metal, it is also a measure of the distance over which the pairs can give rise to 'non-local' effects; e.g., an electric field applied to one member of a pair will affect its partner a distance $\sim \xi_0$ away. It turns out that ξ_0 depends exponentially on the strength of the interaction binding the pairs:

$$\xi_0 \sim a \exp(1/V_0 \rho_0) \qquad (12)$$

(where the prefactor a is of the order of a few lattice spacings) and hence can be very large on the scale of the inter-electron spacing: e.g., for Al, ξ_0 is of the order of 10^4 Å (1 micron).

A striking feature of the BCS ground-state wave function given by eq. (8) is that not only are all electrons bound in pairs, but they are bound in pairs with *the same* 'molecular' wave function $\varphi(r_1 \sigma_1 : r_2 \sigma_2)$. Let us ask what happens if we relax this feature. First, imagine that we break up a single pair and put the two electrons in question into independent single-particle plane-wave states k_1 and k_2. It turns out that the extra energy necessary to do this is a sum of two independent terms $E(k_1)$ and $E(k_2)$, where the 'single-particle excitation energy' $E(k)$ is given by the expression

$$E(k) = (\varepsilon_k^2 + |\Delta_k|^2)^{1/2}, \qquad (13)$$

where ε_k is the normal-state excitation energy of the one-electron state in question *relative to the Fermi energy*, and Δ_k is a complex quantity which, for reasons we shall see below, is usually called the 'energy gap'. However, an equally important property of Δ_k is that it is related to the Fourier transform F_k of the relative wave function $f(r)$ of the Cooper pairs by the expression (valid at zero temperature)

$$F_k = \Delta_k/2E_k. \qquad (14)$$

In the simple case of $S = 0$, $l = 0$ pairing considered by BCS and with their choice of model interaction, it turns out that Δ_k is just a constant Δ, independent of k. It is then immediately obvious that the minimum value of the single-particle excitation energy (13) is $|\Delta|$, whence the name 'energy gap'. Moreover, it is clear from eq. (14) that the quantity F_k is appreciable only for $|\varepsilon_k| \lesssim |\Delta|$ and falls off for values of $|\varepsilon_k|$ larger than this roughly as $|\varepsilon_k|^{-1}$; it is

---

[*] Note that the term 'coherence length' is used in the theory of mesoscopic systems with a different meaning; see chapter 9.

this behavior which, when Fourier-transformed, gives rise to the behavior (11) of the coordinate-space relative pair wave function $f(r)$, with, moreover, ξ_0 related to $|\Delta|$ by

$$\xi_0 \sim \hbar v_F / |\Delta|, \tag{15}$$

v_F being the Fermi velocity as usually defined. (Thus, from eq. (12), $|\Delta| \sim (\hbar v_F/a)\exp(-1/\rho_0 V_0)$.

Thus, to break up a single pair and put the two electrons into independent plane-wave states costs a minimum energy of $2|\Delta|$ ($|\Delta|$ for each electron). But we might imagine another way of modifying the BCS ground-state wave function (8); namely, we continue to constrain all N electrons to be paired but allow the pair state to be different for different pairs. As an example, let us suppose that we start with the wave function (8), with a form φ_0 of φ which corresponds to the center of mass of the pair being at rest (i.e. φ is independent of the COM variable $R \equiv (r_1 + r_2)/2$): now we take just half of the $N/2$ pairs and replace φ_0 by a function whose COM dependence is $\exp i K \cdot R$, correspond-ing to a COM momentum $\hbar K$. Thus, in words, half the pairs have momentum zero and the other half have momentum $\hbar K$. How much energy does this cost? If we could really regard the Cooper pairs as equivalent to independent 'diatomic molecules', the answer would clearly be simply the number of 'moving' pairs, namely $N/4$, times the kinetic energy of a single moving pair, $\hbar^2 K^2/4m$, where m is the electron mass. Now, however, we confront the crucial difference between Cooper pairs and independent diatomic molecules: because of the indistinguishability of the electrons, by constructing the wave function in this way we lose the potential energy corresponding to scattering between states with COM momentum zero and those with COM momentum $\hbar K$.

This is perhaps the subtlest point in the whole of the pairing theory, so it is worthwhile taking a moment to illustrate it by the analogy with a single molecule. Suppose we have two particles interacting by some attractive potential $V(r)$ with Fourier transform V_k, and let the Fourier transform of the wave function $\psi(r_1 r_2) \equiv \psi(rR)$ with respect to the relative coordinate r and COM coordinate R be denoted ψ_{kK} (i.e. k is the Fourier variable conjugate to r and K that to R). Then it is straightforward to show that the expectation value of the kinetic energy has the form

$$\langle T \rangle = \sum_{k,K} \left(\frac{\hbar^2 k^2}{m} + \frac{\hbar^2 K^2}{4m} \right) |\psi_{k,K}|^2 \tag{16}$$

(note that, whereas the total mass is $2m$, the reduced mass for relative motion is $m/2$!), while the expectation value of the potential energy is

$$\langle V \rangle = \sum_{k,k',K} V_{k-k'} \Psi_{k,K} \Psi_{k',K}. \tag{17}$$

For simplicity in the following argument I will assume that all V_k are < 0 (so that all c_k will have the same phase). The crucial point is that, in expression (17), because of conservation of total momentum, *components $\Psi_{k,K}$ with different COM momentum K are not coupled*. Suppose now that the wave function of the ground state with the COM at rest has the form

$$\Psi_{k,K} = c_k \delta_{K,0} \tag{18}$$

and divide the possible relative momenta k into two roughly equal sets C_1 and C_2 in an arbitrary way. We now consider the two possible excited-state wave functions:

$$\text{(a)} \ \Psi_{k,K} = c_k \delta_{K,K_0}. \tag{19}$$

(this is just eq. (18) given a Galilean boost, and it is clear from eqs. (16) and (17) that its excitation energy is just $\hbar K_0^2 / 4m$ and tends to zero with K_0);

$$\text{(b)} \ \Psi_{k,K} = \begin{cases} c_k \delta_{K,0} & \text{if } k \in C_1 \\ c_k \delta_{K,K_0} & \text{if } k \in C_2 \end{cases}. \tag{20}$$

What is the excitation energy of the state (20)? It has a contribution from eq. (16) which is proportional to K_0^2. However, it also has a contribution from eq. (17),

$$\langle V \rangle = - \sum_{\substack{k \in C_1 \\ k' \in C_2}} V_{k-k'} c_k^* c_{k'} > 0, \tag{21}$$

indicating the fact that we have *lost* the (attractive) potential energy associated with scattering from states in C_1 to those in C_2. The excitation energy (21) does not tend to zero with K_0; hence, there is a *finite energy gap* against deforming the molecular wave function according to eq. (20).

Because of the indistinguishability of the electrons, an essentially similar phenomenon occurs for Cooper pairs: if we start from $K = 0$ and try to put some fraction (less than unity) of the pairs (e.g.) in a state with finite COM momentum K_0, we get a term in the excitation energy corresponding to eq. (21) which does not tend to zero with K_0. Although we have gone through the argument explicitly for the case of a Galilean boost, it applies equally to other kinds of possible deformation: e.g., if it should happen that the ground-state 'molecular' wave function φ for $K = 0$ is two-fold degenerate, e.g. due to some broken rotational symmetry, it would still cost a finite energy to put half the Cooper pairs in one possible state and the rest in the other, compared with putting all of them in a single state (Yip 1984).

The result just obtained is rather remarkable. It says, first of all, that the

Cooper pairs indeed undergo a sort of Bose condensation, but for reasons which are, in essence, dynamical (i.e. based on energy considerations) rather than statistical as in the case of the free Bose gas. But it says more than this: unlike the case of the free Bose gas,* the energy of the state obtained by putting a fraction α of the pairs in an excited state (e.g. a state with finite K) and the rest in the ground state is *not a monotonic function of α*. Thus, Bose condensation is dynamically enforced even in circumstances where it would not be (statistically) enforced for the free Bose gas.

One final point should be noted: at least for the simple BCS case ($l = 0$, $S = 0$), the function $\varphi(r_1 - r_2)$ (or $F(r_1 - r_2)$), which describes the relative wave function of the Cooper pair with COM at rest, is unique, and any deformation of it either in shape *or in overall magnitude* will lead to an increase in energy. It is intuitively rather obvious (and can be shown by a microscopic calculation) that this is a 'local' result, i.e. that (e.g.) an increase in the overall magnitude of F at one value of the COM cannot be compensated by a decrease in another.

It is possible to put the various considerations discussed above together to form a rather simple statement, which is often taken as a starting point in phenomenological discussions of superconductivity. Let us define a so-called 'order parameter' $\Psi(R)$ which is a measure of the 'strength of condensation' at value R of the COM variable. The precise definition of $\Psi(R)$ in terms of the microscopic wave function is somewhat arbitrary, but a conventional definition relates it to the quantity $F(r_1\sigma_1 : r_2\sigma_2)$ (for the simple BCS case) by the relation

$$\Psi(R) = F(r_1\sigma_1 : r_2\sigma_2), \quad \sigma_1 = \uparrow, \sigma_2 = \downarrow, \quad r_1 = r_2 = R, \qquad (22)$$

i.e., in words, it is the probability amplitude of finding two (paired) electrons with opposite spins at the point R. Being a quantum-mechanical probability amplitude, $\Psi(R)$ is a complex quantity, and, for a uniform situation, physical energies etc. can depend only on its modulus squared. On the basis of the above considerations, we can then state that the total energy at $T = 0$ has a contribution of the form

$$E = \int f(|\Psi(R)|^2)\, dR, \qquad (23)$$

where the function f has a minimum at some non-zero value of Ψ. In a situation where $\Psi(R)$ varies appreciably in space, we would expect also terms depending on the gradient; the simplest form of such a term which respects the obvious symmetry requirements is

---

* The interacting Bose gas actually turns out to be similar to superconductors in this respect.

$$E_{\text{grad}} = \int \gamma(|\nabla \Psi(\boldsymbol{R})|^2) \, d\boldsymbol{r}, \tag{24}$$

and this is often used for qualitative calculations (although the form which actually follows, at $T = 0$, from a microscopic calculation is a bit more complicated). Thus, the energy as a functional of Ψ can be taken as the sum of eqs. (23) and (24), and, as we shall see below, this immediately leads us to the explanation of many of the fundamental phenomena of superconductivity. (In qualitative arguments below, I shall often not distinguish explicitly between the quantities φ, F and Ψ; the 'interesting' properties are common to all of them.)

I want to stress that, although the notion of an 'order parameter' which is defined above is often introduced in the modern literature as if it were a quite trivial consequence of the pairing hypothesis, this is very far from the truth; in particular, the very definition of $F(\boldsymbol{r}_1\sigma_1 : \boldsymbol{r}_2\sigma_2)$ and hence of $\Psi(\boldsymbol{R})$ rests implicitly on the assertion that 'Bose condensation' takes place into only a *single* pair state, which, as we have seen, is guaranteed only by a rather subtle argument in which the indistinguishability of the electrons is an essential ingredient. In fact, the moment that one has to deal with any situation more complicated than the simple BCS one (and in particular cases where there are different groups of paired particles, distinguishable e.g. by their spins, as in the case of superfluid ^{3}He), the whole question of the 'uniqueness' of the pairing state becomes highly non-trivial; see, e.g., Leggett (1995).

7.6 BCS theory at finite temperature

The generalization of the BCS theory to finite temperature raises few new conceptual problems. All the arguments concerning the uniqueness of the pairing state (at any given T, P etc.) go through just as at $T = 0$; the main difference, now, is that not all the N electrons form Cooper pairs, since entropic considerations now make it favorable to 'break' some of the pairs and put the relevant electrons into independent single-particle states. These 'normal' (unpaired) electrons behave qualitatively much like the electrons in a normal metal and are said to form the 'normal component' of the system; all characteristically superconducting effects come from the Cooper pairs as at $T = 0$. It turns out that the energy gap $|\Delta|$ depends on T and tends to zero at some temperature T_c; as a result of eq. (14),[*] the same is true of the pair wave function F, and thus T_c is the point at which the number of Cooper pairs vanishes and the system ceases to be superconducting. The quantity $k_B T_c$ is of

---

[*] At finite T, the right-hand side of eq. (14) has to be multiplied by $\tanh(\beta E_k/2)$ $(\beta \equiv 1/k_B T)$, but this does not affect the results qualitatively.

the order of the zero-temperature gap $\Delta(0)$ (in the simple BCS model, $\Delta(0) = 1.75 k_B T_c$). It is interesting that, although the amplitude of the pair relative wave function F tends to zero as $T \to T_c$, the *shape* of the function is not strongly temperature dependent, and, in particular, the 'pair radius'* ξ_0 near T_c is only slightly different from its zero-temperature value.

The expression for the Gibbs free energy G as a functional of the (temperature-dependent) order parameter $\Psi(\mathbf{R})$ turns out to have a particularly simple form in the limit $T \to T_c$:

$$G\{\Psi(\mathbf{R})\} = G_0(T) + \int d\mathbf{r}\{\alpha(T)|\Psi(\mathbf{R})|^2 + \tfrac{1}{2}\beta(T)|\Psi(\mathbf{R})|^4 + \gamma|\nabla\Psi(\mathbf{R})|^2\},$$

(25)

where $\alpha(T) \sim (T - T_c)$ (so changes sign at T_c), $\beta(T) \sim \gamma(T) \sim$ constant. The form (25), which can be justified from a microscopic calculation based on the pairing theory, was actually written down on phenomenological grounds several years prior to the work of BCS by Ginzburg and Landau (1950). It is clear from eq. (25) that, in homogeneous equilibrium, the value of Ψ is zero for $T > T_c$, as of course it must be, while for $T < T_c$ it is non-zero and proportional to $(T_c - T)^{1/2}$ (a temperature dependence which is shared by the energy gap $|\Delta|(T)$).

7.7 How the pairing theory explains the phenomenon of superconductivity

As we have seen, the two fundamental features of superconductors which needed (in 1957) to be explained were the Meissner effect and the stability of supercurrents. The explanation of the Meissner effect follows along the general lines sketched earlier in this chapter for atomic diamagnetism: When a magnetic vector potential $\mathbf{A}$ is applied to a single electron, the expression for the current (with $c = 1$) is

$$\mathbf{j} = \frac{e}{m}(\mathbf{p} - e\mathbf{A}),$$

(26)

where $\mathbf{p}$ is the 'canonical' momentum, which in operator form is $-i\hbar\nabla$. Correspondingly for a Cooper pair (a two-electron complex) the current is, intuitively speaking,

$$\mathbf{j}_{\text{pair}} = \frac{e}{m}\psi^*(\mathbf{p}_1 + \mathbf{p}_2 - 2e\mathbf{A})\psi = \frac{e}{m}\psi^*(-i\hbar\nabla_R - 2e\mathbf{A})\psi - \text{c.c.},$$

(27)

where ∇_R is the gradient of the pair wave function ψ with respect to the COM coordinate $\mathbf{R}$. Now, if we consider a simply connected geometry, the tangential component of the gradient has to be zero since the wave function must be

---

* ξ_0 should not be confused with the 'correlation length' $\xi(T)$, which diverges as $T \to T_c$.

single-valued and no nodes are allowed (they would cost too much energy; see below). Thus, only the term proportional to A survives in eq. (27); and since there are $n_s/2$ Cooper pairs per unit volume, where n_s is the number density of superconducting electrons, we find for the electric current J the result

$$J = \frac{-n_s(T)}{m} e^2 A, \tag{28}$$

in agreement with the Londons' phenomenological result, eq. (2). Note that the factor of 2 has canceled out.

This derivation of the Meissner effect immediately suggests that, in a multiply connected geometry, e.g. a superconducting ring or hollow cylinder, more interesting things can happen. Consider the geometry shown in Fig. 7.1, where the dashed path C lies considerably more than a penetration depth λ_L from the surface. Then the Meissner effect results in the current density being zero on the path C, so we can integrate the right-hand side of eq. (27) around the ring to obtain

$$2e \oint A \cdot dl = \oint \hbar \nabla \varphi \cdot dl, \tag{29}$$

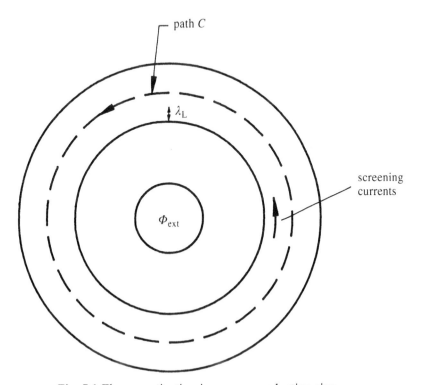

Fig. 7.1 Flux quantization in a superconducting ring.

where φ is the phase of Ψ. But the phase must be single-valued mod 2π, so we obtain

$$\oint A \cdot dl = nh/2e \tag{30}$$

or, since by Stokes's theorem the left-hand side is the flux Φ trapped within the contour C,

$$\Phi = n\Phi_0, \quad \Phi_0 \equiv h/2e, \quad n = 0, 1, 2, \ldots. \tag{31}$$

This is the celebrated phenomenon of *flux quantization*, which was predicted by Fritz London (with a quantization unit h/e) in 1948, and re-derived by Byers and Yang (1961) and by Onsager (1961) with the correct unit in the light of the pairing theory, simultaneously with its experimental discovery. What happens is that, if we apply through the ring an external flux Φ_{ext} which is not an exact multiple of the flux quantum Φ_0, it will induce surface currents to flow on the inside surface of the ring to just the right extent to cancel the non-integral part.

The explanation of the stability of supercurrents in the pairing theory is a little more subtle. Consider a ring geometry, for simplicity thin compared with the London penetration depth, so that the fields set up by the currents may be neglected (i.e. we can set $A = 0$), and suppose we manage to set up a circulating current which involves, at least partially, the Cooper pairs. From eq. (27), we see that this must mean that the integral of the phase gradient around the ring is non-zero, i.e., schematically,

$$\Psi(R) \sim \exp(in\theta), \quad n \neq 0, \tag{32}$$

where θ is the angle around the ring. Now the critical question is: Can the order parameter relax to its thermodynamic equilibrium form (namely $\Psi(R) \sim$ const., corresponding to zero circulating current) by a process which is energetically 'downhill all the way'? One possibility would be to let the Cooper pairs drop back individually to the ground state so that, at intermediate times, some of them would be in the original state, $n \neq 0$, and some in the ground state ($n = 0$) (and possibly some in intermediate states). However, we have already seen that this process is *not* 'downhill all the way'. A second possibility is to keep all the Cooper pairs in the same state but to modify this state continuously from $\Psi_n \sim \exp(in\theta)$ to $\Psi_0 \sim$ const.: say for definiteness

$$\Psi(t) = \alpha(t)\Psi_n(\theta) + \beta(t)\Psi_0(\theta)$$
$$\alpha(-\infty) = \beta(+\infty) = 1, \quad |\alpha(t)|^2 + |\beta(t)|^2 = 1. \tag{33}$$

This is precisely how a single electron in the excited p-state of an atom would make the transition (in the approximation that the EM field is treated as classical). However, it is easy to see on topological grounds that in the process described by eq. (33) the wave function must develop at least one node for some value of θ and some t. In the case of the single atomic electron this does not matter because the equation for the energy (Schrödinger's equation) is linear, and we have, quite generally for the energy expectation value, irrespective of the presence of nodes,

$$\langle E(t) \rangle = E_\mathrm{p}|\alpha(t)|^2 + E_\mathrm{s}|\beta(t)|^2, \tag{34}$$

which is monotonically decreasing as a function of time if $|\alpha(t)|^2$ is. On the other hand, for the Cooper-pair system, the energy is a non-linear function of $|\Psi(R)|^2$ (cf. eqs. (23) and (25)), and, in particular, to make a node in the wave function turns out to cost so much energy that the barrier against doing so is virtually insuperable (except extremely close to T_c when the pair density is very small). Consequently, on these essentially topological grounds, the supercurrent is metastable for astronomical times, except extremely close to T_c where a slow decay can be observed.

(It should be noted, by the way, that this argument for the stability of the supercurrent works, at least in the simple form given, only because of the very simple form – a complex scalar – of the order parameter of an ordinary superconductor. Other superfluid Fermi systems, such as liquid ^{3}He, are believed to have more complicated forms of pairing and hence of the order parameter, and in some cases at least the question of stability of supercurrents can become extremely delicate: cf. Mermin (1978). Fortunately, for superconductors even of the 'exotic' type (see below), these complications are usually removed by the fact that the order parameter is pinned by the crystal lattice.)

There is one more complication that we need to address: in the above argument, it is implicitly assumed that there are no *pre-existing* nodes in the pair wave function. In the so-called type I superconductors (which include most of the elemental ones), this is indeed justified. However, there exists a wide class of superconductors known as type II (including all the 'new types' discussed below) in which the application of a magnetic field of appropriate strength results in the formation of *vortices* – line nodes of the pair wave function around which a screening current circulates, rather as it does around the hole in the flux quantization set-up discussed above. Under certain circumstances, these vortices can move across the current, thereby decreasing the number n in eq. (32) and giving rise to a finite electrical resistance. This effect is particularly important in the high-temperature (cuprate) superconductors; see below.

7.8 The Josephson effect: SQUIDS

The Meissner effect, the persistence of supercurrents and flux quantization are all examples of what have become known as 'macroscopic quantum effects (phenomena)' in superconductors (or superfluids). The word 'macroscopic' can be understood here in several senses: the effects occur because a macroscopic number of Cooper pairs are condensed into the same 'molecular' state, they are manifested by macroscopic values of variables such as the electric current, and they extend over macroscopic dimensions. (It should be emphasized, however, that, despite a widespread misconception, none of these effects in themselves manifest a superposition of macroscopically distinct states of the kind which is interesting from the standpoint of the quantum measurement paradox. See Leggett (1980).) There exist other 'macroscopic quantum effects' in super-conductors apart from those just mentioned; probably the most spectacular, and certainly the most useful for various practical applications, is the Josephson effect (Josephson 1962).

The Josephson effect occurs when two bulk superconductors are connected by a so-called 'weak link' (or 'Josephson junction'), i.e. a region which permits electrons to pass but with much greater difficulty than in the bulk. The classic example of such a weak link is a thin ($\sim$10–20 Å) insulating oxide barrier, but other types are also possible. Because of the difficulty of transferring electrons (and hence *a fortiori* Cooper pairs) through the weak link, we can regard the order parameters Ψ_1, Ψ_2 of the two bulk superconductors as independent in the first approximation. However, they are then coupled by the process of tunneling through the junction. One way of visualizing this situation is to ignore the dependence of the 'molecular' wave function $\varphi(r_1\sigma_1 : r_2\sigma_2)$ on spins and on relative coordinates and to regard it as a function only of the COM coordinate $R \equiv (1/2)(r_1 + r_2)$. Then, very schematically, we can write $\varphi(R)$ in the form

$$\varphi(R) \sim a\varphi_{\mathrm{L}} + b\varphi_{\mathrm{R}}, \tag{35}$$

where φ_{L} and φ_{R} are functions which are finite only in the left and right bulk superconductors, respectively, and a and b are complex constants; then we have schematically $\Psi_1 \sim a$, $\Psi_2 \sim b$. On the basis of an analogy with the case of a single particle whose wave function is similarly 'shared' between two bulk regions as indicated in eq. (35), we should expect the energy to depend *inter alia* on the relative phase of a and b, i.e. Ψ_1 and Ψ_2: in fact, gauge invariance implies that the lowest-order term has to be of the form

$$E(\Psi_1, \Psi_2) \sim \mathrm{const.} - \tilde{E}_{\mathrm{J}}(\Psi_1^* \cdot \Psi_2 + \mathrm{c.c.}) \tag{36}$$

(where the constant may depend on the magnitude of Ψ_1, Ψ_2 but not on their relative or absolute phase). Defining $E_J \equiv \tilde{E}_J |\Psi_1| \cdot |\Psi_2|$, we can rewrite eq. (36) in the form

$$E = \text{const.} - E_J \cos \varphi, \tag{37}$$

where $\varphi \equiv \arg(\Psi_1^* \cdot \Psi_2)$. So the energy is a minimum (in this simple model) when $\varphi = 0$; i.e. the phase of the pair wave function is the same in the two bulk superconductors – as we would expect from experience with, e.g., the tight-binding model of electrons in a crystal lattice.

It turns out that the existence of an energy of the form (37) gives rise to the possibility of a non-dissipative supercurrent across the junction, which itself depends on the relative phase

$$I_s = I_c \sin \varphi, \quad I_c \equiv (2e/\hbar)E_J. \tag{38}$$

Moreover, the time dependence of the phase is related to the potential difference V applied across the junction by

$$\frac{\mathrm{d}\varphi}{\mathrm{d}t} = \frac{2eV}{\hbar}. \tag{39}$$

These relations, originally discovered by Josephson (1962), form the basis of a rich variety of applications of the Josephson effect in metrology, infrared detection and elsewhere: see, e.g., Barone and Paterno (1982).

One particularly interesting application which combines the Josephson effect with concepts related to flux quantization is the device known as a dc SQUID (superconducting quantum interference device): see Fig. 7.2. Any current in the device flows, because of the Meissner effect, along the edges, and thus the current on the dashed contour is zero. Thus from eq. (29) we find for the phase of the order parameter

$$\varphi_A - \varphi_C = \frac{2e}{\hbar} \int_A^C A \cdot \mathrm{d}l, \tag{40}$$

etc. Since the total change of phase on going around the ring (which of course includes the discontinuous changes across the junctions themselves) must be $2n\pi$, we find that the *difference* $\Delta\varphi$ of the phase differences φ_1, φ_2 across the two junctions (taken in the same (parallel) direction) is given by

$$\Delta\varphi = 2n\pi + \frac{2e}{\hbar} \oint A \cdot \mathrm{d}l = 2n\pi + 2\pi(\Phi/\Phi_0), \tag{41}$$

where $\Phi_0 \equiv h/2e$ is the flux quantum as above, and Φ is the total flux through the ring (which may have an 'induced' component but is not now necessarily an integral multiple of Φ_0). It now follows from eq. (38) that the maximum value of current which the two junctions in parallel can carry without dissipation is the maximum with respect to φ of the sum of $I_c \sin \varphi$ and

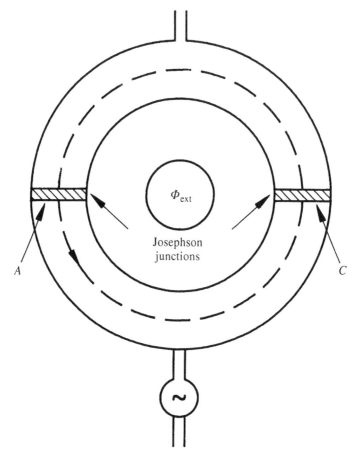

Fig. 7.2 A dc SQUID magnetometer (schematic).

$I_c \sin(\varphi + \Delta\varphi)$ (where I assume for simplicity that the critical currents of the two junctions are equal). The result is

$$I_c^{tot} = 2I_c \cos\left(\frac{\pi\Phi}{\Phi_0}\right),$$ (42)

which in particular is a maximum ($2I_c$) when Φ is an integral number of flux quanta and zero when it is a half-odd-integral number. A remarkable feature of this result is that the flux trapped through the ring can affect a *macroscopic* quantity, the critical current, even though there may be no electrical or magnetic field anywhere in the region actually occupied by the electrons carrying the current! This is a particularly dramatic example of the Aharonov–Bohm effect (see chapter 9), and is the basis of a type of very accurate magnetometer. (See, e.g., Barone and Paterno (1982).)

7.9 The mechanism of attraction: new types of superconductor

In the discussion of the pairing theory so far, I have assumed that, in the presence of an attractive interaction between electrons near the Fermi surface, the many-body ground state will have the generic form given in eq. (8); however, I have said nothing about the origin of the necessary attraction nor about the ways in which it may constrain the form of $\varphi(r\sigma: r'\sigma')$, or equivalently of the pair wave function $F(r\sigma: r'\sigma')$. This is deliberate, since, generally speaking, the arguments given above about how the pairing hypothesis explains the phenomena of superconductivity are insensitive to the form of F and hence *a fortiori* to the origin of the attraction responsible for pairing. However, in order to discuss meaningfully the new classes of superconductor discovered in the last two decades, it is necessary to address these issues.

In the original BCS model, which has served as an indispensable starting point for studying the superconductors known before 1979, the attraction which binds the pairs is visualized as due to the exchange of virtual phonons: one electron polarizes the ionic lattice, and the positive polarization charge then attracts the second electron to the track left by the first. (Meanwhile, the direct Coulomb repulsion between the two electrons is strongly reduced because of collective electronic screening effects.) In the original BCS calculation, this phonon-induced interaction was modeled by a constant attraction operating only between electron states within an energy $\hbar\omega_D$ of the Fermi energy, where ω_D is the phonon Debye frequency (maximum frequency); however, subsequent, more sophisticated, calculations have treated the interaction more realistically. One feature which almost invariably emerges from such calculations is that, since the phonon-induced interaction is not very sensitive to momentum transfer, it is always advantageous to form the Cooper pairs in the simplest possible pairing state, namely that with $l = 0$, $S = 0$; we have already used this state repeatedly for illustration. One of its properties is that, for $|\varepsilon_k| \lesssim k_B T_c$, the gap function $|\Delta_k|$ is approximately constant, and this has the consequence that the number of normal electrons excited by breaking the pairs falls off exponentially at low temperatures, giving rise to an exponential dependence in those properties (such as the specific heat) which are associated with the normal component. Such a dependence is indeed generally observed in the superconductors known prior to 1979. It is also noteworthy that none of those 'old-fashioned' superconductors had a transition temperature above 25 K, and indeed there was a general feeling that a temperature around 30 K would be the ultimate limit for any phonon mechanism of superconductivity.

In the last two decades, however, three quite new classes of superconductors have been discovered for which the mechanism has certainly or possibly nothing to do with phonon exchange, and in some of which, at least, the pairing may be of a type quite different from the BCS model. I deal with them in non-historical order:[*] the alkali-doped fullerenes, the heavy-fermion superconductors and, in the next section, the high-temperature (cuprate) superconductors. It should be remarked that, in the analysis of possible scenarios for these materials, it has often been useful to refer to the one example of a superfluid Fermi system which is almost universally believed to form Cooper pairs with an 'exotic' symmetry, namely the superfluid phases of liquid ^{3}He (see, e.g., Leggett (1975)): since in this system the pairs are formed of neutral atoms rather than electrons, it does not form part of the subject matter of this chapter, but it has been a very useful test-bed for some general ideas concerning 'exotic' pairing.

The first, and in the present context perhaps least exciting, class of new superconducting materials is the alkali-doped fullerenes. It has been discovered in the last few years that certain large molecules of carbon, in particular C_{60} and C_{70}, are exceptionally stable and will form regular molecular crystals under appropriate conditions. In the undoped state these crystals are insulating, but when doped with an appropriate number of alkali atoms per unit cell, as, e.g., in the compound K_3C_{60}, they can become superconducting with transition temperatures as high as 33 K (which, prior to 1986, would have been the world record!). Despite these high transition temperatures and the fact that the general nature of the crystals is very different from that of a traditional superconductor, the properties of the superconducting state seem remarkably similar to the classic textbook examples, and, in particular, the low-temperature properties seem entirely consistent with an energy gap $|\Delta_k|$ which is close to constant over the Fermi surface. Thus, there are strong indications that the Cooper pairs formed in these materials are of the simple BCS ($l = 0, S = 0$) type. That of course does not answer the question whether the mechanism of formation of the pairs is the traditional phonon exchange, and indeed some very interesting mechanisms which are purely electronic in nature have been suggested (see, e.g., Chakravarty *et al.* (1991)); but the prevailing majority opinion is probably that, despite the fact that T_c can go somewhat above what used to be thought of as the ultimate limit for the phonon mechanism, that mechanism is the likeliest explanation of superconductivity in these materials.

The situation is different with those members of the class of heavy-fermion metals which become superconducting. As will be explained in chapter 8, the heavy-fermion systems are a class of metals containing either Ce, U or some

---

[*] Historically, the class of heavy-fermion superconductors was discovered in 1979, that of the cuprates in 1986 and the doped fullerenes in 1991.

other actinide, which have anomalous properties at low temperatures; in particular, the specific heat, if interpreted in terms of an effective mass m^*, can correspond to a value of m^* more than 1000 times the bare electron mass. Since 1979, it has been found that some, though not all, of the heavy-fermion metals become superconducting, though only at temperatures below 1 K. Now, according to eq. (15) and the fact that $\Delta(0) \sim k_B T_c$, the pair radius ξ_0 is related to T_c (in the model) by

$$\xi_0 \sim \hbar v_F / k_B T_c, \tag{43}$$

where v_F is the Fermi velocity, and thus, other things being equal, low transition temperatures imply large pair radii. In the case of the heavy-fermion systems, however, other things are not equal: in fact, the very large value of m^*, coupled with a 'normal' value of the Fermi wave vector k_F, implies a very small value of v_F, and, as a result, the size of the Cooper pairs in the heavy-fermion systems is much *less* than in a typical 'old-fashioned' superconductor, typically of the order $50-100$ Å. It is interesting that, in contrast to the high-temperature (cuprate) superconductors (see below), the occurrence of super-conductivity in the heavy-fermion systems does not seem tied to a particular lattice structure; e.g., both UBe_{13} (cubic) and UPt_3 (hexagonal) become super-conducting.

There is a widespread belief that the Cooper pairs formed in some or all of the heavy-fermion superconductors are *not* of the simple BCS-model type ($l = 0, S = 0$) but that they correspond to a more sophisticated pair wave function which does not have the same symmetry as the crystal lattice (and thus is somewhat analogous to the pairs formed in superfluid ^{3}He). The evidence for this hypothesis is somewhat circumstantial, but nonetheless strong. In the first place, the phase diagram of UPt_3 in the $B-T$ (magnetic field–temperature) plane seems to indicate the presence of at least three different superconducting phases, while the compound $U_{1-x}Th_xBe_{13}$ appears, for a certain range of x, to have a 'split' transition (i.e. the usual peak in the specific heat which occurs at T_c appears to be split). While it is difficult to accommodate these features within the simple BCS model, they can be given a natural explanation if the pairing is 'exotic'. Secondly, various properties associated with the normal excitations, such as the specific heat, appear to vanish as $T \to 0$ not exponentially as one would expect for the BCS model (see above) but rather as a power law, which is exactly what one would expect if the energy gap $|\Delta_k|$ has nodes – a feature which is a common, though not universally necessary, consequence of 'exotic' pairing. Because of the various different crystal symmetries involved and the possibility of strong spin–orbit coupling, the analysis of the various possible pairing states is quite complicated and I will not discuss it here (the problem is reviewed for the case of UPt$_3$ by Sauls (1994): for a more general discussion at a less technical level, see Cox and Maple (1995)). At any rate, it seems highly probable that the heavy-fermion superconductors provide the first example of 'exotic' pairing in a system of electrons. This would suggest (and the suggestion is bolstered by the apparent absence of any appreciable isotope effect in these materials) that the mechanism of attraction cannot have much to do with phonons but might be purely or mainly electronic in origin. However, the detailed nature of such a mechanism remains controversial, and, as emphasized in chapter 8, is likely to remain so as long as a generally agreed theory of the normal state is lacking.

7.10 High-temperature (cuprate) superconductivity

Although the discovery of superconductivity in the heavy-fermion compounds showed that the simple BCS model might not be universally applicable, it did little to shake the confidence of most people working in the area that the problem of superconductivity was essentially understood and that there were likely to be few qualitative surprises; in particular, I suspect that most people, if asked in the summer of 1986 to bet on the possibility of occurrence of the phenomenon at any temperature above 30 K, would have given very long odds indeed against it. The discovery late that year of the first member of the class of cuprate ('high-temperature') superconductors, and the subsequent rapid expansion of that class to include materials which become superconducting at temperatures up to over 150 K, was therefore a severe shock to the low-temperature community.

The nature and origin of superconductivity in the cuprates are still highly controversial and contentious issues; I will confine myself here to giving a summary of the salient experimental facts and some brief comments on our current understanding of them.

Without exception, every compound which has been reproducibly shown (at the time of writing) to exhibit the phenomenon of superconductivity at any temperature above 35 K is characterized by a striking structural feature, namely the presence of planes containing Cu and O atoms in the ratio 1:2, and arranged approximately in the square (tetragonal) structure shown in Fig. 7.3. (In most cases, the lattice is not exactly tetragonal but slightly orthorhombic, and a small degree of buckling may occur.) The class of materials having this property is known generically as the cuprates: not all cuprates are superconducting, but many are, and one interesting observation is that, while the presence of the CuO$_2$ planes seems essential, very little else is: in fact, the residual (off-plane) atoms can be of various types and numbers, the only necessary requirement

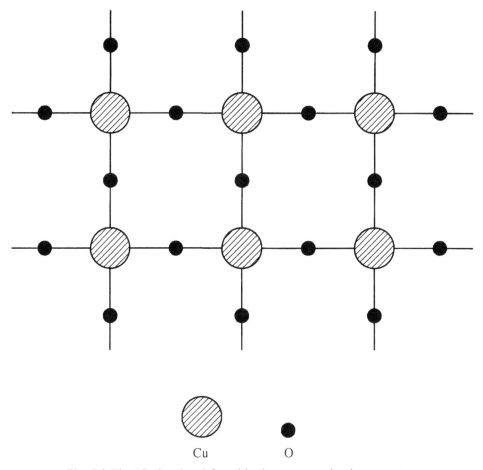

Fig. 7.3 The 'CuO$_2$ plane' found in the superconducting cuprates.

apparently being that they can donate electrons (or more usually holes) to the CuO$_2$ planes. In particular, the number of CuO$_2$ planes per unit cell (1, 2, or 3), and the maximum spacing between the planes, does not seem particularly critical: e.g., while the so-called 'infinite-layer' compound Sr$_{1-x}$Ca$_x$CuO$_2$, in which each successive pair of CuO$_2$ planes is 'spaced' only by a single Sr or Ca atom and thus has a separation of ~3.5 Å, is superconducting up to around 115 K, the compound HgBa$_2$CuO$_4$, with an inter-plane spacing of ~9.5 Å, already has a transition temperature of above 95 K. Apart from their high critical temperatures, many of the cuprate superconductors are also characterized by critical magnetic fields for the destruction of superconductivity which are, by traditional standards, so enormous that at present it is impossible to drive these materials normal at low temperatures!

The normal state of the cuprate superconductors (i.e. the phase above T_c) is itself highly anomalous: very few experimental properties are even qualitatively similar to those of the traditional textbook metals such as Al or Pb. A particularly striking example is the dc electrical resistivity, which for the materials in each group with the highest T_c appears to be linear in temperature at all temperatures above T_c, with no indication of the drop which is seen in 'textbook' metals at temperatures of the order of the Debye temperature Θ_D. It is, moreover, striking that if this resistivity is interpreted in terms of the traditional Drude model with a phenomenological scattering time τ, then τ appears always to be of order $\hbar/k_B T$ (again, this is true for textbook metals at $T \gtrsim \Theta_D$, but not more generally). Other properties, such as the Hall effect and the Pauli spin susceptibility, also behave in a highly unusual way.

In the light of this, it is all the more striking that the microscopic properties of the superconducting state in these materials actually appear, at the qualitative level, rather surprisingly conventional. In fact, if one was given a graph of the behavior of some typical experimental property (specific heat, spin susceptibility, London penetration depth) against reduced temperature T/T_c, and did not look too closely at the details of the behavior in the limit $T \to 0$, it would be quite difficult to tell that one is not looking at a traditional BCS-model superconductor. However, if one indeed looks closely at the low-temperature behavior, it turns out that the temperature dependence is usually not exponential as predicted in the BCS model, but corresponds more closely to a power law (cf. the heavy-fermion case). Moreover, if one assumes a 'pairing' model and tries to extract a zero-temperature energy gap $\Delta(0)$ from the data, it turns out that its ratio to $k_B T_c$ is substantially larger than the BCS value of 1.75.

While the 'microscopic' properties appear, then, to be quite similar on a crude scale to those of traditional superconductors, the difference in the macroscopic electromagnetic behavior is almost qualitative. The experimental magnetic behavior of the cuprate superconductors corresponds to the extreme type-II limit, and one result of this is that, at fields well below the critical field, vortices are very plentiful and moreover appear to be very mobile compared with those in the traditional superconductors; consequently, in magnetic fields of the order of a few tesla the sample can develop a substantial resistance due to vortex slippage (cf. above), even when in other respects its state appears to be fully 'superconducting'. Indeed, the whole question of whether the definition of 'superconductivity' in the cuprates is unique is at present a controversial one. (It should be stressed, however, that the difficulties in this connection are specific to the case of substantial external magnetic fields; for zero field there is no important difference from traditional superconductors.)

A crucial question which arose in the earliest days of research on high-temperature superconductivity is: Is the mechanism of superconductivity in those materials indeed the formation of Cooper pairs, as in the traditional superconductors, or could it be something completely different? Initially the answer to this question was not at all clear, and there was much speculation on mechanisms which *prima facie* had nothing to do with the BCS theory. One idea which at one time seemed very appealing, particularly in view of the strongly layered nature of the cuprates, was connected with the concept of 'anyons' (Leinaas and Myrheim 1977; Wilczek 1990); these are objects which can exist only in a two-dimensional system, obey statistics intermediate between the familiar Bose and Fermi types, and (allegedly) are superfluid even without the need for attractive interaction (see, e.g. Wilczek (1990)). A severe constraint was soon put on these and similar speculations by the experimental observation that the cuprate superconductors (or at least the 'workhorse' material $YBa_2Cu_3O_{7-\delta}$ (YBCO)) show the phenomenon of flux quantization with the standard unit $h/2e$ – a circumstance which suggests very strongly that it must be possible, even if it is not necessary, to describe what is going on in the language of Cooper pairing. (In fact, it has been shown by Rokhsar (1993) that the most popular variant of 'anyon' theory is indeed simply an alternative language for describing a certain kind of Cooper-paired state.) I will therefore assume from now on that the fundamental mechanism of superconductivity in the cuprates is indeed the formation of Cooper pairs as described above.

Given this, what do we know about the nature of the pairs? The first question relates to their size. If one believes that some kind of BCS-like model is quantitatively adequate, one can infer this from the magnitude of the zero-temperature gap $\Delta(0)$ (cf. eq. (15)); alternatively, there is a rather more indirect way of trying to infer it from the critical magnetic field. While these methods do not give the same numerical value, both place the pair radius[*] in the range 10–30 Å. This is considerably smaller than even in the heavy-fermion systems, indeed it may be only two or three times the typical distance between electrons; thus, the pairs in the cuprates are much closer to being something like 'dielectronic molecules' than in any other known system. One interesting consequence of this is that it may be legitimate to think of the pairs as, in some sense, existing even above the transition temperature T_c, which then has the significance of the point at which they undergo (a kind of) Bose condensation; and indeed there are some phenomena in the normal phase which it is tempting to interpret according to this scenario.

A question which has attracted considerable interest since almost the earliest

---

[*] This refers to the 'in-plane' radius. It is a much more delicate question whether it makes sense to define a 'size' of the pairs normal to the planes, since they are usually regarded as primarily defined within a single plane.

days is: Is the nature of the relative (internal) wave function of the pairs
'conventional', i.e., roughly speaking, of the $l = 0$, $S = 0$ variety considered in
the original BCS calculation, or is it something more 'exotic', as in superfluid
^{3}He and, probably, some or all of the heavy-fermion systems? Actually, one
needs to phrase this question a little more precisely. It is believed that spin–
orbit coupling is quite weak in the cuprates (as distinct, possibly, from the
heavy-fermion systems); moreover, all the evidence from nuclear magnetic
resonance suggests that the spin state of the pair is a singlet ($S = 0$). So writing
for COM at rest

$$F(r_1\sigma_1 : r_2\sigma_2) = 1(\uparrow\downarrow - \uparrow\downarrow)F(r) \quad (r \equiv r_1 - r_2) \tag{44}$$

(cf. eq. (10) for the notation), let us consider the behavior of the relative
coordinate wave function $F(r)$ which, by the Pauli principle, must have even
parity. The interesting question is not so much the detailed form of this
function, but how it transforms under the operations of the relevant crystal
symmetry group, which, if we neglect the small orthorhombic anisotropy, is the
group of the square, C_{4v}. Now the fundamental operations of this group can be
taken as $\pi/2$ rotations and reflections in a crystal axis, and it turns out that
there are exactly four irreducible even-parity representations, which are
denoted in technical group-theoretic notation, respectively, A_{1g}, A_{2g}, B_{1g} and
B_{2g}, or more informally s^+ (or simply s), s^- (or g), $d_{x^2-y^2}$ and d_{xy}. I will
concentrate here on the two candidates which are generally thought to be of
most interest, namely s and $d_{x^2-y^2}$ (for a more complete discussion see, e.g.,
Annett et al. (1996)). The 's-wave' state has the unique property that it is
invariant under all operations of the crystal symmetry group, and is therefore
the analog of the $l = 0$ state for three-dimensional space; it is thus in some
sense the most 'BCS-like' candidate. However, it is important to realize that it
need not look at all 'isotropic', indeed it may even have nodes, provided that
the $\pi/2$ and reflection symmetry is preserved. A typical example of such a
state is shown in Fig. 7.4(a). By contrast, the so-called $d_{x^2-y^2}$ state changes
sign on $\pi/2$ rotation; a typical example is shown in Fig. 7.4(b). It can be seen
from Figs. 7.4(a) and (b) that, except close to the 45° axis, the *magnitude* of
the pair wave function (order parameter) as a function of angle may look quite
similar for the two cases; the major difference is the relative sign of the various
'lobes'.

Why is the question of the symmetry of the pair wave function regarded as
so important? Apart from its intrinsic interest, a major reason is the hope that it
may act as at least a partial discriminant between competing microscopic
theories. At present there exists a vast array of theories of high-temperature
superconductivity, ranging from phonon-based models of the traditional BCS

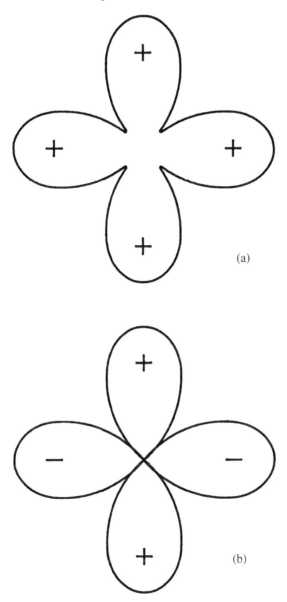

Fig. 7.4 Schematic representation of (a) the s^+ and (b) the $d_{x^2-y^2}$ states.

variety to sophisticated approaches which start from an assumed stronger repulsive interaction between the electrons, and in some cases may superficially appear not to involve pairing at all (see above). Generally speaking, phonon-based theories, and also those which attribute a crucial role to inter-plane coupling (e.g. Chakravarty *et al.* (1993)) have predicted with some confidence that the pairing state is of the simple s-wave type, while theories which start from a strong coupling scenario tend with equal confidence to predict the $d_{x^2-y^2}$ state.

So what is the answer? A detailed discussion of the various experiments performed over the last few years to resolve this question is given in Annett *et al.* (1990) and updated in Annett *et al.* (1996) (see also Scalapino (1995)); I will briefly summarize it here. The earlier (pre-1993) experiments mostly looked either for evidence of nodes in the gap (manifested by power-law dependencies of experimental quantities associated with the normal excitations), or more directly for evidence of gap anisotropy (e.g. in angularly resolved photo-emission spectroscopy (ARPES)). These experiments indicated that the gap must be quite strongly anisotropic, and in particular must be small or zero near the 45° axis; however, as we see from Figs. 7.4(a) and (b), such behavior could be reconciled not only with a $d_{x^2-y^2}$ state but with a sufficiently anisotropic s-state. In the last three years, a number of experiments have been done which exploit the Josephson effect, mostly in the 'dc SQUID' geometry of Fig. 7.2 or something related. These have the potential to determine the relative *sign* of the various lobes of the pair wave function and hence to resolve the issue definitively (see Annett *et al.* (1996) for a detailed discussion). There are by now about a dozen experiments of this type in the literature, all but one of them on YBCO, and with two puzzling exceptions they all seem to indicate that the pairing state is at least predominantly of the $d_{x^2-y^2}$ type. This conclusion, incidentally, is of some importance not only in a fundamental but also in an applied context: if the state is indeed of $d_{x^2-y^2}$ type, it has nodes and hence a non-zero (non-exponentially small) number of excitations at any temperature, however low, and the resulting inevitable dissipation may be a highly relevant factor in the design of devices based on these materials.

7.11 Conclusion

In this chapter we have seen that the occurrence of pairing of itinerant electrons, and the resulting onset of the complex of phenomena we label superconductivity, is very much more widespread than would have been guessed 20 years ago. Is the end in sight? Personally I think not: once we have freed ourselves from the notion that all instances of pairing must necessarily conform to the original simple BCS model (spectacularly successful as that has been!), I believe there is a high probability that by sufficiently inspired trial and error we will find superconductivity in yet other classes of materials, including perhaps some which currently do not even exist, and a reasonable chance that some of the readers of this chapter will live to see the 'Holy Grail', super-conductivity at room temperature.

Acknowledgement

This work was supported by the National Science Foundation under grant DMR 92-14236.

8

The heavy electron

MICHAEL SPRINGFORD

University of Bristol

The study of electrons in metals has contributed richly to the development of our understanding of the physics of condensed matter. Sometimes this has taken the form of the discovery of a new electronic ground state of the system, such as occurred with superconductivity, while at other times we are alerted to the need for better models and calculational procedures, as in the case of the electronic and magnetic properties of the transition metals. As we shall outline in this chapter, the discovery of heavy electrons in metals continues this tradition, both by signalling the presence of a remarkable property of matter and by creating the need for a new theoretical framework to deal with it.

Until the 1970s, many aspects of the behaviour of electrons in a typical non-magnetic metal seemed to be reasonably well explained by a simple model based on Landau's model of a Fermi liquid (chapter 5). In this, the net result of the strong interactions experienced by the electrons was, in summary, to cause them to behave as though they possessed an effective mass m^* somewhat different from that of free particles, together with only weak residual inter-actions between them. The masses of these renormalised electrons, for the majority of metals known at the time, were found to be in the range $0.1-10\ m$.

The first hint of something new came in 1975 with experiments by Andres, Graebner and Ott (Andres *et al.* 1975) on the intermetallic compound CeAl₃. Their measurements at low temperatures yielded two results which differed dramatically from anything previously encountered in metals. Below about 0.1 K, the electronic specific heat and the electrical resistivity were observed to vary with temperature according to $C = \gamma T$ and $\rho = AT^2$, but with values of γ and A that were, respectively, $\sim 10^3$ and $\sim 10^6$ times greater than values typical of normal metals! While these two results are consistent with the presence of a Fermi liquid characterised by renormalised electrons with huge masses, of order 1000 times greater than the free electron mass, it seemed improbable at the time that such an interpretation could be correct. Further results, however, were soon added to the debate by Steglich *et al.* (1979), who showed that the compound $CeCu_2Si_2$ also possessed a giant specific heat at low temperatures. Moreover in this case, below 0.5 K, a large fraction of the material was observed to undergo a superconducting transition accompanied by a jump in the specific heat, which strongly suggested that Cooper pairs are formed out of the heavy Fermi liquid. If uncertainty still surrounded the interpretation of these results, perhaps in view of the difficulty of preparing single-phase $CeCu_2Si_2$, it was surely laid to rest with the discovery of both a giant specific heat and superconductivity in the actinide compound UBe_{13} by Ott *et al.* (1983). The stage was set for an explosion of interest in this extraordinary group of solids, and there followed the discovery and exploration of many new compounds, a process which continues today; some popular examples being $CeCu_6$, CeB_6, $CeRu_2Si_2$, UPt_3, U_2Zn_{17} and UBe_{13}. Particularly stimulating was the discovery that superconductivity should occur in an environment of magnetic ions, which appeared to hold the exciting prospect that superconductivity in this class of materials may be of an unconventional kind; an aspect of their behaviour which is addressed in chapter 7. Containing either rare earth or actinide elements, each of these materials has in common the presence of 4f- or 5f-electrons in unfilled shells which, at low temperatures, would seem to be itinerant. As we shall discuss in this chapter, it is the very strong electron–electron interaction acting within the f-shell which ultimately leads to the presence of renormalised electrons with masses which may indeed be as large as 1000 *m*! As a consequence, these materials are frequently designated, after their most remarkable common feature, as *heavy-electron* compounds or, given the extent of the renormalisation and the expectation that such particles must bear little resemblance to free electrons, as heavy fermions.

8.1 Two key experiments

Although heavy-electron systems are not universal in their behaviour, they nevertheless possess recognisable patterns of behaviour which characterise them as a group. The compound $CeCu_6$, which has been widely studied, is notable for exhibiting the properties of a very heavy Fermi liquid, remaining paramagnetic without the onset of an ordered magnetic state or superconductivity, at least down to temperatures of a few millikelvin. We shall use this compound to illustrate the importance of two key experiments in the historical development of this field. Turning first to the specific heat, and having in mind the variation typical of normal metals,

$$C = \gamma T + \beta T^3,\tag{1}$$

in which γ and β represent, respectively, the magnitudes of the electronic and phonon contributions, Fig. 8.1 shows the experimental result for CeCu$_6$. The low-temperature limiting value of C/T yields $\gamma = 1670$ mJ/mole K^2, which is strongly enhanced over the value encountered in ordinary metals, for which $\gamma \approx 1$ mJ/mole K^2 is more typical. A second key experiment was the measurement of the magnetic susceptibility for which an interesting and informative crossover occurs as a function of temperature. Around room temperature, the susceptibility of CeCu$_6$ obeys a Curie–Weiss law characteristic of a concentration N of local magnetic moments μ_e,

$$\chi = \frac{N\mu_e^2}{3k_B}\left(\frac{1}{T-\theta}\right), \tag{2}$$

in which μ_e is close to the value of 2.54 μ_B expected for the f^1 electronic configuration of cerium on the basis of Hund's rules. However, at low temperatures, the situation is quite different. The susceptibility then depends

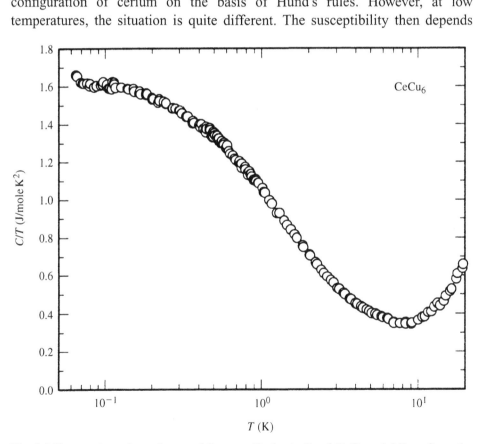

Fig. 8.1 Temperature dependence of the specific heat, C, of CeCu$_6$ yielding, from the low-temperature limiting value of C/T, a giant value for the electronic heat capacity, $\gamma = 1.67$ J/mole K^2 (Amato *et al.* 1987).

only weakly on temperature and therefore resembles a Pauli spin term, but with a magnitude for the volume susceptibility of 3.5×10^{-3}, again enhanced over that encountered in normal metals by a factor of between 10^2 and 10^3. The combined effect of these two experiments, which for a non-interacting Fermi gas are both measures of the density of electronic states at the Fermi energy, is to suggest an abnormally large value for this quantity at low temperatures or, equivalently, the presence of an abnormally narrow band of carriers at the Fermi energy.

It is interesting at this stage to express these results somewhat differently in order to demonstrate the remarkable nature of what has been discovered. For a non-interacting Fermi system, the specific heat coefficient is a direct measure of the density of states at the Fermi energy ε_F and hence, with a dispersion relation $\varepsilon_k = \hbar^2 k^2 / 2m^*$ for fermions of mass m^*,

$$\gamma = \frac{\pi^2 k_B^2}{3} \sum_k \delta(\varepsilon_F - \varepsilon_k) = \frac{k_B^2}{3\hbar^2} k_F m^*. \tag{3}$$

In an interacting Fermi system, we know from Fermi liquid theory (see chapter 5) that the specific heat measures the density of 'quasiparticle' states at the renormalised Fermi energy E_F; an exact expression following from Luttinger (1960),

$$\gamma = \frac{\pi^2 k_B^2}{3} \sum_k \delta(E_F - E_k), \tag{4}$$

δ being the Dirac delta function. We can now choose to write for the quasiparticle bands near E_F, $E_k = \hbar^2 k^2 / 2m^*$, so that eq. (3) remains valid but with $k_F m^*$ now interpreted as relating to the quasiparticles. By means of this dispersion law, we have introduced a fictitious Fermi energy E_F' whose main function is to signal the presence of a new (and much reduced) energy scale. However, a key result of Luttinger's analysis was to show that the *volume* of the Fermi surface is unaffected by the interactions, even for an anisotropic system, so that changes in $k_F m^*$ must reflect changes in the effective mass. We conclude therefore that the specific heat measurements are consistent with the presence of a Fermi liquid in which the heavy electrons are characterised by masses of $\sim 1000\ m$. Further support for this idea comes from the observed giant value of the Pauli spin susceptibility, which also reflects the density of states,

$$\chi_P = 2\mu_B^2 \sum_k \zeta(k)\delta(E_F - E_k). \tag{5}$$

In this expression, which is valid for spin-$1/2$ fermions and isotropic spin-independent interactions (Luttinger 1960), $\zeta(k)$ measures the modification of the effective moment due to the interactions. In much of what follows in this chapter, we shall often refer to the heavy electrons as 'quasiparticles'. This is solely for convenience, and should not be taken to imply the existence of an acceptable theoretical framework for their description.

8.2 Transport properties and the onset of coherence

The transport properties of solids often yield complementary information to that obtained from thermodynamic measurements, and a striking example of this is seen in the temperature dependence of the electrical resistivity of our illustrative material $CeCu_6$. Indeed, of all the transport properties, it is the electrical resistivity which most clearly reflects the underlying physics of the heavy-electron phenomenon. The resistivity in Fig. 8.2 is seen to fall from its relatively high value, for a metal at room temperature, of ~ 80 $\mu\Omega$ cm to a shallow minimum in the region of 200 K before rising again with falling temperature. Such behaviour is reminiscent of the Kondo effect associated with magnetic impurities in metals (Kondo 1964), a point to which we shall return. At the lowest temperatures the resistivity is observed to fall to a small value

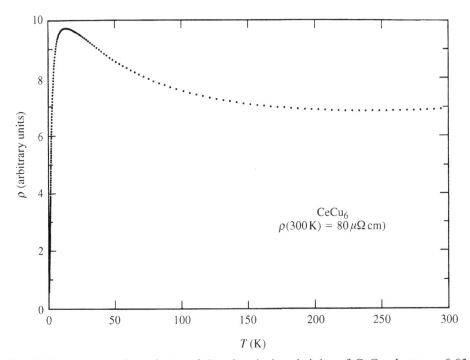

Fig. 8.2 Temperature dependence of the electrical resistivity of $CeCu_6$ between 0.03 and 300 K as reported by Ott *et al.* (1985). The resistivity at 300 K is 80 $\mu\Omega$ cm.

consistent with the expectation that, in a perfectly periodic metal, the resistivity will vanish as the temperature approaches zero. Of particular interest, however, is that for temperatures $\leqslant 200$ mK the resistivity variation shows the T^2 dependence expected of a Fermi liquid. These observations lead naturally to a consideration of the important characteristic energy scales in the problem. It is customary to define a temperature T^* for each system, in terms of which many of its thermal properties can be expressed as a function of the dimensionless quantity, T/T^*. For $T/T^* \ll 1$ a low resistance or *coherent* state is established, a region whose onset is often defined in terms of a coherence temperature T_{coh}, in which the elastic scattering associated with disorder has become vanishingly small and the resistivity is determined by the inelastic scattering processes characteristic of a Fermi liquid.[*] For example, in $CeCu_6$, the characteristic and coherence temperatures are, respectively, $T^* \approx 4$ K and $T_{\mathrm{coh}} \approx 0.15$ K.

Transport measurements in a magnetic field, such as the Hall effect, are expected to provide a useful probe of energy scales in systems which incorporate magnetic ions. In ordinary metals under appropriate experimental conditions, the Hall effect provides a measure of the carrier concentration and charge. For heavy-electron systems, the Hall effect differs from that in ordinary metals both in its large magnitude and in its strong temperature dependence. In $CeCu_6$, for example, the Hall effect shows considerable structure at temperatures below 1 K (Milliken *et al.* 1988). The physical origin of these low energy scales is at the heart of the heavy-electron phenomenon, but, before considering this, it is instructive to examine first the effect of disorder brought about by the presence of impurities.

8.3 The effect of impurities

The metal $LaCu_6$, with no f-electrons, is isomorphic with $CeCu_6$, and the solid solution series $Ce_xLa_{1-x}Cu_6$ therefore provides a convenient laboratory for studying the effects of f-electron concentration and of disorder on the heavy-electron state. Measurements of the thermodynamic properties yield an interesting general result; namely that the effects measured are essentially single-ion effects (Onuki and Komatsubara 1987). This is exemplified clearly by the magnetic susceptibility, which, when expressed per mole of cerium, is almost independent of the concentration x. The same conclusion follows from experiments on the heat capacity at temperatures below ~ 10 K, which also scale closely with the cerium concentration. In contrast, transport measurements,

---

[*] Note that 'coherence' here has a different meaning from 'coherence length' defined for superconductors (chapter 7) or mesoscopic systems (chapter 9).

such as the electrical resistivity per mole of cerium, while showing a convergence at the higher temperatures, reveal a dramatic departure from scaling and loss of coherence as the cerium concentration departs from $x = 1$. The effect of alloying on the Hall effect broadly reflects that on the electrical resistivity. As x departs from unity, the negative Hall coefficient at low temperatures, which is evidently associated with coherence, changes sign to reflect the presence of disordered local magnetic moments, and this positive effect decreases progressively with decreasing x to the limit $x = 0$ when the Hall effect of $LaCu_6$ is relatively small, negative and temperature independent as expected for a normal metal.

8.4 The Kondo effect

The foregoing discussion of some key properties of heavy-electron compounds, particularly the results of the thermodynamic experiments which appear to emphasise their single-ion properties, suggests a model for the description of these systems which has its genesis in ideas developed and refined over the past few decades for dealing with magnetic impurities in metals (see, for example, Hewson (1993)). Historically this field is associated with the experimental observation, made originally in the 1930s (de Haas *et al.* 1934), that certain impurities in metals, such as, for example, a dilute concentration of iron in gold, lead to the appearance of a minimum in the electrical resistance when expressed as a function of temperature. The phenomenon is widely referred to as the Kondo effect after the author who first suggested the physical origin of the effect (Kondo 1964).

Kondo's approach to this problem was influenced by the observation that a correlation exists between the presence of a resistance minimum and the existence of a Curie–Weiss term in the magnetic susceptibility at higher temperatures, signalling the presence of local moments. He therefore assumed that a local magnetic moment with spin S associated with an isolated impurity already existed in a metal, and that it was coupled to the conduction electron spin via an exchange interaction J. An interaction of this form had been proposed earlier by Zener (1951) and is widely referred to as the s-d model. Kondo's Hamiltonian therefore had the form

$$H = H_c + H_v + H_{sd}, \tag{6}$$

in which $H_c = \sum_{k,\sigma} \varepsilon_k n_{k,\sigma}$ describes non-interacting conduction electrons, H_v is the potential scattering term of the isolated impurity and H_{sd} is the magnetic interaction, which can be written in the compact form

$$H_{sd} = J\boldsymbol{S} \cdot \sigma. \tag{7}$$

In the absence of H_{sd}, the potential scattering leads to a finite residual value of the electrical resistivity which is essentially independent of temperature. Furthermore, the inclusion of phonon scattering only adds a further term to the resistivity, which increases monotonically with temperature and so cannot of itself account for the existence of a minimum. This then was the problem. When the full Hamiltonian (6) is used, calculation of the scattering by the impurity to second order yields a result similar to that for potential scattering alone, which is independent of temperature. Kondo's pioneering contribution was to extend the calculation for the first time to *third* order in the coupling constant J. Using the techniques of many-body perturbation theory, new terms now enter, the most important of which are those in which the spin of a conduction electron and an impurity spin are flipped in the scattering process, and these lead to a third-order term involving $\ln(T)$:

$$\rho = \frac{3\pi mJ^2 S(S+1)}{2\hbar e^2 \varepsilon_F}\left(1 - 4JN(\varepsilon_F)\ln\left(\frac{k_B T}{D}\right)\right). \tag{8}$$

This was the breakthrough needed, for with J negative, representing antiferromagnetic coupling of the conduction electrons to the local magnetic moment, and noting that $k_B T/D \ll 1$, D being the conduction electron bandwidth, the third-order term *decreases* with increasing T. Combined with the usual residual and phonon resistivities that we have already mentioned, this gave the possibility of a minimum in the measured resistivity as a function of the temperature. The agreement with the experiments at the time, which are reproduced in Fig. 8.3, is convincing.

It is clear, however, that the perturbation result (8) is unsatisfactory in the sense that it diverges at $T = 0$. Summing the higher-order terms only increases the problem by revealing that the divergence occurs at a finite temperature, the so-called Kondo temperature T_K, where

$$k_B T_K \approx D e^{-1/2JN(\varepsilon_F)} \tag{9}$$

and that the perturbation approach is only valid for $T \gg T_K$. The search for a solution valid in the region $T \ll T_K$ became known as the Kondo problem, and was to remain centre stage in condensed matter physics for more than a decade. It may be thought of as culminating in K. G. Wilson's (1975) seminal paper in which he devised the non-perturbative method of the 'numerical renormalisation group', a contribution for which he received the Nobel prize in 1982. A full discussion of the many developments that took place during this period may be found elsewhere (Hewson 1993).

The fundamental question of how a local impurity magnetic moment can survive in a metallic environment was addressed by Anderson (1961), who introduced a model which, in its later periodic form, has subsequently

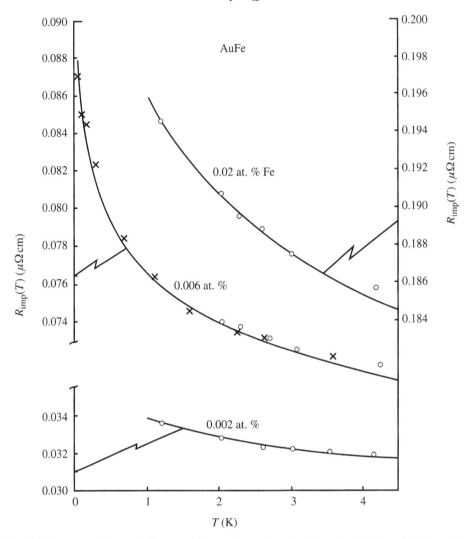

Fig. 8.3 A comparison of the experimental results for the electrical resistivity of a dilute concentration of iron impurities in gold with the logarithmic form in eq. (8); reproduced from Kondo (1964).

contributed richly to our understanding of the heavy-electron problem. Continuing for the moment with the impurity case, however, the Anderson Hamiltonian has the form

$$H = H_c + H_i + H_{ci} + H_U, \tag{10}$$

in which $H_c = \sum_{k,\sigma} \varepsilon_k n_{k,\sigma}$ and $H_i = \sum_{\sigma} \varepsilon_i n_{i,\sigma}$ refer, respectively, to the conduction electrons and the impurity, H_{ci} represents a hybridisation which gives rise to a finite matrix element V_{ci} between conduction and impurity states and

$H_U = U \sum_\sigma n_{i,\sigma} n_{i,\sigma'}$ results from the Coulomb interaction U between two electrons situated at the *same* impurity site. The introduction of a finite U is essential in order to establish a degenerate ground state and hence the possibility of a local magnetic moment. A full examination of the problem shows that this possibility is retained even in the presence of hybridisation for the parameter regime $\varepsilon_i < \varepsilon_F$ and $\varepsilon_i + U > \varepsilon_F$, and for scattering near the Fermi energy. Furthermore, under these conditions, the interaction is antiferromagnetic, as required by Kondo.

The non-perturbative approaches to the single magnetic impurity problem developed by Wilson and others, and the confirmation afforded by exact solutions of the s-d and Anderson models using the so-called Bethe ansatz (Bethe 1931) to give results over the full temperature range, have undoubtedly endowed this field with a sense of completeness. While recognising that particular systems may not be understood in detail, and that aspects of the theory remain to be resolved, there is a consensus that the underlying physics of the isolated magnetic impurity in a metal is essentially understood. In the low-temperature regime, $T \ll T_K$, the picture that emerges is that of a Fermi liquid of weakly interacting quasiparticles (Nozières 1974, 1975; Yamada 1975a, b). In this limit, the impurity is screened by conduction electrons which are effectively bound so as to compensate the local moment to form a singlet state. The characteristic screening length is $\xi \approx h v_F / k_B T_K$ which, for the heavier electrons, may not greatly exceed a typical lattice parameter. The net result is that the fully screened magnetic impurity scatters conduction electrons like a non-magnetic impurity so that, as $T \to 0$, the electrical resistivity approaches the unitarity limit and varies with temperature according to

$$\rho = \rho_0(1 - AT^2 + O(T^4) + \cdots). \tag{11}$$

The density of quasiparticle states shows a resonance at the Fermi level of width $\sim k_B T_K$, giving rise to an enhancement of both the specific heat and the spin susceptibility, but with their ratio R (often referred to as the Wilson or Sommerfeld ratio) differing from unity, as expected for interacting particles and in accordance with eqs. (3)–(5) of our discussion above:

$$R = \frac{\chi_i/\chi_c}{\gamma_i/\gamma_c} \neq 1. \tag{12}$$

The subscripts i and c refer here to the values with and without the impurity. It is important to distinguish between the resonance at the Fermi level (referred to as the Kondo or Abrikosov–Suhl resonance) and the original impurity level, which is situated ε_i below the Fermi energy ε_F. The resonance is a many-body effect resulting from the combined effects of U and V in the Anderson Hamiltonian.

8.5 The Kondo lattice

This brief discussion of isolated magnetic impurities in metals has introduced several themes which are reminiscent of the properties of heavy-electron compounds that were described in the introduction. There is the occurrence in both of a minimum in the electrical resistivity as a function of temperature, and both show Fermi liquid behaviour in a low-temperature regime with enhanced values of the specific heat and spin susceptibility. Furthermore, both systems incorporate ions which are magnetic in the free state; theories of the Kondo effect having been developed to deal primarily with 3d transition metal impurities, while heavy-electron compounds, as we have seen, involve transition metals from the 4f and 5f series. Given these observations, together with the result from thermodynamic experiments in heavy-electron systems that they largely reflect single-ion behaviour, it is natural to enquire if the analogy between magnetic impurities and concentrated systems can be usefully developed.

The idea was explored by Doniach (1977), who referred to a system of conduction electrons exchange-coupled to a regular array of local spins, such as we find in heavy-electron compounds, as a 'Kondo lattice'. His model consisted of a one-dimensional analogue, or 'Kondo necklace' with Hamiltonian

$$H = H_J + H_W, \tag{13}$$

in which a competition is set up between Kondo (like) compensation to form a singlet ground state, as expressed by H_J, and the effect of the long-range coupling between the local spins, which is mediated by the conduction electrons, the so-called RKKY interaction (Ruderman and Kittel 1954; Kasuya 1956; Yosida 1957), as expressed by H_W, which, acting alone, would give rise to magnetic order. Both interactions scale with the exchange coupling J such that the binding energy of the Kondo singlet in the weak coupling regime is

$$W_K \approx \frac{1}{N(\varepsilon_F)} \, e^{-1/N(\varepsilon_F)J}, \tag{14}$$

while that of the RKKY antiferromagnetic state is

$$W_{AF} \approx CJ^2 N(\varepsilon_F), \tag{15}$$

C being a dimensionless constant of order unity which depends on the electronic structure. As the exchange coupling is increased at zero temperature, the analysis reveals the presence of a second-order transition at a critical value J_c from an antiferromagnetic state to a Kondo singlet, as shown in Fig. 8.4. Although Doniach's approach cannot be generalised to three dimensions, the

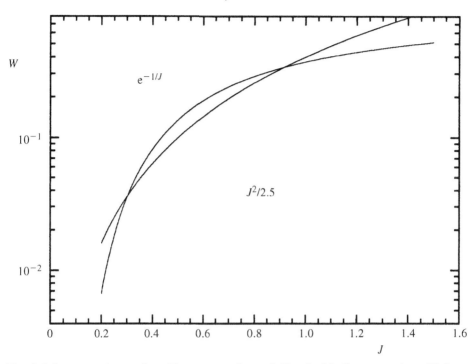

Fig. 8.4 A comparison of antiferromagnetic and Kondo binding energies, W, for a one-dimensional Kondo 'necklace' model (Doniach 1977). A crossover occurs with increasing exchange, J, from a magnetically ordered state to a spin compensated state at $J \sim 0.3$. The second crossover at $J \sim 0.9$ lies outside the range of validity of the theory.

idea suggests that heavy-electron systems may be thought to originate in this way, and that in systems which lie very close to J_c a weak lattice magnetisation may result from incomplete compensation. Weak ordered moments of $\sim 10^{-2}\,\mu_B$ are indeed commonly observed in heavy-electron systems (Grewe and Steglich 1991). Nevertheless, in extending these ideas developed for isolated magnetic impurities to the lattice case, it is solutions based on the Anderson model which have proved most fruitful, and it is to this, as a model for heavy-electron systems, that we now turn.

8.6 The periodic Anderson model

While, as we have seen, there are similarities between dilute magnetic alloys and heavy-electron systems at the level of observation, there are several reasons to doubt that this analogy can be developed to advantage at the microscopic level. In the Kondo lattice, the separation between magnetic ions (for example

Ce ions in CeCu$_6$) is comparable with the Kondo screening length in the impurity problem, so that the idea that we might regard the lattice simply as an assembly of individually screened ions is clearly unacceptable. If screening is an appropriate concept in the lattice case, then it must be a collective and macroscopic phenomenon. Additionally, as the dramatic fall in the electrical resistivity as $T \rightarrow 0$ in Fig. 8.2 reveals, there is the presence of coherence in the lattice case. The coherent propagation of f-electrons in Bloch states in well-ordered Kondo lattices is evidently associated with translational invariance and therefore lies outside the magnetic impurity problem. These are good reasons for doubting that a simple connection exists between the heavy-electron problem and the Kondo effect, yet, rather remarkably, when the Anderson impurity model (10) is extended to the case of a periodic lattice the new physics emerges.

The natural extension of eq. (10) to the case of a periodic lattice of magnetic ions is

$$H = \sum_{k,\sigma} \varepsilon_k n^c_{k,\sigma} + \sum_{i,m} \varepsilon_f n^f_{i,m} + U \sum_{i=1}^{N} \sum_{m,m'} n^f_{i,m} n^f_{i,m'} + H_{cf}. \tag{16}$$

The first two terms refer to the conduction and f-electrons, respectively, but now we have introduced a lattice site index, i, and the angular momentum state of the f-electron, m. The third term expresses, as before, the energy penalty U due to Coulomb repulsion, which occurs when the f-site is doubly occupied, while the hybridisation term H_{cf} yields matrix elements, V_{cf}, between the initially localised f-states, $|\phi_f\rangle$, and itinerant conduction states, $|\varphi_c\rangle$:

$$V_{cf} = \langle \phi_f | H_{cf} | \varphi_c \rangle. \tag{17}$$

It is the introduction of the Coulomb term U which causes the motion of the f-electrons to be correlated, which in turn has caused these solids to be referred to generically as *strongly correlated electron systems*. In spite of the very considerable effort expended on the theory of this lattice problem since the discovery of heavy-electron systems, no exact solution has yet been found. However, if we assume that U is infinitely large, a reasonable approximation to the situation prevailing in many cerium-based compounds at low temperatures where the 'bare' f-level lies several electron-volts below the Fermi energy and the excited (doubly occupied) state as far above, a *mean field solution* to eq. (16) may be obtained which yields physically appealing results (Tesanovic and Valls 1986; Newns and Read 1987). An elegant method for achieving this, the so-called 'slave boson' method devised by Coleman (1984), introduces a scalar Bose field, b_i, which, by establishing the occupancy of the f-site i, essentially expresses its availability for hybridisation. The method was first used for the impurity problem, under the conditions $U \to \infty$, where the mean field solution of eq. (10) is obtained by introducing a mean value for the magnitude of the Bose field, b_0:

$$\langle b_i \rangle = |b_0| \, e^{i\phi_i} \equiv b_0. \tag{18}$$

For the lattice case, Tesanovic and Valls (1986) proposed an additional constraint, namely that the phases ϕ_i at the different sites, which are uncorrelated in the random and dilute impurity problem, should be phase locked in the lattice case. By this stratagem the motions of the conduction and f-electrons are made coherent and the Kondo lattice may then be visualised as a *macroscopic* singlet state.

The mean field solution of eq. (16), for the simple case of one conduction electron band and an f-orbital degeneracy of 2 ($J = 1/2$), is illustrated schematically in Fig. 8.5. In the absence of both the Coulomb repulsion U and hybridisation V terms in the Hamiltonian, the unperturbed f-level is situated at an energy $-\varepsilon_f$ below the chemical potential μ, which, for the metallic case, is crossed by a broad (spd) conduction band. For $U = 0$ but $V \neq 0$, the solution is also straightforward; the f-level hybridises with the conduction band but the f spectral weight remains centred at $-\varepsilon_f$. Similarly, for $V = 0$ but $U \neq 0$ the solution to eq. (16) is intuitively clear; the ground and first excited 'atomic' levels of the f-electron system remain sharply defined either side of μ and are separated by an energy U. However, the effect of U and V acting together introduces new physics. Under these conditions, the f-levels are renormalised to a new position, E_f, a little *above* μ:

$$E_f = \mu + k_B T^*, \tag{19}$$

where $T^* > 0$, and they hybridise with the conduction band to give two renormalised quasiparticle bands, E_k^+ and E_k^-. The chemical potential adjusts to accommodate the f-electrons, and the quasiparticle Fermi surface at k_F lies in the region of flat bands. We may refer to the excitations of the system as quasiparticles as the complex many-body problem has been mapped onto a single-particle picture. At the Fermi level the renormalised density of states is

$$N(E_F) \approx N(\varepsilon_F) \frac{D}{k_B T^*}, \tag{20}$$

which, because $D/k_B T^* \gg 1$, is much enhanced over its bare value, and the quasiparticle Fermi velocity is similarly reduced by the same factor. Thus, heavy quasiparticles are slow quasiparticles. The characteristic temperature T^*, which is given by

$$kT^* = \mu \exp\left(-\frac{|\varepsilon_f|}{2N(\varepsilon_F)V^2}\right), \tag{21}$$

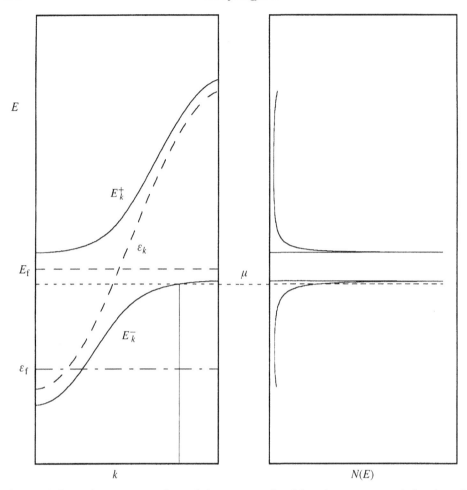

Fig. 8.5 Schematic representation of the renormalised band structure and density of states for a heavy-electron metal, according to the mean field solution of the periodic Anderson model. Interactions transform the 'bare' f-level at ε_f to form a many-body band at E_f near the chemical potential μ. Hybridisation gives rise to two bands E_k^+ and E_k^-. The Fermi surface is defined by $E_{k_f}^- = \mu$.

plays a similar role to the Kondo temperature in the impurity case, to which, experimentally, it is often found to be similar in magnitude. Thus, in summary, for temperatures $T < T^*$, the mean field solution to eq. (16) yields an excitation spectrum consisting of heavy quasiparticles. In this approximation, in which fluctuations around the mean field are not included, the quasiparticles are non-interacting. Their inclusion leads to interactions and to a decoupling of the underlying f-electron and conduction electron subsystems for $T > T^*$. Finally, we note that the mean field approximation corresponds to the leading term in a $1/N$ expansion (Newns and Read 1987), where N is the degeneracy $(2J+1)$ of the ground state. When $J=1/2$, as has been assumed here, this justifiably gives concerns regarding convergence of the model. Nevertheless, mean field theory gives a plausible and intuitively appealing description of the origin of heavy electrons.

8.7 The heavy-electron liquid

The evidence presented so far in support of the existence of quasiparticles with unusually high masses poses a number of intriguing and fundamental questions. It is perhaps worth stressing the remarkable nature of the discovery that renormalisation of electrons can take place in a solid on a scale of 10^2–10^3 without a transition to either a localised or magnetic state. Indeed, so great is the effect that the concept of renormalisation may be inappropriate. Electrons are evidently interacting with local magnetic moments to produce a 'soup' in which the elementary excitations are itinerant quasiparticles with giant masses. This is a novel Fermi liquid of which it is pertinent to ask whether a Fermi surface exists and, if so, to what extent it can be accounted for by the well established methods of electronic structure calculation. It is further appropriate to ask what experimental methods might be used to verify the existence of the anticipated heavy-mass quasiparticles, rather more directly than via transport or thermodynamic measurements, with a view to investigating their temperature and magnetic field dependences and even anisotropies. Ultimately the aim is to develop a microscopic description of the heavy-electron state which is material specific and which permits a satisfactory physical description of such exotic excitations. At the time of writing, only a modest start has been made towards achieving these goals, to describe which a convenient starting point is Luttinger's analysis (Luttinger 1960).

For non-interacting fermions, the Fermi surface is trivially defined by $\varepsilon_k = \mu_0$, in which ε_k are the unperturbed single-particle (electron) energies and μ_0 is the unperturbed chemical potential. In this case, the Fermi surface marks the boundary between occupied and unoccupied momentum states in the ground state of the system at $T=0$. However, in the presence of interactions, this definition is meaningless. Luttinger (1960) showed that, under certain conditions, a Fermi surface can still be rigorously defined, even in the presence of interactions. In fact, Migdal (1957) was the first to point out that the momentum distribution of quasiparticles characterising interacting Fermi particles *could* still possess a discontinuity, although this statement cannot generally be true for arbitrary interactions. This point was also emphasised by Landau (1956), who cited the properties of an assembly of deuterium atoms, which, as we know, interact in such a manner that they form molecules, as a result of which liquid deuterium possesses an energy spectrum of the Bose type. In Landau's words, '... the presence of a Fermi energy spectrum is connected not only with the properties of the particles, but also with the properties of their interactions'. The condition imposed by Luttinger is that the forces between the particles are such that a power series expansion in their strength gives an excellent asymptotic representation. It is an interesting question whether such an expansion is a good representation for real systems, such as electrons in metals.

Using this approach, Luttinger defines the new (renormalised) single-particle energies E_k by the expression

$$E_k - \varepsilon_k - K_k(E_k) = 0, \tag{22}$$

and hence the *Fermi surface of the interacting system* by

$$\mu - \varepsilon_k - K_k(\mu) = 0, \tag{23}$$

in which K_k expresses the interaction between the particles. The new Fermi surface divides momentum space into two parts according as to whether the left hand side of the last equation is positive or negative. The former region (of k-space) is defined as the interior of the new Fermi surface. Because the interaction is not assumed to be isotropic, the new Fermi surface will not in general be spherical in shape. On the basis of Luttinger's analysis, two statements may be made which are of rather general validity:

(1) The momentum distribution function in an interacting Fermi system shows a discontinuous change on crossing the new Fermi surface.
(2) The interactions leave the volume in k-space of the Fermi surface unchanged.

Of interest is that a further generalisation of this approach may be made to the case of interacting particles moving in the periodic potential of the lattice, the band case (Luttinger 1960), including particles of arbitrary spin as well as the effect of a magnetic field on the orbital motion of the particles. Thus, the above two powerful statements remain valid even under these conditions. Having now established that a Fermi surface, in the sense described above, should exist, even in such a strongly interacting system as a heavy-electron compound, we turn our attention to experiments that might be used to validate these ideas.

8.8 The de Haas–van Alphen effect

Of the many experiments that may be used to investigate electrons in metals, the de Haas–van Alphen (dHvA) effect has played a seminal role. It derives its name from the discovery by de Haas and van Alphen (1930a,b, 1932) of an oscillatory variation of the magnetisation of a bismuth single crystal held at the temperature of liquid hydrogen, as a function of the applied magnetic field. Quite independently of this discovery, Landau (1930) predicted in the same year that such oscillations might be expected as a result of the quantisation of the motion of conduction electrons, which, in the presence of a magnetic field, are constrained to move in helical orbits. Some years later, on the basis of rather general semiclassical arguments, Onsager (1952) concluded that the frequency of the oscillations, which are periodic in the reciprocal of the magnetic field, would be related to the extremal cross-sectional area of the Fermi surface measured in a plane perpendicular to the applied field. Within a few years, Lifshitz and Kosevich (1955) had obtained a complete semiclassical theory, which still provides the basis for the interpretation of experiments in normal metals.

The starting point for the Lifshitz–Kosevich (LK) theory is the standard expression for the free energy per unit volume of a non-interacting electron gas,

$$\Omega_0 = -\frac{1}{\beta} \sum_k \ln\left(1 + e^{\beta(\mu_0 - \varepsilon_k)}\right), \tag{24}$$

in which the sum is over the energy levels k, including spin, and $\beta = 1/k_B T$. With the aid of the Onsager relation,

$$A_n(k) = \frac{2\pi e B}{\hbar}(n + \gamma), \tag{25}$$

which imposes a quantum condition on the allowable areas $A_n(k)$ of the cyclotron orbits in k-space, n being a positive integer, an expression is derived for the free energy in a magnetic field and hence, using $\tilde{M} = -(\nabla_B \Omega_0)_T$, for the oscillatory component of the magnetisation. Rather surprisingly it turns out that the *quantum oscillatory* diamagnetisation is rather easier to calculate than the steady part of the orbital diamagnetism, a situation that persists in the case of an interacting Fermi system, as we shall see. In the non-interacting case, one finds that the oscillatory part is given by a Fourier sum,

$$\tilde{M} = \sum_{r=1}^{\infty} \alpha(r, B, T) \cos\left(\frac{2\pi r F}{B} + \gamma\right), \tag{26}$$

in which the dHvA frequency, F, is related directly to the extremal cross-sectional area of the Fermi surface in a plane normal to the applied field,

$$F = \frac{\hbar A_{\text{ext}}}{2\pi e}. \tag{27}$$

That this simple and elegant result is valid for Fermi surfaces of arbitrary shape makes the dHvA effect a powerful spectrometer for the study of itinerant electrons in metals. The amplitude $\alpha(r, B, T)$ contains information on the electron effective mass, scattering rate and g-factor, all of which may therefore be measured, averaged around a well defined 'orbit' on the Fermi surface. Such experiments, performed in the majority of metals throughout the 1960s and 1970s, underpin much of our present-day knowledge of their electronic properties.

Returning to heavy-electron systems, the dHvA effect would seem to offer an ideal tool for their investigation. The observation of quantum oscillations would itself indicate the presence of a Fermi surface, and their amplitudes could be analysed to determine the quasiparticle masses. Although, in this case, experiments preceded a detailed theory, it is helpful to reflect on the influence of interactions on the dHvA effect when applied to highly correlated electron systems. Such an analysis was first undertaken by Luttinger (1961), whose aim it was to use the assumed analytic property of the particle interactions mentioned above, to investigate the response of an interacting fermion system to a uniform applied magnetic field and, in particular, to calculate the dHvA effect. In so far as one only wishes to calculate *oscillatory* effects, Luttinger showed that the thermodynamic potential is given exactly as for the non-interacting case, except that the single-particle energies, ε_k, in eq. (24) are replaced by their renormalised values E_k, characteristic of the quasiparticles, the elementary excitations of the interacting system. The problem therefore resolves into one of determining Ω, exactly as was done originally by Lifshitz and Kosevich for the non-interacting case. It follows that the amplitude and frequency of equilibrium quantum oscillatory properties, such as the dHvA effect, are determined by the quasiparticle excitations and their Fermi surface. Only the phase of the oscillations is not given correctly by this procedure. An interesting result of the manner in which Luttinger applies gauge invariance, by connecting it with that of the original Hamiltonian, is that it resolves the question of the charge of the renormalised particles in their response to a magnetic field, which is shown to be the same as that of the original particles, that is $-e$ and not some other value (Falicov 1960; Stern 1960).

It is now possible to combine the lattice mean field model above with Luttinger's result to provide a theory of the dHvA effect for the case of heavy-electron systems (Rasul 1989; Wasserman *et al.* 1989). The structure of the resulting theory is identical to eq. (26) but with two important changes; namely that both the phase, γ, and the amplitude factor, $\alpha(r, B, T)$, are renormalised. In the context of heavy electrons, the amplitude change is particularly interesting. As in the LK theory, the temperature dependence of the amplitude has a functional form determined by the Fourier transform of the derivative of the Fermi–Dirac function, leading to

$$\alpha(T)_B = \frac{X}{\sinh X}, \tag{28}$$

where $X = 2\pi^2 r k_B T / \hbar \omega_c$ and $\omega_c = eB/m^*$. But now, as we have shown, $m^* \approx D/k_B T^* \gg 1$. Thus the mean field result indicates that the heavy-electron mass measured in a dHvA experiment should reflect the same giant renormalisation as that inferred from the low-temperature electronic specific heat. We therefore have a method for determining the masses of heavy electrons rather directly from their cyclotron frequencies, thereby providing a microscopic test of the quasiparticle description. Although we have presented these theoretical ideas as a prelude to the experiments, historically it was the experiments that came first.

8.9 Quantum oscillation experiments

Two fundamental requirements must be satisfied for the observation of quantum oscillations in metals. The first, $\omega_c \tau \geqslant 1$, effectively places an upper limit on the magnitude of the scattering of carriers arising from the presence of various defects in the lattice. It is for this reason that such experiments are normally only possible in single crystals of high purity and perfection. The second, $\hbar \omega_c \gg k_B T$, requires that the thermal broadening of states at the Fermi energy is not so great as to smear out the quantum effects which arise from the underlying Landau level structure. Applied to heavy-electron compounds, in which the magnetic energy level separation, $\hbar \omega_c$, may be reduced by a factor of $\sim 10^3$ from its value in ordinary metals, this means that experiments normally performed at temperatures of a few kelvin must now be made at temperatures of a few millikelvin, that is in the temperature range of a dilution refrigerator.

The first Landau quantum oscillation experiments in heavy-electron systems were performed in 1986 in CeCu$_6$ (Reinders *et al.* 1986, 1987) and in UPt$_3$ (Taillefer *et al.* 1987; Taillefer and Lonzarich 1988). In both of these it was the dHvA effect that was studied, and the general conclusions of the two investigations were rather similar. The most striking result was undoubtedly the mere observation of quantum oscillations, implying the presence of a well defined Fermi surface, even in the strongly interacting Fermi liquid that comprises the coherent heavy-electron state. Furthermore, in both cases, around eight dHvA frequencies were detected in the initial experiments, reflecting the evident complexity of the Fermi surfaces in these two heavy-electron compounds. Turning to the temperature dependence of the amplitudes of the quantum oscillations, as expressed by eq. (28), the quasiparticle masses were measured to be in the range 6 m–80 m in CeCu$_6$ and 25 m–90 m in UPt$_3$. Thus, for the first time, the presence of heavy electrons was demonstrated by directly probing their dynamical properties. Given that the mass is determined by an orbital integral of the Fermi velocity, $m^* \propto 1/v_F^*$, a value for $v_F^* \sim 10^4$ m s^{-1} was inferred for CeCu$_6$, in line with the discussion of eq. (20) that heavy electrons are slow electrons. It is interesting to reflect that, for the heaviest electrons, one can envisage a situation where v_F^* is comparable with the velocity of sound, so presaging a breakdown of the Born–Oppenheimer approximation which under-pins most electronic structure computations. Finally, from the evolution of the dHvA amplitude with magnetic field, the magnitude of $\omega_c \tau$ was measured, and hence an estimate could be made of the quasiparticle mean free path. In both sets of experiments, this was found to be ~ 100 nm.

An important question that follows from these experiments is the extent to which the results can be accounted for by conventional electronic structure calculations based on the local (spin) density approximation for the exchange-correlation potential, as described in chapter 4. We shall return to this question later, but for now we note that the situation in UPt$_3$ is rather clearer than that for CeCu$_6$. As shown by Wang *et al.* (1987), in the former 5f-compound, the Fermi surface dimensions appear to be in reasonable agreement with the local density calculations. On the other hand, the measured masses are typically a factor ~ 20 times greater than the calculations predict. This is roughly the same factor as that by which the calculated density of states needs to be enhanced in order to account for the low-temperature linear coefficient of specific heat γ; an observation which is consistent therefore with the idea that the anomalously large heat capacity in UPt$_3$ is indeed to be identified with the presence of heavy electrons. In the 4f-compound CeCu$_6$, the Fermi surface dimensions do not agree with the electronic structure calculations (Norman and Koelling 1993). Although this may be a result of the loss of some symmetry which takes place upon cooling the crystal, as CeCu$_6$ undergoes a weak monoclinic distortion from an orthorhombic structure which is not included in the calculations, it may equally herald the breakdown of the theory. It raises the important and fundamental question of whether the local density approximation can be used as a first-principles basis for the description of the heavy-electron state. Heavy electrons, while itinerant, might be expected to have rather spatially localised wavefunctions so that the local density approximation, which presupposes the existence of extended Bloch functions, may break down. The compound CeCu$_6$ is perhaps the first example of this.

Since these first two experiments, a growing number of quantum oscillatory studies have been made in heavy-electron materials employing not only the dHvA effect but also the Shubnikov–de Haas effect (quantum oscillations of the electrical resistance) and magnetoacoustic properties (Onuki *et al.* 1991). A picture emerges of a class of solids whose low-energy excitations, at sufficiently low temperatures ($T \ll T^*$), consist of charged fermions with unusually large masses. Only in a rather few cases, however, is the information regarded as sufficiently complete that a clear picture of the complete Fermi surface has emerged. One of these is the tetragonal compound $CeRu_2Si_2$, for which a more meaningful comparison is made below between the thermodynamic and quantum oscillation experiments. Such a comparison is crucial to our understanding of the origin of the heaviness as it can confirm or otherwise the contention that the origin of the giant specific heat coefficient, γ, is solely attributable to the presence of heavy, charged quasiparticles. This view has sometimes been challenged for reasons that we shall now discuss.

8.10 An alternative view

Intermetallic compounds incorporating 4f- and 5f-atoms constitute a large class of solids in which heavy-electron behaviour is the exception rather than the rule. In the majority of these compounds, the f-electrons are situated close to the Fermi level so that they exhibit a non-integer valence and participate in the metallic bonding. The heavy-electron compounds, however, are (very nearly) integer valent compounds in which the unrenormalised f-level is situated well below the Fermi level, consistent with our discussion of them in terms of the periodic Anderson model. For these, the large Coulomb interaction, expressed by U in eq. (16), suppresses charge fluctuations, and the low-energy excitations are entirely due to the spin degrees of freedom. With this in mind, Kagan *et al.* (1992a, b) proposed a different mechanism to account for the mass enhancement, which leads to different kinds of excitations. Starting with the Kondo Hamiltonian (6) generalised to a Kondo lattice, and motivated by Abrikosov's treatment of the impurity problem, in which the spin operators $S_i \cdot \sigma$ are represented by pseudo-fermion operators, they developed a description of a *spin Fermi liquid*. The elementary excitations of the spin liquid, spinons, interact with conduction electrons, giving rise to a renormalisation of their mass. The picture that emerges is of two interpenetrating Fermi liquids, one charged (heavy electrons) and one neutral (spinons), both of which contribute to the low-temperature thermodynamics, but only one of which, the charged one, contributes to the dHvA effect. Thus the density of states at the Fermi energy, as measured in the two experiments, would, according to this model, be different. For example, given a knowledge of the Fermi surface, together with the variation of the quasiparticle velocity v_F^* over it, one may calculate the (charged quasiparticle) density of states by means of the Fermi surface integral,

$$N_{\mathrm{qp}}(E_{\mathrm{F}}) = \frac{1}{4\pi^3\hbar} \int_S \frac{\mathrm{d}S}{|v_{\mathrm{F}}^*|}, \qquad (29)$$

for comparison with the (total) density of states obtained from the specific heat,

$$N_{\mathrm{total}}(E_{\mathrm{F}}) = \frac{3\gamma}{\pi^2 k_{\mathrm{B}}^2}. \qquad (30)$$

A difference could flag the presence of neutral quasiparticles and clearly have a profound impact on our perception of the underlying physics. In the few cases where this has been attempted to date, in CeB_6, $CeRu_2Si_2$ and UPt_3, there is little evidence for the presence of a neutral Fermi liquid, although the experiments and calculated band structures are not without uncertainties, so that a small contribution of this type cannot be entirely ruled out.

8.11 The band structure of heavy electrons

We have seen in the foregoing that the existence of heavy electrons in certain rare earth and actinide compounds arises as a result of the presence of many-body interactions whose combined effect is to cause the motion of the underlying electrons to be strongly correlated. The net result is the creation of excitations of the system whose properties reflect giant renormalisations which can be as large as 10^2-10^3. Furthermore, there is a convergence of experimental evidence that, in the low-temperature regime, these heavy electrons constitute a heavy Fermi liquid with an attendant Fermi surface, without necessarily precipitating a transition to either a superconducting or magnetic ground state. A question of some interest, therefore, is whether the standard methods for electronic band structure calculations are valid in the presence of such strong renormalisations.

The standard calculational procedure for dealing with interacting electrons in metals is based on density functional theory, in which the basic variable is the charge density, as is discussed in some detail in chapter 4. As normally applied, this is a ground-state theory which would not be expected to describe adequately low-energy excitations from the ground state, such as are measured experimentally by the electronic specific heat coefficient γ. In the practical realisation of the density functional method, several approximations are made which raise interesting questions about its validity in our present problem (see, for example, Norman and Koelling (1993)). Calculation of the ground-state properties is reduced to the problem of non-interacting electrons moving in some effective potential. As a by-product in this process, one obtains one-electron eigenvalues as auxiliary quantities, which are then used to obtain the charge density. Although these eigenvalues are without meaning in the formalism, the Kohn–Sham ansatz (Kohn and Sham 1965) asserts that they may be used to obtain the (Kohn–Sham) Fermi surface. However, what is measured in an experiment, as we have discussed above, is the quasiparticle Fermi surface as defined by eq. (26). Except in the trivial case of Jellium, Schonhammer and Gunnarsson (1988) have shown that these two Fermi surfaces will differ, although the difference for many metals appears to be small, even for such cases as the f-electron metals, where correlations are known to be significant. A further approximation is the local density approximation, in which the real exchange correlation functional is approximated at each point in space by a local value which is based on the result appropriate to a homogeneous electron gas of the same density. Although crude, this simple approximation to a complex many-body interaction appears to work remarkably well, even in strongly inhomogeneous systems (see chapter 4). Perhaps most surprisingly, the agreement even extends to certain heavy-electron systems, such as UPt_3, although here, as we have already noted, while the Fermi surface is reasonably well determined, the predicted quasiparticle masses are greatly in error.

It was to address this shortcoming that a different approach to calculating the band structure of strongly correlated systems was devised, which combines the results of material-specific *ab initio* methods with some phenomenological considerations. The renormalised band method (Fulde 1991; Zwicknagl 1992) rests upon an ansatz which asserts that a microscopic model calculation such as we have just described, based on the local density approximation to describe weakly correlated conduction electrons, may subsequently be supplemented by additional information representing the existence of strong correlations, to arrive at a realistic expression not only of the Fermi surface, but also of the dynamical properties of the quasiparticles. The charge density in the ground state is assumed to be given correctly by the local density approximation, even for the strongly correlated electrons, and the Fermi energy is assumed to be unchanged. In the scattering formulation of this approach, in which the band structure problem is formulated in terms of scattering theory, the characteristic properties of a given material are expressed in terms of the phase shifts, $\eta^i_\nu(E)$, for a wave of energy E and symmetry ν incident at the lattice site i. The Fermi surface and velocities follow from the set of values of the phase shifts and their derivatives evaluated at the Fermi energy. In its application to heavy electrons, the supplementary information takes the form of a renormalised f-phase-shift at the lanthanide or actinide site, whose value is set by fitting to experimental data. The renormalised band method is therefore essentially a semiphenomeno-logical one-parameter theory.

The method has met with some success, particularly in the case of $CeRu_2Si_2$, for which the experimental data derived from the dHvA effect are sufficiently complete to permit a searching comparison with the theory (Zwicknagl 1992). It has also shed light on the following question: Under what conditions will conventional band calculations based on the local density approximation be expected to yield a Fermi surface in agreement with the quasiparticle Fermi surface as measured by the dHvA effect? This question is evidently equivalent to asking about the effect of the correlations upon the phase shifts at the Fermi energy. These, however, are not independent of each other, linked as they are to the volume enclosed within the Fermi surface, which, as we have seen, is unaffected by the interactions. The fixed number of electrons establishes a connection between the phase shifts of the f- and the non-f-states which is unaffected by local correlations, and the Fermi surface is essentially determined by the non-f-conduction bands and the geometry of the lattice. It follows from this that a comparison between the calculated and measured Fermi surfaces is not a valid test of the description of the f-states. These considerations also suggest that the conditions for the validity of a local density approximation calculation of the Fermi surface will depend upon the value of T^* relative to the spin–orbit T_{so} energy scale (Zwicknagl 1992), such that the requirement that the f-shells are associated with a single degree of freedom translates to the requirement that $T_{so} \gg T^*$ and that the crystalline electric field effects are negligible.

A recurrent theme prompted by the comparison between Fermi surface measurements and electronic structure calculations is the vexed question of whether or when the f-electrons behave as itinerant or localised in the ground state. Certainly, compounds can be found in which the f-electrons are unambiguously localised (e.g. UPd_3) or itinerant (e.g. UPt_3). The question here is whether the restricted class of heavy-electron compounds can be described as falling into one or other of these two categories. From the perspective of the periodic Anderson model, the f-spectral weight appears largely in the narrow many-body band located close to the chemical potential, which adjusts to accommodate the f-electrons. This picture is not inconsistent with the renormalised band approach, in which the f-charge density is obtained from a local density approximation calculation with the f-electrons treated as band states. From these considerations, we are led to a view in which the f-electrons are itinerant and are to be included within the volume of the Fermi surface. However, appearing to be in conflict with this conclusion are reports of measured Fermi surfaces in heavy-electron systems (Onuki *et al.* 1991) which appear to resemble closely those in the f-electron free analogue compounds in which the cerium atoms are replaced by lanthanum. One explanation for this visualises the f-electrons in such materials as being localised and so regarded as part of the atomic core; a situation resembling that in the element praeseodymium. We must also recognise, however, that the apparent similarity between certain Ce and La compounds might be the result of an incomplete knowledge of the Fermi surface, particularly as it is the heaviest sheets whose presence it will be most difficult to establish experimentally. Until this matter is resolved by more comprehensive experiments, the question of the itineracy of f-electrons as a universal feature of the heavy-electron state cannot be answered with assurance.

8.12 Final remarks

Many intriguing properties of heavy electrons have been omitted in this brief essay. We have, for example, said little of their magnetic behaviour, the ubiquitous antiferromagnetic correlations, their propensity for magnetic order with unusually small moments or the phenomenon of metamagnetism. Nor have the semiconducting heavy-electron systems received the attention they would merit in a complete survey. Perhaps the difficulty we have in discerning universal patterns of behaviour is little more than a restatement of our ignorance. In the absence of an agreed theory of heavy electrons, such diverse properties are not easy to accommodate within a general framework.

On the superconductivity in the heavy-electron liquid, whose presence has been the stimulus for so much of this work, the reader's attention is directed to chapter 7. There is little doubt, however, that the key to the understanding of the origin and nature of the unconventional superconductivity lies in a better appreciation of the physics of the normal state; that is where the new physics is. Finally, although it is pertinent to ask about the nature of the heavy-electron quasiparticles, if that label is valid, we have to confess to much ignorance. Quantum oscillation experiments permit us to follow their coherent propagation, measure their effective masses, Fermi velocities and mean free paths, but such information falls woefully short of a microscopic description. That, coupled with the development of an appropriate language for the description of such states of condensed matter in which many-body interactions play a decisive role, is surely a task for the coming years.

Acknowledgement

It is a pleasure to acknowledge many illuminating discussions on these matters with Allen Wasserman.

9

The coherent electron

Y. IMRY

The Weizmann Institute of Science

and

M. PESHKIN

Argonne National Laboratory

9.1 Introduction

In this chapter, we explore the concept of coherence as it applies to electrons and we discuss some examples of the fundamental physics that has been learned with the help of coherent electrons.

'Coherence' and 'coherence length' are commonly defined very differently for different purposes. We shall have in mind a loose definition that seems to us to capture the essence of the coherent electron in its most typically quantum mechanical behaviours. A coherent state of one or more electrons is a state which exhibits quantum mechanical effects on length scales much larger than atomic or molecular dimensions. The coherence length measures the distance over which the electrons in question display interference effects. Quantum effects on scales that are in some sense large have proved to be fascinating in their own right, and are powerful as tools to study the physical world.

First we digress to mention some other definitions of coherence that appear frequently in the literature. (1) In superconductivity theory (chapter 7), 'coherence length' is sometimes used to mean the typical distance between the paired electrons in a Cooper pair. (2) An electron which is localized in the sense that its wave function is a wave packet whose spatial extent is and remains small compared with the scale of the electron's orbit, so that the electron's motion and interactions are approximately describable by classical physics, is sometimes said to be in a 'coherent state'. That definition, which is sometimes applied to electrons in orbits as small as a highly excited atom or as large as tens of metres in a synchrotron, is almost opposite to the one we use in that it applies to the classical limit, where quantum effects become irrelevant. (3) Some authors use 'coherent state' as a synonym for 'pure state', that is a state described by a wave function in contrast to a statistical mixture of such states (see section 9.3). (4) Finally, photons can exhibit an additional, entirely different, kind of 'coherent state' in which the wave function contains a large indefinite number of photons in the same state, with definite phase relations between the components with different number of photons. The limiting case is the classical field, as in a radio wave. Electrons cannot do that because of the Pauli principle, which forbids more than one electron per state.

Interference effects in classical physics, for instance in light waves, were known long before the advent of quantum mechanics. Quantum mechanics, with its wave–particle duality, implies that interference effects should also be exhibited by electrons. The experiment of Davisson and Germer (1927a, b), which demonstrated interference of electron waves, gave one of the important proofs of the validity of the quantum picture. Davisson and Germer verified that a beam of electrons can be diffracted in exactly the same way as a beam of light that strikes a diffraction grating, and that the electron's wavelength is given by

$$\lambda = \frac{h}{p}, \tag{1}$$

as predicted by quantum mechanics. In eq. (1), h is Planck's constant and p is the momentum of the electron. Their electrons had wavelengths around 0.1 nm, too short for the ruled gratings used in optical diffraction, but they were able to observe diffraction at large angles by using a thin nickel crystal as their grating. Electron diffraction has since become an important tool for analysing the structure of solid surfaces (Pendry 1974), including various chemical changes that they undergo.

We shall concentrate mainly on two kinds of electron coherence phenomena that have especially interested us in the course of our own research:

(1) interferometry using electron beams in a vacuum, which has recently enabled the development of electron holography and thereby the study of a surprising new quantum mechanical phenomenon called the Aharonov–Bohm effect that has cast a new light on the meaning of electromagnetic fields and potentials, and
(2) electron-interference phenomena in normal-metal circuits having mesoscopic (sub-micron) dimensions (Landauer 1970; Bergmann 1984; Webb *et al.* 1985; Webb and Washburn 1989; Levy *et al.* 1990; Imry 1996).

We will also speak briefly about superconductivity, which is also a coherent electron phenomenon, albeit a more subtle one.

A single electron in a pure state is one that has a wave function:

$$\psi(\mathbf{x}, t) = A(\mathbf{x}, t)\, e^{i\varphi(\mathbf{x}, t)} \tag{2}$$

The amplitude $A(\mathbf{x}, t)$ and phase $\varphi(\mathbf{x}, t)$ are real. By contrast, an electron in a mixed state is represented by a statistical mixture of pure states. Measurable quantities are always quadratic in the wave functions, so when a pure-state wave function is a sum of two parts,

$$\psi = \psi_1 + \psi_2,$$

the results of measurements depend upon quantities such as

$$\psi^*\psi = A_1^2 + A_2^2 + 2A_1 A_2 \cos(\varphi_1 - \varphi_2). \tag{3}$$

The last term in eq. (3) is a typical interference term. It depends upon the relative phase of the two parts of the wave function, and it looks like the interference terms that appear in classical optics and other classical wave phenomena. However, there is an important difference. Pure states are usually confined to atoms and molecules. At larger distances, one is most often dealing with some statistical mixture of pure states so that the relative phases become randomized, and interference effects disappear. In condensed matter, that phase randomization comes about through the thermal spread of the electrons and through interactions of the electrons with each other and with thermal agitations of the atoms. In electron beams, the phases become randomized by averages over a spread of frequencies and wave numbers. In either case, quantum mechanical interference effects are washed out by the randomization of the relative phases.

Nevertheless, there are special situations where something very like a pure wave function survives at laboratory-scale distances, and typical quantum mechanical interference phenomena may be restored. Electron beams may be made sufficiently monochromatic so that stable phase differences persist over useful distances. Normal-metal circuits may be made small enough so that the wave-function information is not completely lost. Then, if the relative phases persist over some coherence length long enough in comparison with the size of the conductor, quantum mechanical interference phenomena appear. These issues are addressed in sections 9.2 and 9.3. Superconductivity is discussed briefly in section 9.4 and in more detail in chapter 7 of this book. In sections 9.5 and 9.6, we discuss the physics that has been learned from the Aharonov–Bohm effect and from experiments with mesoscopic normal-metal circuits, respectively, while section 9.7 presents off-diagonal long-range order, the quantum theoretical concept which underlies the kind of coherence exhibited by superconductors.

9.2 Interference phenomena, holography and coherence length

One may produce a photograph by causing a light wave to fall on some subject and allowing the wave emerging from the subject to strike a photographic film. The emerging wave may be reflected from the object under study, as in ordinary photography with visible light, or it may be transmitted, perhaps refracted or partially absorbed. Either way, the emerging wave is characterized by a phase and an amplitude at each point. The film records the time-averaged square of the amplitude at each point on the film, and the phase information is lost. Lost with it is the depth information contained in the relative phases of waves reflected at different distances from the film. The scale of the structures to be studied is limited by the resolution, which, with special exceptions,* is not better than about one wavelength, even with ideal optical elements.

For electron beams, similar general considerations apply, but the details are very different. At practical energies for elecron microscopy, in the 100 keV range, the electron wavelength is a few times 10^{-3} nm, that is up to 10^6 times shorter than optical wavelengths. On the other hand, electron 'lenses', in reality electric and magnetic fields, have more serious limitations than the analogous optical lenses. Spherical and chromatic aberrations cannot be eliminated, and the achievable resolutions are on the order of 100 wavelengths.

Using interferometry, one can record phase, and hence also depth, information, and one can improve resolutions. As discussed in the Introduction, the interference term in eq. (3) contains the information on the phase difference between two waves. In the simplest case, a monochromatic beam may go through two slits and the two partial waves may be made to converge on a film as in Fig. 9.1 to form the interference pattern illustrated in Fig. 9.2. The interval d between the dark bars in Fig. 9.2 is given by

$$d = \frac{\lambda}{\sin \beta}, \tag{4}$$

where β is the angle between the two beams. In practice, electron optics may be used to make β as small as about 10^{-6} radian. The record of an interference pattern on a photographic emulsion is called an interferogram, or hologram. If one beam is shifted in phase, the interference pattern is shifted by the same phase angle. If one of the beams is partially intercepted by a small transparent object, the resulting image can be magnified by electron optical methods and focused on the film. Then the interference pattern may resemble the one in Fig. 9.3, and the shift of the interference pattern measures the phase shift of the transmitted wave. More generally, when one beam has any elastic interaction with an object, the interferogram may be more complicated, but at every point it will record both the amplitude of the emergent beam and the phase relative to the reference beam, which did not interact with any object. In Fig. 9.3, the interferogram was made by passing one beam through samples

---

* For example, better resolutions are achievable with near-field microscopy

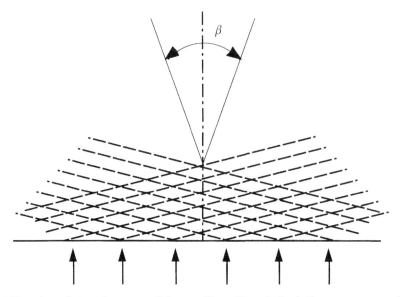

Fig. 9.1 Two interfering beams striking a film. The dashed lines are wave fronts. Arrows indicate the points of maximum intensity.

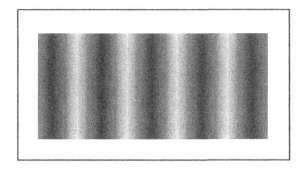

Fig. 9.2 Interference pattern on the film from the beams of Fig. 9.1.

of size about 100 nm. Note the well-resolved phase shifts of a few times 2π, which corresponds to a few wavelengths.

In holography, one uses the interferogram/hologram as a transmission grating for visible light. The diffracted light wave contains the same phase information as does the hologram, so one has in effect reconstructed the electron wave as a light wave without loss of any information. (That is why the familiar optical holograms, created with visible light, are seen by our eyes as three-dimensional pictures. Ideally, the wave transmitted by the hologram is identical to the wave that came from the original object.) Electron holography has the great advantage that the unavoidable aberrations of the electron optics can be

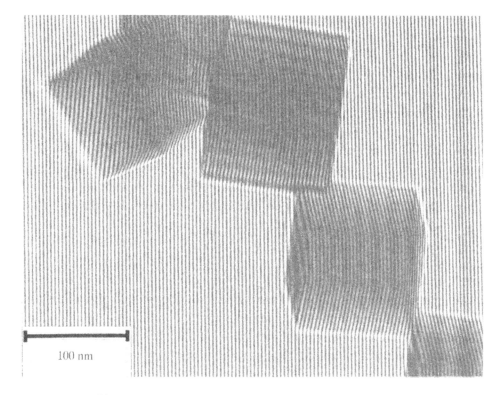

Fig. 9.3 Interferogram crystals of MgO (Tonomura 1993).

compensated by the optical tricks which are available for visible light. It also provides an angular magnification in the ratio of the wavelength of the light to the wavelength of the electron, as can be seen from eq. (4). Phase shift resolutions down to 0.01 times 2π have been achieved by electron holography.

The essence of holography can be seen by considering how one reconstructs the full information in the object wave ψ_0 from the hologram. ψ_0 can be written as a superposition of plane waves moving in all directions, with weights equal to $f(k_x, k_y)$. (For simplicity, we consider a monochromatic wave, so the third component k_z is determined by the first two.) Just shining ψ_0 on a screen gives an intensity which is the absolute square of the wave's amplitude, and the phase information is lost. In forming a hologram, one circumvents that loss by 'beating' the object wave ψ_0 against a reference wave ψ_r. Taking the absolute square of the sum of these two waves produces, in addition to the sum of the intensities, two interference terms, one proportional to the ψ_0 and one to its complex conjugate ψ_0^*. In the simplest case, the reference wave is a plane wave $\exp(ik_z z)$ and the emulsion lies in the $z = 0$ plane. Then the pattern on the hologram contains the Fourier transform of ψ_0. That transform is inverted by the diffraction of the light wave incident upon the hologram. One has to disentangle ψ_0 from ψ_0^*. A good introduction to that process, as well as to other techniques of holography, is given in the book *Electron Holography* (Tonomura 1993). An overview of the state of the art in 1994 and its application to diverse research problems requiring high-resolution imaging is also available (Tonomura *et al.* 1994).

All of electron holography is an application of the coherent electron. Holography was invented conceptually in 1949 by Gabor (1949, 1951) precisely as a way of overcoming the resolution limit imposed on electron microscopes by their unavoidable aberrations. However, Gabor's concept was not brought to fruition until 30 years later, when Tonomura was able to make sufficiently coherent electron beams in an electron microscope.

A real electron beam is not in a plane wave state. It consists of a stream of wave packets, typically far apart compared with the dimensions of any interference experiment and compared with their own spatial extent. Each wave packet represents a different electron and interferes only with itself when it has been split by a beam splitter and then reassembled at the photographic film after one of the two partial waves interacted with the sample under study. For the interference to be observed, for example in the geometry of Fig. 9.1, the relative phase of the two beams' contributions to each point must be stable throughout the time of exposure to the wave packet. Stated in terms of wave numbers, that says the packet's width Δk must be much less than the mean wave number k, so that the dark and light bands on the interferogram in Fig. 9.2, spaced proportionately to k^{-1}, will not be washed out by the average over k. The quantity Δx defined by

$$\Delta x = \frac{1}{\Delta k} \tag{5}$$

defines a coherence length for the electron beam. Two contributions to the wave function, $A \exp\{ikx\}$ and $B \exp\{i(k + \Delta k)x\}$, maintain their relative phase only over a range of x values differing by less than Δx. Outside that range of x, the contributions of different k to the wave packet cancel, and the wave function goes to zero.

For electron beams, the quantity defined in eq. (5) is referred to as the temporal coherence length or the longitudinal coherence length, or simply as the coherence length. For holography, one also requires phase coherence across the wave front. That is measured by the spatial or transverse coherence length given by $\lambda/\sin\beta$, where β is the angle between the two beams. In practical effects, the distances between points on an object that can contribute to measurable interference effects are limited by these coherence lengths, and of course the object beam must be coherent with the reference beam on the same length scales.

In practice, there is always a trade-off between intensity and coherence length. For useful intensities, longitudinal coherence lengths up to about 1 μm and transverse coherence lengths up to some tens of micrometres represent the state of the art in 1995. Those achievable coherence lengths limit the sizes of the structures that can be investigated by electron holographic methods. Large transverse coherence lengths are especially important for investigating domain structure in solids.

Electric fields, either within the transparent sample, or more generally in the electron microscope, will shift the phase of the electrons' waves. That, however, is not the only way to shift the relative phase of two coherent electron beams. Magnetic fields can also create a phase shift, and electron holograms can be used to explore the effects of magnetic fluxes on the scale of 10^{-7}G cm^2. Consider an interference measurement in which a magnetic flux Φ threads the region between the two beams, as in Fig. 9.4. The wave function at a point on the film is a sum of Feynman amplitudes

$$\exp\left\{\frac{i}{\hbar}\int L\,\mathrm{d}t\right\} \tag{6}$$

over all classical paths between the space-time point where the electron originated before the beam was split and the space-time point on the film where the wave function is being evaluated. The classical Lagrangian L contains the kinetic energy, and any potential energy due to applied fields or to matter being traversed, and a contribution from the magnetic flux equal to $ev \cdot A$, where v

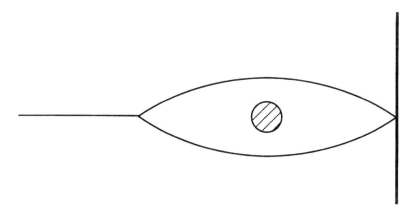

Fig. 9.4 Magnetic flux between the two halves of a coherently split electron beam. The striped circle represents a flux-bearing solenoid.

is the classical velocity and A is the vector potential. Then every Feynman amplitude is multiplied in the presence of the magnetic flux by the factor

$$\exp\left\{\frac{ie}{\hbar}\int A \cdot v \, dt\right\} = \exp\left\{\frac{ie}{\hbar}\int A \cdot dx\right\}. \tag{7}$$

The relative phase of the two beams at any point on the film is shifted by the amount

$$\frac{e}{\hbar}\oint A \cdot dx = \frac{e\Phi}{\hbar}, \tag{8}$$

where the integral is taken on the closed path that goes from the source of the electrons to a point on the film along one path and back along the other path, and Φ is the magnetic flux between the two paths. So interferometry and holography can record magnetic flux structures in the same way as they record matter that interacts with electron beams. Since a 2π phase shift is equivalent to no phase shift, all observable effects of the flux must be periodic in the flux with period equal to London's unit,

$$\Phi_0 = 2\pi\frac{\hbar}{e}, \tag{9}$$

about 10^{-7}G cm^2. This universal flux quantum sets the scale for fluxes observable by holography. The response to a quantity of flux greater than London's unit is ambiguous because of the periodicity, and the response to a quantity of flux much less than London's unit will result in a phase shift much less than 2π. Fig. 9.5 shows the structure of a fluxon near the edge of a type-II superconductor, imaged by electron holography. The total flux in this case is one-half of London's unit.

For magnetic structures, just as for material objects, the coherence length needs to be at least equal to the size of the region being investigated so that a well-defined and stable phase shift between the two beams can exist.

The coherence lengths discussed in this section are the ones for electron beams. When the interfering waves interact with other degrees of freedom, as is the case for electrons in solids, phase coherence can be lost, as we shall now discuss.

9.3 Coherence in normal metals

The coherence discussed above for electron beams is associated with the interfering electron beam having a narrow energy width. Otherwise, different electrons in the beam have different wavelengths and produce different interference patterns. In solid-state systems, where there is no literal beam, the energy width is equal to the thermal energy at finite temperatures and very

vacuum

Pb

2 μm

Fig. 9.5 Holographic reconstruction of magnetic flux entering and leaving a superconducting Pb sample (Matsuda *et al.* 1989). The arch at the left connects two flux lines with fluxes in opposite directions.

small voltages. There is an important distinction between this 'coherence' and the more fundamental process of true dephasing (Stern *et al.* 1990). In the latter, each electron wave loses its phase memory and the ability to interfere. The exception is the case of superconductors, where the charge carriers (electron pairs for all known superconductors) are coherent over arbitrarily large distances. This is due to a particular 'long-range order', which exists in such systems and is discussed in section 9.7. In this section, we consider only normal conductors.

For beams, interference can be achieved by reducing the energy-width of the beam using, for example, an energy filter,* provided, of course, that there is sufficient intensity. As long as the interaction of the electron under discussion with all the other degrees of freedom (such as the translation state of a dust particle, other electrons, lattice vibrations, etc.) can be neglected, the electron has a definite wave function and remains coherent. However, when the electron interacts with other degrees of freedom, and if those degrees of freedom are then excited, the ability of the electron to interfere with itself is lost. In the case of two-wave interference, the condition to lose the interference is that those degrees of freedom (which will be referred to as 'the environment') have recorded which of the two paths the electron took. Therefore, interference is lost. The condition for recording this information is that the two partial waves leave the environment in different states, as explained below. Condensed matter systems are a particularly useful arena to study these issues, due to the known interactions of electrons among themselves and with the lattice vibrations. In fact, these studies have led to a better understanding of the interacting electrons in conducting media.

To get phase randomization, the scattering should be inelastic, changing the state of the environment. Elastic scattering by an arbitrary static potential, such as that due to defects and impurities in a solid, just gives the electron a definite, however complicated, phase. The realization that elastic scattering retains phase memory has played an important role in the inception of mesoscopic physics. Since it is known that inelastic scattering in solids decreases with decreasing temperature (or with decreasing excess energy of the electron), the coherence length can be made large enough at realistically low temperatures to enable coherence to exist in small but attainable samples.

It is interesting that the above concept of the elimination of interference by the environment registering which path the particle took is equivalent (Stern *et al.* 1990) to a seemingly different way of describing the loss of interference. As

---

* Exceptions are special interference phenomena, called 'weak localization' or 'coherent back scattering', in which the energy spread of the beam is irrelevant (Bergman 1984; Webb and Washburn 1989; Imry 1996).

seen in eq. (3), intensity hinges on the relative phase $\varphi \equiv \varphi_1 - \varphi_2$ between the two interfering paths. This relative phase is a quantum mechanical quantity that has a quantum uncertainty. Due to interactions with a dynamic environment, the uncertainty in φ increases with the length of the path and becomes of the order of 2π for times at which the environment has a probability of order unity of having changed its state. It is quite straightforward to prove mathematically that the condition for the two partial waves leaving the environment in nearly orthogonal states is equivalent to that for producing a large uncertainty in φ. This follows because both effects are governed by the same interaction between the interfering particle and the environment. The dephasing time τ_φ can thus be defined equivalently either as the time to change the environment in the above sense or as the time to produce a large uncertainty in φ. In the former description, the dephasing is understood from the fact that, in the interference experiment, one is averaging out the variables of the unmeasured environment. In the latter description, it is the averaging out of the interference in the electron's wave function by the large uncertainty in φ.

For electrons in conducting solids, the dominant interaction with the other degrees of freedom is often the one with all the other conduction electrons. The above equivalence between the two ways to describe the dephasing follows from the 'fluctuation-dissipation' theorem, which relates the fluctuations of the environment that produce $\Delta\varphi$ to the ease of exciting it (related to the first description). This can be turned into a rather straightforward way of calculating τ_φ. Interesting temperature dependences of τ_φ have been predicted. They are particularly unusual for very thin wires ('quasi-one-dimensional' systems) in which they have received impressive experimental confirmation. Thus, electron dephasing in disordered solid-state conductors is, by now, quite well understood.

9.4 Coherence and superconductivity

Superconductivity is now known to be a coherent state of electron pairs, and is addressed in chapter 7. Here, we shall speak only of the most general considerations linking superconductivity with the coherent electron.

The first successful theory of superconductivity was Fritz London's phenomenological theory (London 1950), developed in the 1930s and 1940s. That theory was initially based on classical dynamics. Its central postulate was that the conduction electrons in a superconductor are forced by some interaction into a state where the average canonical momentum of the electrons vanishes at every point in the superconductor, regardless of magnetic fields that may be present, be they due to external sources or to the currents in the superconductor itself. The canonical momentum p is not an objective measurable quantity. It depends upon the vector potential A, which itself depends upon the choice of gauge. London's postulate applies to the gauge in which

$$\nabla \cdot A = 0 \tag{10}$$

so that

$$A(x) = \int \frac{x' - x}{|x' - x|^2} \times B(x') \, \mathrm{d}^3 x' \tag{11}$$

where B is the sum of whatever external magnetic field is applied to the conductor plus the magnetic fields created by any currents in the conductor.

Once London's postulate is accepted, it follows that the electrons at a point x move with the persistent velocity

$$v(x) = \frac{1}{m}(p + eA(x)) = \frac{e}{m}A(x), \tag{12}$$

e being the absolute value of the electron's charge. Then there is a persistent current given by

$$j(x) = -\Lambda A(x), \tag{13}$$

where Λ is London's constant,

$$\Lambda = \frac{ne^2}{m}, \tag{14}$$

and n is the density of conduction electrons, assumed to be constant. London did not have a quantum mechanical model to justify the vanishing of the average canonical momentum everywhere, independently of external fields. His speculation toward that goal,

... we may say that the present interpretation offers a remarkable possibility of reducing superconductivity to an apparently very simple model. The long range order of the (canonical) momentum would be a specific quantum effect. It would *not* be due to distinct electrons at separate places having the same momentum. It would arise from wave packets of wide extension in space assigning the same local mean momentum to the entire superconductor ...

is based on the assumption of a coherent state, in his words 'wave packets of wide extension in space'.

London's classical equations are unchanged formally by going over to quantum mechanics. London's phenomenological postulate then says that the wave function is 'frozen', unchanged by the presence of magnetic fields, to make p vanish at every point. Then eqs (12) and (13) are unchanged except that v, p and j have to read as the quantum mechanical expectation of the velocity, canonical momentum, and current at the point x, summed, of course, over all the electrons.

The simplest dynamical model to justify London's postulate is obtained by pretending that the conduction electrons are bosons contained only by the edges of the conductor and interacting only weakly with each other and with the atoms of the conductor. Then all the conduction electrons would occupy the single-electron ground state, and a coherent state wave function is constant throughout the conductor. In fact, the modern BCS theory does justify that model except that the effective charge carriers are pairs of electrons that act approximately as bosons and occupy a coherent state (chapter 7).

9.5 The Aharonov–Bohm effect

Research using coherent electron beams has produced many achievements in the past ten years (Tonomura 1993), one of the most dramatic of these being the experimental verification of the Aharonov–Bohm, or AB, effect (Aharonov and Bohm 1959; Peshkin and Tonomura 1989), which has profound implications for our fundamental understanding of electromagnetism in quantum mechanics as well as having practical applications.

We saw in section 9.2 that the relative phase of the two beams in an interferometer can depend upon a magnetic field confined to the region between the two beams. According both to conventional quantum mechanics, and more recently to experiment, a magnetic field can cause observable effects on the motion of an electron that never enters the region where the field exists. When that was first pointed out on theoretical grounds in 1959 by Aharonov and Bohm, most physicists were taken completely by surprise. Many did not believe it because our whole understanding of classical electromagnetism had been based on a notion of locality, explained below, which underlies the work of Faraday and Maxwell and which was now being challenged. (It is interesting that there were hints in earlier works directed to other subjects (Dirac 1931; Ehrenberg and Siday 1949; London 1950), and in fact Dirac's and London's work was widely studied, but the consequences of those hints were not pursued before 1959.)

In classical physics, electromagnetic forces between charged particles are mediated by the Maxwell fields E (electric) and B (magnetic) according to the Lorentz force law

$$F = q[E(x) + v \times B(x)], \tag{15}$$

where q, x and v are the particle's charge, position and velocity. The charged particle is in turn the source of fields according to the Maxwell equations. The Lorentz force is local in the sense that it depends only on the field at the position of the particle. The theory is also local in a more general sense. The force on the particle delivers energy and momentum to the particle. The Maxwell field contains energy and momentum, their densities given, respectively, by

$$H = \frac{1}{8\pi}(E^2 + B^2); \quad \boldsymbol{P} = \frac{1}{4\pi}(\boldsymbol{E} \times \boldsymbol{B}). \tag{16}$$

The field, of which the particle is the source, when added to the field acting on the particle, increments the total field energy and field momentum by an amount that precisely compensates for the energy and momentum delivered to the particle by the Lorentz force. This incremental field energy and momentum are initially localized at the position of the particle. The same is true of angular momentum. Those physical quantities then move with the speed of light, following the Maxwell equations, and the energy, momentum and angular momentum may ultimately be delivered to other particles. By endowing the electric and magnetic fields with physical reality in this sense of Faraday and Maxwell, the energy and momentum are conserved at every instant and can be tracked as they move from place to place. What might have seemed a nonlocal, delayed interaction between two distant charged particles has been replaced by a local, instantaneous interaction between the particle and a physical object called the Maxwell field, and the reward is the restoration of the conservation principles.

There is another way to proceed in classical theory. Instead of using Newton's laws, one can start from the Hamiltonian or the Lagrangian. For that purpose one has to introduce quantities called gauge fields, the scalar potential *V* and the vector potential *A*, defined to obey

$$E = -\nabla V - \frac{\partial A}{\partial t}$$
$$B = \nabla \times A. \tag{17}$$

In the Hamiltonain form of the theory, one then defines the Hamiltonian,

$$H = \frac{1}{2m}(\boldsymbol{p} - q\boldsymbol{A})^2 + V, \tag{18}$$

where $\boldsymbol{p}$ is the canonical momentum,

$$m\boldsymbol{v} + q\boldsymbol{A}. \tag{19}$$

Now, the motion of the particles is given by Hamilton's equations,

$$\frac{d\boldsymbol{x}}{dt} = \frac{\partial H}{\partial \boldsymbol{p}}$$
$$\frac{d\boldsymbol{p}}{dt} = -\frac{\partial H}{\partial \boldsymbol{x}} \tag{20}$$

in terms of $\boldsymbol{x}, \boldsymbol{p}, V$ and $\boldsymbol{A}$ instead of $\boldsymbol{x}, \boldsymbol{v}, \boldsymbol{E}$ and $\boldsymbol{B}$.

In this description, it may appear that the physics is somehow contained in the gauge fields instead of the Maxwell fields, but there are problems. The gauge fields are not unique. Starting from any V and A obeying eqs. (17), one can make a gauge transformation to other choices of V and A that represent the same Maxwell fields and the same physics. Also, the local gauge fields depend in part upon the Maxwell fields at distant points and earlier times. The notions of locality and conservation, which for Faraday and Maxwell endowed the Maxwell fields with physical reality, have been obscured.

In classical physics, there is an easy way out. The solutions of Hamilton's equations are demonstrably identical to those of Newton's equations, so nothing has objectively been changed. The gauge fields are some auxiliary quantities introduced for mathematical convenience, but the physics resides in the underlying Maxwell fields.

In quantum mechanics, there are no Newton's equations to help us reformulate the theory in terms of the Maxwell fields. We have only the Hamiltonian and Lagrangian forms of the theory, and those are based on the gauge fields, not on the Maxwell fields. The Schrödinger equation,

$$i\hbar \frac{\partial \psi}{\partial t} = \mathrm{H}\psi = \frac{1}{2m}[-i\hbar\nabla - qA(x)]^2\psi + V(x)\psi, \tag{21}$$

cannot be formulated in terms of the local Maxwell fields. Nevertheless, it was usually thought that a remote magnetic field could not influence the motion of a particle in quantum mechanics, just as it could not in classical theory. That belief was based on the existence of a quantum mechanical gauge transformation to make the local gauge field due to the remote Maxwell field vanish in the domain of the particle.

To see this, suppose some magnetic field B is confined to a region far from the domain of the particle. It can easily happen that some choice of the vector potential A has non-vanishing values within the domain of the particle, so that $A(x)$ in the Schrödinger equation does not vanish. However, because B vanishes where the Schrödinger equation is obeyed, it is possible to introduce a quantum mechanical gauge transformation

$$\psi' = U(x)\psi \tag{22}$$

that will remove the vector potential from the Hamiltonian and the Schrödinger equation but leave all physical consequences of the Schrödinger equation unchanged. Thus, the remote magnetic field can have no effect upon the motion of the particle. In other words, this quantum mechanical gauge transformation has in common with the analogous classical gauge transformation the fact that it preserves all observable quantities, namely the Maxwell fields and the position and velocity of the particle, while removing A from the Hamiltonian and from the equations of motion of the observables.

The required gauge transformation U in eq. (22) is given by

$$U(\mathbf{x}) = \exp\left\{\frac{iq}{\hbar}\int^{\mathbf{x}} A(\mathbf{x})\cdot d\mathbf{x}\right\},\tag{23}$$

where the line integral on the right-hand side is carried from any fixed starting point in the domain of the particle to the point $\mathbf{x}$ along any path that is entirely within the domain of the particle. For $U(\mathbf{x})$ to exist and be a function of $\mathbf{x}$ alone, the integral on the right-hand side has to be independent of the path of integration from the fixed starting point to $\mathbf{x}$. That condition requires that $\nabla \times A$ vanishes everywhere in the domain of the particle, which says that $\mathbf{B}$ always vanishes at the position of the particle. Thus, the effects of a magnetic field *within* the domain of the particle cannot be removed from the Schrödinger equation by this trick or any other.

All this has been standard physics since the early days of quantum mechanics. However, it was usually overlooked that the proof may fail if the domain of the charged particle is a multiply-connected region, one with at least one hole in it. In Fig. 9.6 a magnetic flux Φ threads the hole through the domain of the charged particle. The magnetic field $\mathbf{B}$ vanishes everywhere in the domain of the particle. Nevertheless, the vector potential A cannot be removed by a gauge transformation because the line integral of A once around the dashed line must obey

$$\oint A(\xi)\cdot d\xi = \Phi.\tag{24}$$

In that case, the gauge transformation U of eq. (23) cannot depend only upon $\mathbf{x}$, independently of the path of integration; one can always add an extra loop

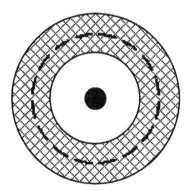

Fig. 9.6 Multiply-connected domain (cross-hatched area) of a charged particle. The black circle in the centre contains magnetic flux coming out of the page. The dashed line is the path of integration in eq. (24).

around the hole to any path from the starting point to **x**, and that will change the integral by Φ. For the gauge transformation to exist, it is not enough that **B** vanishes in the domain of the charged particle if that domain is multiply connected and a magnetic flux threads the hole through it.

To demonstrate that a magnetic field confined to a hole in the domain of the charged particle can affect the motion of the particle, Aharonov and Bohm simply solved the Schrödinger equation for a charged particle scattered by an infinite solenoid containing a magnetic flux Φ and found that the scattering does indeed depend upon the magnetic flux in the region forbidden to the charged particle.* The physics of how that comes about can be seen most simply by considering a bound-state problem (Peshkin *et al.* 1960) instead of the scattering problem. Consider a charged particle confined to the shaded region in Fig. 9.6 by some scalar potential, symmetric about the z axis. In the simplest gauge, the vector potential is given, in cylindrical coordinates, by

$$A_\varphi = \frac{\Phi}{2\pi\rho}; \qquad A_z = A_\rho = 0. \tag{25}$$

When this is put into the stationary Schrödinger equation and the wave function is written as

$$\psi(z, \rho, \varphi) = \psi(z, \rho) \exp\{im\varphi\}, \tag{26}$$

the result is

$$-\frac{\hbar^2}{2m}\left\{\frac{\partial^2\psi}{\partial z^2} + \frac{1}{\rho}\frac{\partial}{\partial\rho}\left(\rho\frac{\partial\psi}{\partial\rho}\right)\right\} + \frac{\left(n\hbar - \frac{q\Phi}{2\pi}\right)^2}{2m\rho^2}\psi = E\psi. \tag{27}$$

Then the allowed values of the centrifugal barrier strength Q, given by

$$Q = \left(n\hbar - \frac{q\Phi}{2\pi}\right)^2 \tag{28}$$

with integer n, depend upon the magnetic flux Φ in the region forbidden to the charged particle. It follows that the observable bound-state energy spectrum depends upon the magnetic flux which the charge particle has never entered. This dependence of the spectrum on the excluded flux underlies the equilibrium persistent currents in mesoscopic circuits discussed in section 9.6. Scattering also depends upon the strength of the centrifugal barrier and therefore upon the excluded flux.

This result, that an electromagnetic field in a region from which a charged particle has been forever excluded can influence the motion of the particle, is

---

* To be precise, they considered the limiting case where the magnetic field is a delta function and the hole has vanishing radius on the scale of the wavelength of the charged particle. However, their results remain valid for a hole of finite radius as well.

called the Aharonov–Bohm effect, and it was vigorously resisted by many physicists for the reasons discussed above. (The version discussed here is the magnetic AB effect. There is also an electric version, but that has been pursued less because it is more complicated theoretically and less accessible experimentally.) Dozens of attempts were made to show that the AB effect is not part of standard quantum mechanics or to introduce variants of the theory that did not contain the AB effect. (Reviews are given by Olariu and Popescu (1985) and Peshkin and Tonomura (1989).) None of these attempts really developed any acceptable theoretical structure, but the decisive test had, of course, to be experimental. Early experiments (Chambers 1960; Mollenstedt and Bayh 1962a, b) apparently confirmed the standard theory, but their total exclusion of the charged particles from the magnetic field was questioned. Finally, the issue was settled by the electron holography experiments of Tonomura, mentioned above.

Fig. 9.7 shows an electron hologram made with a split electron beam, one side of which passed through an iron loop impenetrable to electrons. In Fig. 9.7(a), the iron is not magnetized and the interference fringes are straight lines except where they are cut off by the shadow of the iron loop. In Fig. 9.7(b), a magnetic flux Φ equal to $\pi\hbar/2e$ circulates in the iron and the interference fringes are shifted, demonstrating a phase shift of half a wavelength between those rays which go through the centre of the loop and those rays which do not. (For an electron, $q = -e$.) The AB effect is confirmed in those experiments with a sensitivity of about 1% of the predicted phase shift.

What has been learned from this application of coherent electron beams about the fundamental issues raised above? First, the gauge principle has, for the first time, been directly confirmed in an experimental test. The gauge principle (Yang and Mills 1954) is the main foundation of current quantum field theories. It asserts (Wu and Yang 1975) that the physics of the interaction between particles and fields is carried, not by the local Maxwell fields, and not by the gauge fields, but by the quantities

$$W = \exp\left\{\frac{iq}{\hbar}\int_x^{x'} A(\xi) \cdot d\xi\right\} \tag{29}$$

for all possible paths of integration between all pairs of end points.* The physics cannot be carried by the local Maxwell fields because the AB effect experiments prove that remote fields can affect the motion of the electrons. It cannot be carried by the gauge fields because of the gauge invariance; different gauge fields describe the same physics. Only the quantities W, which enter the

---

* W is more general than the quantity U that appears in the gauge transformation of eq. (23). W may depend upon the path of integration from x to x', and it exists in the general case where a local magnetic field may be present. U can be a gauge transformation only when there is no local magnetic field.

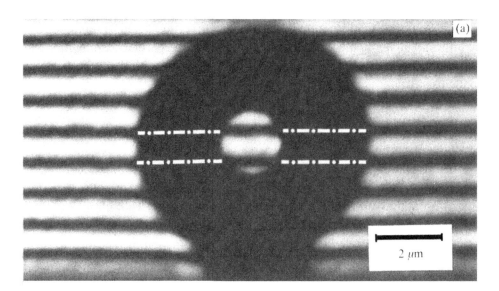

electron phase distribution

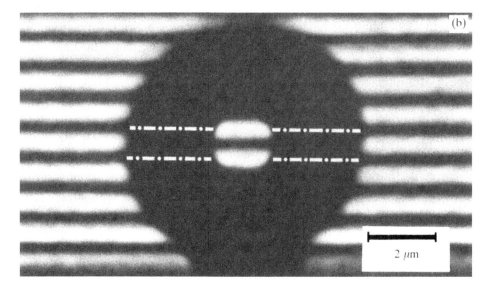

electron phase distribution

Fig. 9.7 Holograms obtained when an iron ring intercepts one beam. In part (a), the iron is not magnetized and the interference pattern is like that of Fig. 9.2. In part (b), magnetic flux equal to one-half of London's unit circulates around the hole, within the iron. (Tonomura *et al.* 1986.)

amplitude for every Feynman path, contain all the physical information and no more. In a simply-connected region, the totality of quantities W is equivalent to the local Maxwell field. In a multiply-connected region, some of them contain the information about the flux through any holes in the particle's domain.

It is noteworthy that a flux through a hole equal to an integral multiple of London's unit,

$$\Phi_0 = \frac{2\pi\hbar}{e}, \tag{30}$$

leaves the quantities W unchanged when a loop around the hole is added to any path of integration. Also for such a flux, the gauge transformation U remains independent of the path. The integral equation (23) depends upon the path, but only by an additive quantity which leaves U unchanged. This implies that all observable phenomena are periodic in the excluded flux, with period equal to London's unit. This periodicity underlies the phenomenon of flux quantization in superconductors.*

What of the local conservation laws? We will not go into details here, but simply state some of the main lessons about the laws from the AB effect. Surprisingly, the scattering experiment involves no force and no exchange of momentum between the charged particle and the Maxwell field. The particle's change in momentum in the scattering process comes entirely from its interaction with the barrier that prevents it from entering the magnetic field. The particle strikes the barrier differently in the presence of the excluded flux because of a phenomenon that happened in the past when the magnetic field was turned on, or more realistically when the particle came through the inevitable return flux to enter the field-free region. (For simplicity, we stick to cylindrical symmetry.) In the former case, the induced electric field created a torque on the particle and changed its kinetic angular momentum $x \times mv$ by an amount equal to $q\Phi/2\pi$, not necessarily an integral multiple of $\hbar$ but just the amount needed to account for the centrifugal barrier strength Q in eq. (28). In the latter case, the Lorentz force on the particle as it traversed the return flux provided the same shift in kinetic angular momentum. The long-cherished conservation laws are, in fact, preserved.

There is, however, a typically quantum mechanical kind of nonlocality that has no classical analogue. Imagine that an electron–positron pair is created in the field-free region and that the electron is scattered from a barrier containing

---

* Flux is not quantized in the same sense as angular momentum, for which only certain values exist in principle. All values of flux exist, but a superconductor has macroscopically lower energy for certain values of the flux, and these are the thermodynamically stable states. The flux quantum for superconductors is one-half of London's unit because the effective charge carriers are pairs of electrons and have charge $2e$.

magnetic flux. That electron also had to be created with kinetic angular momentum values different from the integral multiples of $\hbar$ by $e\Phi/2\pi$, and that condition on the creation seems to be nonlocal by any reasonable definition. In other words, turning on the flux, in addition to causing a torque on existing electrons, changed the Hilbert space of states available to electrons yet to be born, and changed it with observable consequences.

In addition to illuminating these fundamental issues, electron holography has been able to explore magnetic fields with fluxes on the scale of London's unit, around 10^{-7} G cm^2, in situations where the electron does enter the magnetic field. The vector potential A, by causing phase shifts, acts much like an index of refraction, causing magnetic fluxes to become visible in interferograms (Bonevich *et al.* 1995). This phenomenon has been valuable for studying the dynamics of magnetic flux lines in type-II superconductors. Fig. 9.5 shows such a flux line. The amount of flux in the flux line is one-half of London's unit.

9.6 Mesoscopic normal conductors

Interference phenomena in normal metals are observed primarily at temperatures of a few kelvin or below, where the electrons may be coherent over the sizes of experimentally practical samples, typically in the micron or sub-micron range.* In real samples there are always some impurities and defects which make the motion of the electrons quite complicated. However, it is usually an excellent approximation to regard these defects as static over the time scales relevant for the electrons. The defects just present an electron with a complicated, random-looking, potential. The electron has, in spite of the complicated potential, a well-defined wave function in the sense of section 9.3, and it is capable of exhibiting interference.[†]

An interesting case occurs when the sample has the shape of a ring or cylinder. The interference of the electronic waves around it will cause a number of quantum effects associated with the sensitivity of this interference to phase shifts due to an AB flux through the ring. This leads to the following:

(a) A dependence of the energy levels of the ring and hence of all its thermodynamic properties upon the AB flux. The energy levels, and therefore all the thermodynamic functions, of the ring are influenced by the disorder, but still depend upon

---

* Since coherence lengths generally decrease with increasing temperature, higher temperatures can be employed for smaller samples. In fact, for sizes on the order of a nanometre or less, the relevant range should reach room temperature.

[†] There are situations in which the disorder can cause the wave functions of the electrons to be 'localized'. This localization allows the wave function to stretch only over well-defined, finite distances, and it vanishes over larger distances over which interference is then impossible.

the AB flux with the periodicity required by eq. (28). This causes currents, which result in measurable magnetic moments, to flow in these systems. These currents do not decay; they are an *equilibrium phenomenon*. In the presence of magnetic flux, the state with minimum free energy is one with circulating currents.

(b) A dependence of the conductance of the ring, as measured by connecting it to leads and electrical measuring apparatus, upon the AB flux. The simplest way to understand this dependence is by thinking of the conductance (current per unit voltage) of the sample in terms of the transmission coefficient through the ring (Landauer 1970; Webb and Washburn 1989; Imry 1996), to which the conductance is proportional. This picture, due to Landauer, regards the sample as connected to two ideal leads which are in turn connected to two reservoirs maintaining the voltage across the sample. The transmission depends on the flux because of the phase shift between the electron's partial waves travelling along the two arms of the ring and the interference between those two.

Both of the above effects have been observed experimentally. Fig. 9.8 shows the nonlinear magnetic susceptibilities of an array of 10^7 0.5 micron copper (normal metal) rings, measured at very low temperatures. The large number of independent samples, the sum of whose responses is measured, is required to increase the signal-to-noise ratio. This measurement, which requires an excellent sensitivity due to the small signals, demonstrates that such persistent currents exist, do not decay, and are periodic in the AB flux. The period as a function of the flux is $h/2e$, i.e. one-half of the fundamental period, h/e. This is due to the averaging out of the latter in the ensemble of many samples, while the $h/2e$ component survives. This is an example of an ensemble average versus a 'mesoscopic fluctuation' effect, to be discussed briefly later. Fig. 9.9 shows such AB oscillations in the resistance of a micron-size gold ring at low temperature. An electron-microscopic picture of the sample is given in the inset. Here the basic period in the flux is h/e, since there is no averaging over many samples.

The $h/2e$ period in normal rings is different from the $h/2e$ fundamental period in superconductors, which is due to the electron pairs in the super-conductors having charge $2e$. The ring geometry is a simple case where the interference is easy to visualize. Related effects obviously also occur in samples of any shape containing an enclosed flux.* In the Feynman path integral picture, which makes quantum effects easy to visualize, the probability

---

* In all these circuits, the magnetic field is not strictly confined to a region from which the electrons are excluded. However, the small amount of flux intercepted by the metal is insignificant for the h/e oscillations. It is relevant only for the aperiodic conductance fluctuations of the ring's wires, to be discussed later, which occur on a larger field scale. The latter show up as the slow field modulation of the oscillations, or equivalently as the peak at small $1/B$ in the Fourier transform, in Figs 9.9(a) and 9.9(b), respectively.

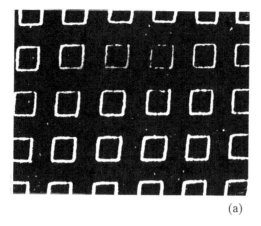

(a)

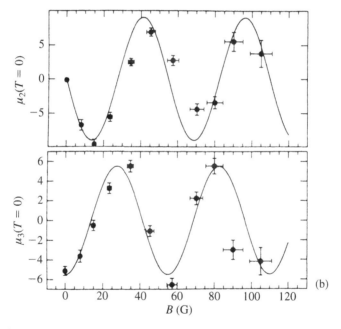

(b)

Fig. 9.8 This figure is taken from Levy *et al.* (1990), who measured the nonlinear response of the ensemble of 10^7 Cu rings, a small part of which is shown in (a), as a function of flux (a value of B of 130 G corresponds to h/e). An ac signal at a low (~ 1 Hz) frequency and an amplitude of 15 G was employed, and the second (μ_2) and third (μ_3) harmonics (double and triple the ac frequency) measured, as shown in (b). This specific nonlinearity arises from the periodic flux dependence. The nonlinear response is used because of the need to discriminate against relatively large background signals (due to the sample holder, etc.) that dominate the linear magnetic response of the system.

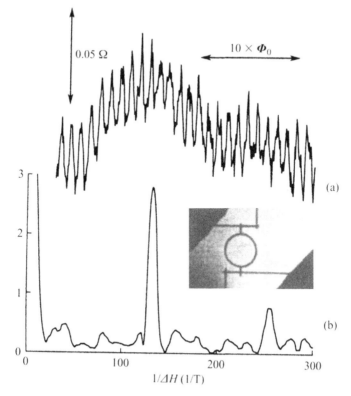

Fig. 9.9 (a) The magnetoresistance of the gold ring shown in the inset, with an inside diameter of ~ 800 nm and a width of ~ 40 nm. The arrow corresponds to ten flux quanta in the hole of the ring. (b) Fourier power spectrum in arbitrary units, showing peaks at h/e and $h/2e$, corresponding to the visible fast oscillations. The low-frequency peak corresponds to the slow modulation, which is a conductance fluctuation due to the flux in the arms of the ring. Taken from Webb *et al.* (1985).

of the electron moving from one point to another is given by the absolute value squared of an amplitude. This amplitude is a complex number which is a sum of many contributions, each related to a possible path between the two points. In a disordered solid, the Feynman paths are the various ways of moving through a terrain full of impurities and defects. The ease with which the electron moves, related to its mobility or conductivity, is influenced in a measurable way by the interference effect. The latter, in turn, is sensitive to external magnetic fields, which provide a useful experimental handle on the interference effects.

A novel concept that has emerged from the above observations is the distinction between properties of the system which are obtained by averaging over the defect configurations and those that are specific to a given sample. The $h/2e$-periodic dependencies of the persistent current or conductance of a ring are examples of the former, whereas the h/e fundamental period is an example of the latter. The h/e-periodic component has a random phase of oscillation as a function of flux (i.e. a minimum or a maximum at zero flux), while that phase is determined, irrespective of the impurity arrangement, for the $h/2e$ component. The impurity-ensemble averaging (as in the experiment shown in Fig. 9.8) has conventionally always been done, and it usually works for bulk properties. However, it turns out that, as long as the sample is coherent, that is smaller than the coherence and dephasing lengths discussed in sections 9.2 and 9.3, the magnitude of these sample-to-sample fluctuations in the conductance is significant. This magnitude can be 'universal', i.e. independent of material, size, purity, etc., and is determined by the electron charge, e, and Planck's constant, h. For example, these fluctuations in the conductance of coherent samples are always on the order of e^2/h, called the 'quantum unit of conductance'. This unit, which first appeared in the quantum Hall effect (von Klitzing *et al.* 1980), is approximately the inverse of 25 kΩ.* This universality is one of the interesting concepts in the new field of *mesoscopic physics*, which grew out of the above.

9.7 Coherence in superconductors as off-diagonal long-range order

The concept of off-diagonal long-range order (ODLRO) provides a proper foundation in theoretical physics for Fritz London's intuitive vision of a superconductor as a giant molecule whose electrons act as if they had a single coherent wave function that represents all the superconducting electrons and extends coherently throughout the conductor. Here we shall define ODLRO and describe some of its main features without proof. A complete discussion of ODLRO has been given by Yang (1962).

A system of N electrons may be in a pure quantum state, one that is described by a wave function

$$\psi(\underline{x}) = \psi(x_1, x_2, \ldots x_N), \tag{31}$$

where the x_j represent the spins as well as the positions of the electrons, $\underline{x}$ stands for all the x_j, and the wave function is constrained to be antisymmetric under exchange of any two electrons. A statistical mixture, for instance a thermal ensemble, of pure states ψ^α with statistical weights P_α such that $\sum_\alpha P_\alpha = 1$, is represented by the N-body density matrix, ρ, defined by

---

* It is easy to understand why this particular combination of natural constants, e^2/h, has the dimension of conductance and why it has this particular value: recall that the impedance of free space divided by 4π, which is about 30 Ω, is given by the inverse of the light velocity, c. Dividing this by the fine structure constant, $\alpha = 2\pi e^2/hc$, yields that h/e^2 is about 25 kΩ.

$$\langle \underline{x}'|\rho|\underline{x}\rangle = \sum_\alpha P_\alpha \psi^\alpha(\underline{x}')\psi^\alpha(\underline{x})^*. \tag{32}$$

For our purposes, it is not necessary to describe the density matrix in any great detail. It contains the complete description of the state of the N conduction electrons, be they in a pure or a mixed state. The expectation of any observable Q is given by the trace of the product ρQ, namely

$$\langle Q\rangle = \mathrm{Tr}\,\{\rho Q\} = \int\int \langle \underline{x}'|\rho|\underline{x}\rangle\langle \underline{x}|Q|\underline{x}'\rangle\,\mathrm{d}^{3N}\underline{x}\,\mathrm{d}^{3N}\underline{x}'$$

$$= \sum_\alpha P_\alpha \langle \psi^\alpha|Q|\psi^\alpha\rangle = \sum_\alpha P_\alpha \int \psi^\alpha(\underline{x})^* Q\psi^\alpha(\underline{x})\,\mathrm{d}^{3N}\underline{x}. \tag{33}$$

Each integral includes summation over all the spin variables as well as integration over all the spatial variables. As the electrons are identical, Q is necessarily symmetric under exchange of any pair of electrons.

The state of the system is said to possess ODLRO if, for some j, the density matrix ρ has a nonzero limit when $|x_j - x_j'|$ approaches infinity. Because of the antisymmetry, the condition is actually independent of j. To be precise, one should also take the limit of an infinite conductor with infinite N and finite density, but it suffices practically to consider $|x_j - x_j'|$ to be macroscopic. To see how ODLRO expresses coherence, consider the reduced one-particle density matrix ρ_1, which is obtained from ρ by integrating out the coordinates of all electrons but one, as is done in evaluating any one-body operator:

$$\langle x_1'|\rho_1|x_1\rangle = \int \langle x_1', x_2, x_3, \ldots, x_N|\rho|x_1, x_2, x_3, \ldots, x_N\rangle\,\mathrm{d}^3 x_2\,\mathrm{d}^3 x_3 \ldots \mathrm{d}^3 x_N. \tag{34}$$

As ρ_1 is hermitian, it can be written in the form

$$\langle x'|\rho_1|x\rangle = \sum_\beta p_\beta \phi^\beta(x')\phi^\beta(x)^*, \tag{35}$$

where the weights p_β are positive numbers whose sum is unity and the ϕ^β are one-body wave functions. Thus, for one-body measurements, the system is equivalent to an ensemble of single-electron states. In the usual case, ρ_1 represents a *mixed* state, i.e. a *statistical* mixture of pure one-body states that have wave functions, even if ρ is a pure state of the N-body system.

It has been proved that for ρ_1 to have ODLRO, p_β must remain finite for some β as N approaches infinity. That is not a surprising result. For any one β, there is a single wave function with definite relative phase between every two points, but averaging over a large number of states washes out that relative phase. In that sense, ODLRO is the precise statement of what is meant by coherence and by London's 'giant molecule'.

For electrons, it has been proved that ODLRO in the one-particle reduced density matrix is impossible. That, too, is not surprising. Electrons are fermions, and no two can have the same wave function, so each electron's chance of being found in a given state cannot be greater than $1/N$, vanishing in the macroscopic limit. Bosons can exhibit ODLRO in their one-body reduced density matrix, but electrons cannot.

Next consider the two-body reduced density matrix

$$\langle x'_1, x'_2 | \rho_2 | x_1, x_2 \rangle,$$

formed by integrating out all the x_j except two. ρ_2 contains the information about two-body correlations, but not higher correlations. It can exhibit ODLRO, and correlated pairs of electrons can therefore be the basic unit of systems exhibiting ODLRO.

The superconducting phase is believed to be one characterized by ODLRO. No proof has been given from any realistic Hamiltonian that the stable phase of any conductor at low temperature has ODLRO. (Neither has it been proved from a Hamiltonian that the stable phase of any class of atoms should be a crystalline solid, which exhibits a different kind of long-range order, diagonal in x.) However, the successful fundamental theory of superconductivity is, in fact, based on electron pairs which exhibit ODLRO in their two-particle reduced density matrix (chapter 7). It has been proved that, when a superconductor surrounds a magnetic field region, quantization of the magnetic flux in units of $\pi\hbar/e$, one-half of London's unit, is a necessary condition for the existence of ODLRO. The circuits of section 9.6 are made of normal metals, which do not have ODLRO and cannot, therefore, have the required coherence of phases at points separated by macroscopic distances. That is why those circuits need to be mesoscopic in scale.

In addition to providing a quantum mechanical definition and description of the coherent state in a conductor, the concept of ODLRO explains the phenomenon of flux quantization in superconductors, in units of $h/2e$, in a fundamental and satisfying way. The history of the interplay between the notion of coherence and the observed phenomena is interesting. Fritz London found from his phenomenological theory in the 1940s that the state of a superconductor surrounding a flux-bearing hole should be periodic in the flux, with period equal to h/e. He concluded that the flux (strictly the fluxoid) was quantized in units of h/e. That notion was not much pursued, partly because the quantum of flux was too small to be observed at that time, and probably also because the whole idea of a coherent single-electron wave function in a superconductor lacked support on the fundamental level. Ginzburg and Landau (1950) greatly improved the phenomenological theory, replacing London's wave function by an order parameter that obeys a nonlinear Schrödinger-type equation obtained by minimizing a properly chosen free energy. When that theory is compared with experimental data, it is found that the fit is improved by assuming that the charge carriers have charge around $2e$. That also was not pursued at the time, perhaps because there was no theoretical foundation for such an assumption. The question of quantized flux in a hole in the super-conductor was not raised. In 1957, the BCS theory (Bardeen *et al.* 1957), based on paired electrons, finally provided a genuine quantum mechanical model, and found that the charge carriers do indeed have charge equal to $2e$. The BCS theory does, in fact, require flux quantization in units of $h/2e$, but again that was not really much noticed until flux quantization, unexpectedly in units of $h/2e$, was discovered experimentally by Deaver and Fairbank (1961). The assumption of ODLRO now tells us, in a completely general and model-independent way, that the basic unit of the coherent state must be at least a pair of electrons simply because electrons are fermions, hence the factor $2e$ in the denominator of the flux quantum.

Acknowledgements

Work at Argonne National Laboratory was supported by the US Department of Energy, Nuclear Physics Division, under contract W-31-109-ENG-38. Work at the Weizmann Institute was partially supported by the Israeli Academy of Sciences and by the German-Israeli Science Foundation (GIF), Jerusalem.

We thank Dr Akira Tonomura and the Hitachi Advanced Research Laboratory for providing copies of their unique holograms.

10

The composite electron

R. J. NICHOLAS

Oxford University

10.1 Introduction

Applying a magnetic field to a charged particle causes circulating cyclotron motion, which leads to a magnetic quantization of the energy levels of a free electron known as Landau levels. When the electrons are in a two-dimensional metal, then this magnetic quantization leads to a total quantization of electron energy, and a whole range of new properties arise. Probably the best known of these are the *quantum Hall effect*, which is an essentially single-particle phenomenon in which the Hall resistance becomes quantized in fundamental units which are simply the ratio of the flux quantum to the electron charge, and the *fractional quantum Hall effect*, where the electron–electron interactions lead to a new quantum fluid which has excitations with an apparently fractional unit of charge. These phenomena have only become observable with the advent of high-quality semiconductor heterostructures, in which the electrons can be confined within a relatively thin layer such that the energy is quantized by amounts which are large compared with the electron Fermi energy and $k_B T$, and the amount of scattering present is reduced enough to allow electron coherence to play a role.

The quantum Hall effect was discovered experimentally in 1980 by Klaus von Klitzing (von Klitzing *et al.* 1980) without any previous theoretical predictions, and has now become established as the international standard to define resistance. It was quickly shown that the value of the resistance unit was simply the ratio of the flux quantum to the charge on the electron, and so, together with the Josephson effect, this allows us to define an electrical system of units based only on the fundamental properties of the electron and $\hbar$. In the near future, we may be able to define a closed system of units through a measurement of current using an electron turnstile device based on simply counting the flow of electrons as a function of time.

The fractional quantum Hall effect is now viewed in terms of an interacting electron gas in which the interactions can be renormalized in terms of electrons bound to a fixed even number of flux quanta. The new particles formed are known as *composite fermions*, which can then be thought of as weakly interacting particles which have their own Fermi energy, and subsequently demonstrate a quantum Hall effect of their own. This picture is relatively recent, having been developed over the period since 1990, and the properties of the composite fermion particles are still the subject of some debate.

In the following, I will introduce in more detail some of the concepts on which these properties are based, and I will look at some of the future directions in which this view of electronic properties is going.

10.2 Quantum Hall effect

10.2.1 Landau levels and the Shubnikov–de Haas effect

The quantum mechanical treatment of an electron in a magnetic field is well known. For a two-dimensional system with a magnetic field applied normal to the two-dimensional plane, cyclotron motion occurs, leading to a quantization of the states in k-space into a series of orbits with areas given by

$$S_p = 2\pi(p + \tfrac{1}{2})eB/\hbar.$$

The electron energies are quantized with

$$E_p = (p + \tfrac{1}{2})\hbar\omega_c = (p + \tfrac{1}{2})\hbar eB/m^*,$$

where ω_c is the cyclotron frequency. The degeneracy of these levels is given by

$$eB/h = B/\Phi_0,$$

where Φ_0 is the magnetic flux quantum. In addition, each level is doubly degenerate due to electron spin, and this degeneracy will be lifted by the relatively small Zeeman energy $g^*\mu_B B$. In the absence of any disorder broadening, this gives a density of states for the two-dimensional system which consists of a set of delta functions separated in energy by $\hbar\omega_c$, each split into two by the Zeeman energy, and with a degeneracy equal to the density of flux quanta per unit area. This total quantization of the density of states in two dimensions was rapidly realized to be very important, and when high-quality two-dimensional systems first became available with the production of silicon MOS (metal oxide semiconductor) transistors, it was found that this led to a very dramatic example of the quantum oscillations of resistivity known as Shubnikov–de Haas oscillations (Fowler *et al.* 1966). Zeroes were observed in the resistivity whenever the occupancy (ν) of the Landau levels was equal to an integer, where the occupancy is defined by

$$v = n_e h/eB = n_e \Phi_0/B,$$

and corresponds to the number of Landau levels filled and also the number of electrons n_e per flux quantum.

Measurements of resistivity and conductivity in high magnetic field are connected by the tensor relations

$$\rho_{xx} = \frac{\sigma_{xx}}{\sigma_{xx}^2 + \sigma_{xy}^2},$$

$$\rho_{xy} = \frac{-\sigma_{xy}}{\sigma_{xx}^2 + \sigma_{xy}^2},$$

where the components ρ_{xx}, ρ_{xy}, σ_{xx}, σ_{xy} represent the diagonal and Hall components of the resistivity and conductivity, respectively. The tensor relationship arises because of the Lorentz force, which acts on the electrons producing a Hall field in the direction perpendicular to the current flow. In the case of any real two-dimensional system, the density of states consists of bands of states corresponding to each of the Landau levels, which are broadened somewhat due to the presence of any residual scattering in the system. Thus, whenever the Fermi energy lies in the gap between any of these bands, we have the equivalent of an insulating system in which the conductivity would be expected to become zero. At the same time, however, the electrons in the filled levels are still contributing to the Hall voltage. The net result of this is that both the resistivity and the conductivity will become zero at the same time due to the tensor relationship between them, and the fact that $\sigma_{xx} \ll \sigma_{xy}$.

10.2.2 Quantum Hall plateaux

The simultaneous tendency of both the resistivity and conductivity to show a series of zeroes at low temperatures, where $k_B T$ is significantly smaller than the separation of the levels $\hbar\omega_c$, was observed in the first experiments. What was not immediately realized was that the relatively weak features that were observed in the Hall voltage ρ_{xy} were also highly significant. This was first demonstrated by von Klitzing *et al.* (1980), who showed that, at the same points where the resistivity went to zero, the Hall voltage became constant, and more importantly that the resistance values measured were quantized in units given by

$$R_{xy} = \frac{1}{p} h/e^2 = \frac{1}{p} \Phi_0/e.$$

This has subsequently been shown to be an extremely precise result, exceeding the abilities of the international standards laboratories to measure it (an absolute accuracy of approximately one part in 10^8). As a result, in 1990 the quantum Hall effect was used to define the standard resistance unit as

$$R_{\text{K-90}} = 25\,812.807\ \Omega.$$

For a suitably chosen sample at low temperatures, as shown in Fig. 10.1, the diagonal resistivity turns into a series of almost delta-function-like peaks, separated by wide regions where the resistivity goes to zero. At the same time, the Hall voltage becomes a staircase-like structure, with sharp transitions between the quantized values each time the diagonal resistivity passes through one of its peaks. This phenomenon is now known as the *integer quantum Hall effect* (IQHE) in order to distinguish it from the features occurring at fractional occupancies discussed below.

The reasons for this remarkable quantization are more complex to explain. At the simplest level, the degeneracy of each Landau level is given by eB/h, and the insertion of this value into the normal definition of the Hall coefficient of $1/n_e e$ leads to the quantum Hall result for an exactly filled level. The significant part of the quantum Hall result lies in the fact that the same value is obtained for a wide range of magnetic fields, even though the Landau levels are not exactly filled. This is due to the presence of localized states, which exist away from the exact Landau level energies, and are caused by the presence of scattering centres in the structure. The extended states, which conduct the whole current in any measurement, carry an additional current which is

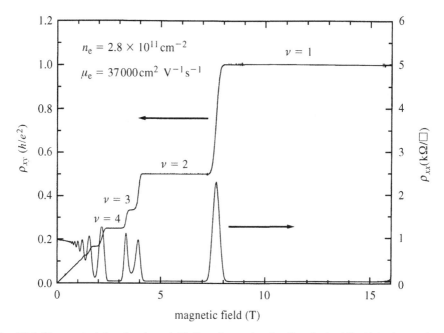

Fig. 10.1 The resistivity (ρ_{xx}) and Hall voltage (ρ_{xy}) of a GaAs/GaAlAs heterojunction at 60 mK, showing quantum Hall plateaux (after Gee (1996)).

sufficient to balance the fact that the localized states act as small insulating regions within the structure. As a result, it becomes irrelevant whether the localized states are occupied or not. The region of the quantum Hall plateaux thus corresponds to the energy range in which the Fermi energy lies within these localized states. For a more detailed discussion, the reader is referred to von Klitzing (1986).

10.3 Composite fermions

10.3.1 The fractional quantum Hall effect

The QHE and resistivity oscillations described so far are all the result of the properties of single electrons in a magnetic field. As the quality of two-dimensional electron systems improved, however, it rapidly became clear that there was a whole new range of properties which were due to collective electron properties caused by the Coulomb interactions between the electrons. This was shown by the observation of the *fractional quantum Hall effect* (FQHE) by Tsui *et al.* (1982), who found that in high-mobility heterojunctions a new quantum Hall plateau occurred (Fig. 10.2) with a fractional level occupancy, $\nu = 1/3$, suggesting a carrier of charge $e/3$. No explanation of this in terms of single-electron properties is possible since there is only a partial population of the lowest Landau level. The picture rapidly became more complex as a whole series of new plateaux and resistivity minima were observed for fractions of the form $\nu = p/(2p + 1)$, where p is an integer (e.g. $\nu = 2/5, 3/7, \ldots$), and at hole conjugates of these fractions given by $\nu = (p + 1)/(2p + 1)$. Fig. 10.3 shows an example of this in a very-high-quality heterojunction in which a large number of different fractions are observed, including further series of features for fractions larger than unity, which correspond to replicas of the original fractional states with a partial population of the next Landau state.

The explanation of this behaviour lies in the formation of an interacting liquid-like state of the electrons, which was first found from numerical studies of the lowest Landau level. A major advance in the understanding of this state was the introduction by Laughlin (1983) of a trial many-particle wavefunction based on a generalization of the exact solution of the Schrödinger equation for three electrons. This is

$$\Psi_m(z_1, \ldots, z_n) = \prod_{(i>j)}^{n} (z_i - z_j)^m \exp\left(-\frac{1}{4}\sum_{1}^{n} |z_l|^2\right),$$

where m is required to be an odd integer in order to preserve the antisymmetry

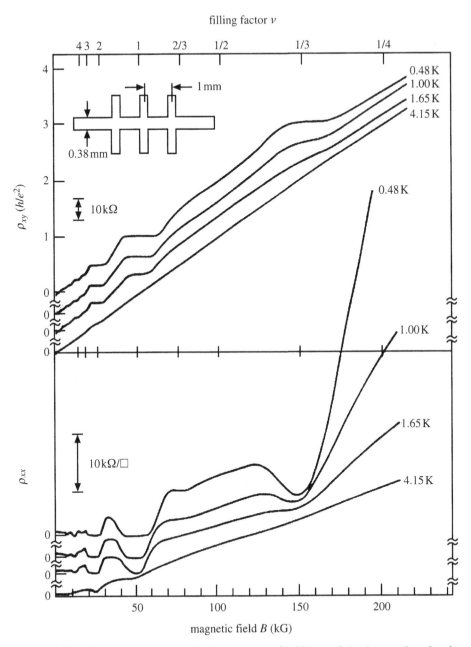

Fig. 10.2 The fractional quantum Hall effect at 1/3 filling of the lowest Landau level as discovered by Tsui *et al.* (1982).

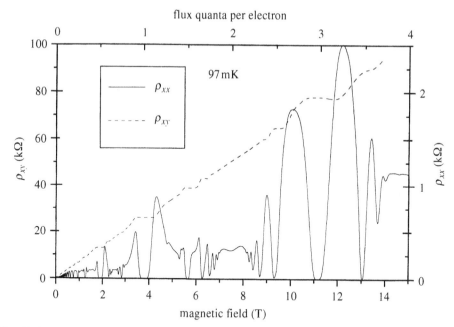

Fig. 10.3 Resistivity and quantum Hall traces for a very-high-mobility GaAs/GaAlAs heterojunction at 97 mK, showing many different fractional occupancies which give rise to resistivity minima. The upper scale shows the magnetic field measured in units of flux quanta per electron.

of the wavefunction. An energy gap arises for the formation of an excited state of this liquid, and this gap causes the resistivity minimum and its accompanying quantum Hall plateau. The quasiparticle excitations of this state are created by piercing the ground state with a single flux quantum, as might happen by changing the magnetic field away from the exact fractional occupancy. The resulting particle has an effective charge of $(1/m)e$, due to partial screening of the induced charge.

The Laughlin picture works well for fractions such as 1/3, where $m = 3$, or for 1/5, where $m = 5$. However, it proved experimentally rather difficult to observe a 1/5 fraction, which required very low temperatures and very pure samples. By contrast, states such as 2/5 and 3/7 were quite strong and easily seen. To describe these states it is necessary to introduce a hierarchical picture (Haldane 1983; Halperin 1984), in which the 2/5 state is a quantum liquid formed from the condensation of the high density of $e/3$ charge quasiparticles which would exist at 2/5, the 3/7 state arises from the condensation of $e/5$ charge quasiparticles, and so on. The picture becomes progressively more complex and less useful as the fractions approach occupancies such as $v = 1/2$, where all of the oscillations disappear.

10.3.2 Composite fermions

The next important step in the understanding of the quantum fluid came with the idea of introducing a statistical or gauge transformation associated with the attachment of flux quanta to the electrons to create a new composite particle. If we consider particle exchange in the presence of a magnetic field, then, in addition to the normal phase factor θ (0 for bosons, π for fermions), there is an additional Aharonov–Bohm, or Berry's, phase, caused by the motion around any magnetic flux present. If we attach q flux quanta to each particle, then the total phase change is

$$\Phi = \pm i(\theta + q\pi).$$

The result of this is that if we attach an odd number of flux quanta to each particle then we change the statistics from fermions to bosons, while for an even number of flux quanta we maintain the fermion statistics. Formally this process of flux attachment is known as a Chern–Simons gauge transformation.

The introduction of the Chern–Simons gauge field was used by Jain (1989) to demonstrate that by choosing an even number for q it was possible to transform the fractional quantum Hall features into an integer QHE of particles consisting of an electron with a fixed number of flux quanta attached to each one, which he called composite fermions. Using this process it is easy to show that the fractional occupancy at $\nu = p/(2p + 1)$ transforms to the integer QHE with occupancy $\nu^* = p$. In this case, the fractional quantum Hall effect becomes the integer effect of the new composite fermions.

The next step in the development of the composite fermion (CF) theory was the ostensibly obvious realization that if we start with the occupancy $\nu = 1/2$ and make the Chern–Simons transformation with $q = 2$, then at this point the gauge field will exactly cancel the external magnetic field. If we therefore make a mean field approximation, the CF particles exist in zero effective magnetic field. This picture and its consequences were discussed in detail by Halperin *et al.* (1993), whose work produced a revolution in the picture of the fractional quantum Hall effect by transforming the emphasis in the description of the phenomenon to even fractions. At $\nu = 1/2$ the ground state of the system will consist of a Fermi sea for the new composite fermions. The significant point about this is that the strong Coulomb interactions between the single-particle electrons turn out to be the mechanism for the binding of the flux quanta to the electrons. Once this has been achieved, the resulting composite fermions behave as only weakly interacting single particles. The choice of $q = 2$ is not unique. We could choose any even number $q = 2m$ and form a further CF particle. For example, with $m = 2$ ($q = 4$), the system can again be transformed to zero effective field at an occupancy of 1/4, and we will see that a whole series of even-occupancy fractions show the characteristic signatures of CF formation.

If we now move away from $v = 1/2m$, the composite fermions move in a residual, or effective, magnetic field (B^*) given by the difference of the applied external field, and the mean field which was used in the formation of the composite fermions. The effective field is thus

$$B^* = B - 2m\Phi_0 n_e.$$

This B^* will lead to a quantization of the CF energies and motion into Landau levels with the same degeneracies as the single-particle states, and energy separations which will be given by some composite fermion cyclotron energy

$$E = \hbar\omega_c^* = \hbar e B^* / M^*,$$

where we now have to define an appropriate effective mass M^* which will describe the properties of the composite fermions. The consequence of this is that we now expect to observe a Shubnikov–de Haas effect (resistivity minima) and a corresponding quantum Hall effect from the composite fermions, but with the difference that the CF features can arise for both positive and negative values of effective field (i.e. as we move both upwards and downwards in total field away from $v = 1/2$). These oscillations occur at magnetic fields given by

$$B^* = \Phi_0 n_e / v^*,$$

where $v^* = p$, an integer. In other words, they occur at the same magnetic field values as seen for the IQHE, with the significant change that the magnetic field is now defined in terms of the effective field B^*. This corresponds to total external fields of

$$B = 2m\Phi_0 n_e \pm B^* = (2m \pm (1/p))n_e \Phi_0$$

and occupancies of

$$v = \frac{p}{(2mp \pm 1)}.$$

In the simplest case ($m = 1$, $p = 1, 2, 3, \ldots$), this corresponds to the series of fractions $1/3, 2/5, 3/7$ for the positive effective field values and $1, 2/3, 3/5, 4/7 \ldots$ for the negative effective fields. These are precisely the occupancy values for the strongest features observed in the fractional quantum Hall effect. We have therefore produced a picture in which it is obvious why the strong FQHE features correspond to the particular series of fractions which are observed. We have a more physical picture of the original prediction by Jain (1989) in which the FQHE features can be transformed into the IQHE by a simple linear translation in magnetic field. The only question mark in this picture is the redundancy of the description of the states at $v = 1$, which is described both as a simple single-particle state and also as a CF state with $p = -1$. In this case, we assume that the simplest (lowest order) picture should be the correct one.

To demonstrate this transformation in practice, we look at the resistivity trace in Fig. 10.3, which shows the FQHE features in a high-mobility hetero-junction at a temperature of 97 mK. The magnetic field units on the upper axis are in units of flux quanta per electron, so that the occupancy $v = 1/2$ corresponds to a field of $2n_e\Phi_0$. We now choose to replot the same data (Fig. 10.4) by transforming $v = 1/2$ and $1/4$ to $B^* = 0$ and plotting the data for both positive and negative values of B^*. We see that the oscillatory features consist simply of multiple repeats of the Shubnikov–de Haas oscillations, with a remarkable degree of self-similarity. The numerator of the fractions also becomes significant. It corresponds very simply to the number of Landau levels occupied, both for single particles and composite fermions.

Careful examination of the uppermost two traces, corresponding to the original single-particle (IQHE) oscillations and the negative B^* oscillations between $v = 1$ and $v = 1/2$, shows that there is even a set of fine structures which also repeat. These correspond to fractions such as 4/3 and 5/3, and 4/5 and 5/7. Such features are described in the CF picture by invoking the ideas of hierarchy first introduced in the original picture of the FQHE, including the idea that the system may simultaneously contain both single-particle electrons and/or combinations of composite fermions of different generations (i.e. different values of m). The simplest example of this idea is for occupancies such as 3/2. Since one Landau level is completely filled at this point, we ignore all electrons in the filled state and attach two flux quanta only to the carriers in the partially (half) filled level. The new composite fermions then move in an effective field generated by any movement away from $v = 3/2$. There is an important difference from the case of $v = 1/2$, however, since the change in the external field also alters the degeneracy of the filled Landau level. Consequently the number of single particles in the unfilled state changes, which alters the number of particles to be treated as composite fermions. In the general case of p occupied single-particle Landau levels, the number of free particles to which we will attach flux quanta is

$$n' = n_e - p\frac{B}{\Phi_0}$$

so that when we attach $2m$ flux quanta to these carriers we have an effective residual field of

$$B^* = B - 2mn'\Phi_0 = B(1 + 2mp) - 2mn_e\Phi_0.$$

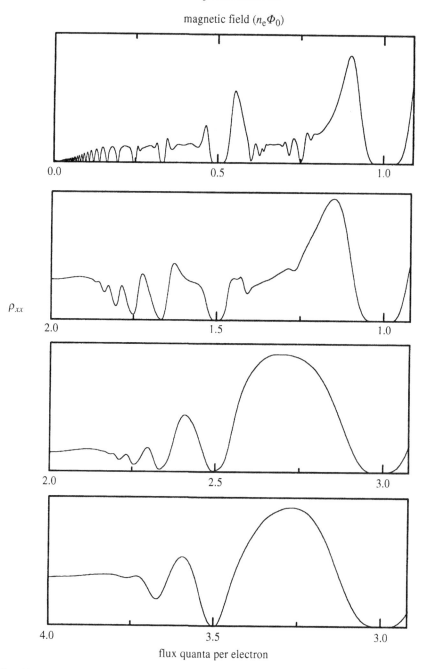

Fig. 10.4 The same data as shown in Fig. 10.3, but now replotted to cover the range of effective magnetic field B^* from 0 to 1.08 starting from an origin of 0, 2, 2 and 4 flux quanta per electron, and plotting the field alternately in the positive and negative directions. These curves thus correspond to the Shubnikov–de Haas oscillations of the single electrons, the $m = 1$ composite fermions for negative and positive effective fields, and the $m = 2$ composite fermions for negative effective fields.

This field then generates a new series of CF Shubnikov–de Haas oscillations at effective field values given by

$$|B^*| = \frac{n'\Phi_0}{p'}.$$

Transforming these values back to the external field B gives

$$\pm\frac{(n_e\Phi_0 - pB)}{p'} = B(1 + 2mp) - 2mn_e\Phi_0$$

or

$$\frac{1}{\nu} = \frac{B}{n_e\Phi_0} = \frac{1 \pm 2mp'}{p \pm p'(1 + 2mp)}.$$

The simplest example of these 'mixed-generation' states is given by the features around $\nu = 3/2$. In this case, $p = 1$ for the single particle (i.e. electron states with $m = 0$), and $m = 1$ for the composite fermions in the partially filled next Landau level. The CF Shubnikov de Haas oscillations now correspond to occupancies such as $4/3$, $7/5$ ($p' = 1, 2$) and 2, $5/3$, $8/5$ ($p' = -1, -2, -3$), representing a mix of electrons and $m = 1$ composite fermions. The analogue of these states for the composite fermions starting around $\nu = 3/4$ comes from putting $p = 0$ for the single particles (i.e. all electrons are in the lowest Landau level), $p' = 1$ for the $m = 1$ CF levels, and then repeating the process described above for a second generation ($m = 2$) of composite fermions with $p'' = \pm1$, 2, 3, generating the fractions $4/5$, $7/9$, $10/13$, and $1/3$ (redundant with the $m = 1$ CF picture), $5/7$, and $8/11$.

In practice, only a few examples of the mixed-generation states have been observed, and never with more than two generations at any one time. The most extensive studies consist of work on the fractions between $\nu = 1$ and $\nu = 2$, in which the two levels which are occupied are for the spin-up and spin-down electron states of the lowest Landau level. There is extensive evidence that the CF states in this region can consist of mixed spin arrangements, with different fractions corresponding to different degrees of spin polarization. It is not at present clear whether the observation of no more than two generations at any one time is a fundamental property of the CF picture, or merely a question of the quality of structures currently available in practice.

10.3.3 Measurement of the CF Fermi surface

We have remarked above that one of the most important features of the composite fermion picture is the idea of the formation of a Fermi surface for the CF particles at even-fractional-occupancy states. The experimental evidence for this was probably the most convincing and unique factor in achieving the rapid acceptance of the CF picture. This came from the observations by Willett *et al*, (1993a, b, 1995), who made measurements of surface acoustic wave (SAW) transmission at very high frequencies (of order 1–10 GHz). These showed that, once the wavelength of the SAWs became comparable with the mean free path of the CF particles, then an enhanced conductivity could be seen due to ballistic transport in the electric field of the SAW. An example of this is shown in Fig. 10.5. Sharp dips were observed in the SAW transmission at the even fractions such as $1/2$ and $1/4$. Peaks were observed because the enhanced conductivity only exists provided that the mean free path traverses more than one SAW wavelength. On moving away from the even-fractional occupancy, corresponding to $B^* = 0$, the composite fermions begin to follow a circular orbit with a semiclassical radius given by

$$R_c^* = \frac{\hbar k_F}{eB^*} = \frac{\hbar (4\pi n_e)^{1/2}}{eB^*}.$$

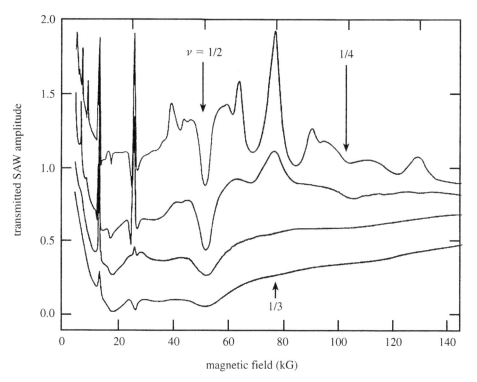

Fig. 10.5 The transmitted surface acoustic wave (SAW) amplitude at 3.4 GHz for a GaAs/GaAlAs heterojunction at temperatures of 0.45, 1.25, 3 and 4 K from the top to the bottom traces. The features at $1/2$ and $1/4$ persist up to considerably higher temperatures than the associated FQHE features (Willet *et al*. 1993a, b).

Once the semiclassical orbit becomes less than the wavelength of the SAW, the enhancement of the conductivity is lost.

If the mean free path of the CF particles is sufficiently long compared with the wavelength of the SAW, then the technique can be used to make a precise measurement of their Fermi wavelength. Since the frequency of the SAW is substantially lower than the cyclotron frequency of the CF particles, we can assume that the SAW presents a static electric field wave to the charge carriers. There will be a geometric resonance when the diameter of the cyclotron orbit is equal to one or more multiples of the SAW wavelength at the condition

$$2R_c^* = n\lambda.$$

At these points there will be a peak in the conductivity provided that the particles complete the orbit without scattering. This was achieved experimentally (Willett *et al.* 1995) by increasing the frequency of the SAW (thus decreasing the wavelength) to a value of order 10 GHz. Fig. 10.6 shows an example of this, in which there is a strong $n = 1$ resonance, with a weaker harmonic. Measurement of the effective field corresponding to the resonance was found to give an excellent agreement with the value of k_F calculated from the number of composite fermions (equal to the number of electrons) in the system at $\nu = 1/2$, thus verifying directly the idea of a Fermi surface formation. There have been some subsequent measurements of ballistic CF motion in patterned structures of antidots and point contacts (Kang *et al.* 1993; Goldman *et al.* 1994), which also provide evidence of semiclassical particle motion, but at the moment the most clear and convincing evidence remains the SAW measurements.

10.3.4 Composite fermion masses

One of the most fundamental properties of any particle is its mass. In the case of composite fermions these are particles which form as a result of the strong electron–electron interactions due to the Coulomb potential. This suggests that the energy scale which we would expect to be associated with the formation of CF energy gaps is the Coulomb energy

$$E_C = \frac{e^2}{4\pi\varepsilon\varepsilon_0 r},$$

where r is the interparticle spacing. For typical two-dimensional electron densities of order 10^{15} m^{-2}, this gives energies of order 100 K.

Measurements of the CF masses have mainly been made by an analysis of the temperature dependence of the strength of the features appearing in the resistivity. This can be done in either the low-temperature limit, in which case

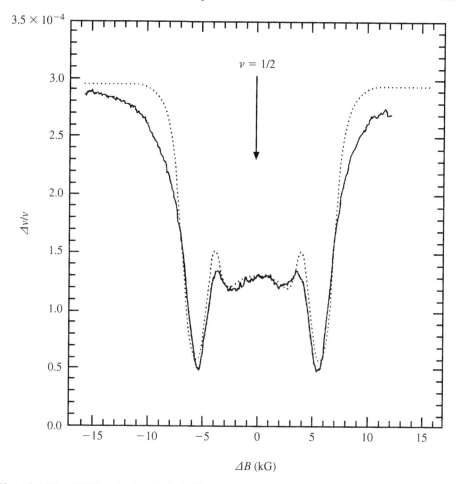

Fig. 10.6 The SAW velocity shift (which gives features in antiphase to the conductivity) versus magnetic field around filling factor 1/2, with SAW frequency of 10.7 GHz at 130 mK. Both principal and secondary geometric resonances of the SAW and the CF cyclotron orbit can be seen (Willett *et al.* 1995).

it is assumed that conduction occurs through activation across an energy gap, whose value can be affected by the amount of disorder broadening in the system, or at relatively higher temperatures by using the temperature dependence of the oscillations (Argyres 1958; Lifshitz and Kosevich 1958; Ando 1974), which gives the effective cyclotron energy of the particles. In very pure samples both methods give similar results. The outcome of one such measurement (Leadley *et al.* 1996) is shown in Fig. 10.7. This shows that the effective CF cyclotron gap increases sublinearly with the magnitude of the effective magnetic field B^*, and the resulting effective mass deduced from the simple relationship $E_g = \hbar \omega_c^*$, also increases with effective magnetic field B^*. In the

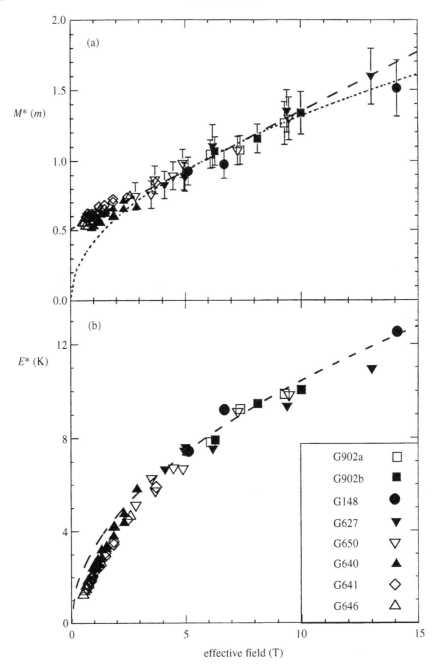

Fig. 10.7 (a) The CF effective mass as measured from the temperature dependence of the Shubnikov–de Haas-like CF resistivity features, as a function of the effective magnetic field B^*. (b) The CF energy gap measured from the same points. The data come from a range of different samples covering densities from 0.6 to 4.8×10^{15} m^{-2}. The dashed lines show a square root behaviour, which might be expected from a dependence related to the Coulomb energy (Leadley *et al.* 1994, 1996).

experiment, two different phenomena are studied. For a given sample, with a fixed carrier density, measurements are made for the series of different fractions approaching $v = 1/2$, with consequently reducing values of B^*. Using a series of different samples, the carrier dependence of the effective mass was studied. The results shown in Fig. 10.7 suggest that the CF effective mass is determined by the single parameter of the effective field B^*, with a dependence proportional to $B^{*1/2}$ at high fields. Such a behaviour is obviously to be expected from the Coulomb energy, since the position in magnetic field of all of the features is proportional to the carrier density, and thus the carrier spacing is proportional to $1/B^{1/2}$. This is also not so surprising for the different fractions, since the number of composite fermions in each Landau level is also proportional to B, and hence to B^*.

An interesting prediction of the theory of Halperin *et al.* (1993) is that, for sufficiently soft electron–electron interactions, there will in fact be a relatively weak (probably logarithmic) divergence of the effective mass at $B^* \to 0$. Some measurements (Du *et al.* 1994; Manoharan *et al.* 1994) report that the mass does increase at the lowest fields for which CF Shubnikov–de Haas oscillations can be seen. The mass increase reported is, in fact, very rapid, and this area of work still remains quite controversial.

10.3.5 Wigner solid

Another interesting prediction of the importance of Coulomb interactions between electrons is that, for sufficiently low densities, the electron will form a solid rather than a liquid ground state. The first prediction of this phenomenon was by Wigner (1934), who pointed out that, for sufficiently low temperatures, the long-range Coulomb interaction would favour the formation of a regular crystalline arrangement of electrons. This situation is favoured at low densities because of the competition between the Fermi energy, which increases proportionally to n_e in two dimensions, and the Coulomb potential, whose strength falls off only as $1/r$, and hence $n^{1/2}$. For high densities, the large value of the Fermi wave vector means that the electrons are delocalized, thus inhibiting the solid formation. Such Wigner crystals have nevertheless been observed in very-low-density two-dimensional electron gases, such as electrons bound to the surface of liquid helium by image charge potentials (Grimes and Adams 1979). For typical semiconductor systems, the densities are too high for a Wigner solid to form, but when a very large magnetic field is applied the electrons become localized within a distance of the size of the cyclotron orbit, and Wigner solid formation has been predicted (Lozovik and Yudson 1975; Yoshioka and Fuku-yama 1979). The occupancies needed for the formation of the magnetically induced Wigner solid (MIWS) are of order $v = 1/6$ (Lam and Girvin 1984; Esfarjani and Chui 1990). There have been a number of reports of the formation of an insulating state with different sorts of transport anomalies at very low occupancies (Andrei *et al.* 1988; Willett *et al.* 1988; Goldman *et al.* 1990; Jiang *et al.* 1990), which have been attributed to the insulating behaviour of the Wigner solid. At the same time, measurements of luminescence have found the appearance of a new line at similarly low occupancies (Buhman *et al.* 1991; Goldberg *et al.* 1992). The temperatures (< 500 mK) and occupancies at which these features occur are consistent with the Wigner solid interpretation; however, it is difficult to give unequivocal proof in semiconductor systems that there is significant long-range order, and that localization does not play an important role. In the presence of disorder, it has been shown (Kivelson *et al.* 1992) that the conduction properties of the electrons can be such as to form an insulating state that is still capable of supporting an essentially classical, metallic Hall voltage (the so-called Hall insulator), and this further complicates the experimental picture.

10.3.6 *Quantum ferromagnets and skyrmions*

Finally we return to what is, in principle, the very simple case of an occupancy of $v = 1$. At this point, we expect that only one spin of the lowest Landau level is occupied and that the system is completely spin polarized. For typical semiconductor systems, the Zeeman splitting of the levels is relatively small, since the effective g-factor of GaAs, which is used for the highest-quality structures, is only 0.4. The Coulomb energy is typically one to two orders of magnitude larger than the Zeeman energy, and this leads to a large exchange contribution to the measured energy gap between the different spin states. We have what could be termed a quantum Hall ferromagnet. Typically this is measured by studying (Usher *et al.* 1990) the activation energy gap for the resistivity minimum at $v = 1$, and gives effective g-factors of order 3–10. It is predicted that there will be no change in the ground state of the system as long as g remains finite (Sondhi *et al.* 1993). The excitations of the quantum Hall ferromagnet are, however, predicted to be rather unusual particles. Several authors (Sondhi *et al.* 1993; Fertig *et al.* 1994) suggest that these may be charged spin-texture excitations, which are the finite Zeeman energy generalization of skyrmions and whose character is very dependent on the size of the Zeeman energy. For large g-values the excitations are single-electron-like, carrying charge $\pm e$ and spin $S_z = 1/2$. In other words, they represent single-spin flips. At small g-values they still carry charge $\pm e$, but now diverge in size and have nontrivial spin order with a divergent spin component S_z and total spin *S*. They could be considered to be something like a two-dimensional equivalent of a domain wall in which the flipping of a single spin propagates outward via the exchange interaction. Thus the existence of only a few spin-flipped electrons may be sufficient to remove all the spin polarization in the system.

Experimental evidence for the existence of such skyrmion-type excitations has come from measurements of the Knight shift of optically pumped NMR by Barrett *et al.* (1994). The Knight shift (shown in Fig. 10.8) is directly proportional to the spin polarization of the system, and therefore would be expected to show a gradual increase as the occupancy falls towards $v = 1$, after which it should saturate at complete polarization. In practice, the data show a completely different behaviour. The system is only close to being fully polarized very near to $v = 1$, and, as soon as the occupancy deviates from this, the polarization falls away rapidly. This is strong evidence that skyrmions are being formed, with the fall in average polarization caused by the additional spin texture introduced.

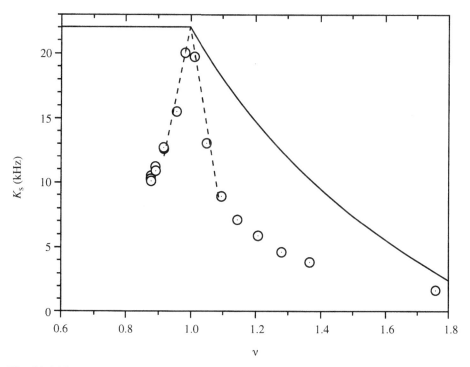

Fig. 10.8 The occupancy dependence of the Knight shift (K_s) measured at 7.05 T and 1.55 K. The solid line shows a dependence that would be expected from single-particle occupation of two spin-split levels, with the system becoming spin polarized for occupancies less than one. The data show a sharp peak in polarization only close to unity, indicating population of skyrmions and skyrmion holes (Barrett *et al.* 1994).

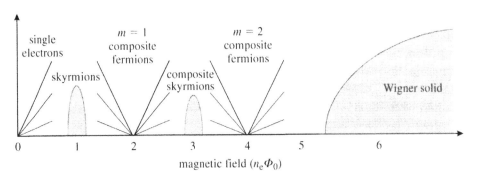

Fig. 10.9 A schematic picture of the different regions of magnetic field in which different new ground states and excitations are thought to occur.

Still to be found are the CF equivalent of the skyrmion excitations. We could expect to see these as excitations when the Fermi energy lies such that exactly one CF Landau level is filled, such as at $\nu = 1/3$, and the next available state comes from the lowest CF Landau level of the opposite spin.

10.4 Summary and conclusions

The combination of a strongly interacting layer of electrons and a high magnetic field produces a whole range of unusual states of matter, with excitations which depend on spin orientation and magnetic flux attachments. A schematic picture of the different regions of magnetic field where these different approaches apply is shown in Fig. 10.9. At low fields, we have the conventional Landau fan for simple electrons, with the ferromagnetic state at $\nu = 1$ with its skyrmion excitations. For fields of $2n_e\Phi_0$ ($\nu = 1/2$), we have the CF Landau fans heading out in both directions of effective field. Around $3n_e\Phi_0$ ($\nu = 1/3$) would be a good place to start looking for spin-textured CF excitations (composite skyrmions), provided that the Zeeman energy is sufficiently small, and at $4n_e\Phi_0$ the second-generation composite fermions have their Landau fan. Finally, for the highest magnetic fields, the Wigner solid forms. The existence of some of these excitations remains unproven, and even some of the possible ground states are controversial, so there may still be several revisions needed of this picture in the future. Some of the significant questions which remain to be answered at this point are what happens to the CF mass as the effective magnetic field approaches zero, and how significant is the introduction of the skyrmion picture for describing the properties of the ($\nu = 1$) quantum Hall ferromagnet?

11

The electron in the cosmos

M. S. LONGAIR
University of Cambridge

11.1 Preamble – physics and astrophysics

The radiation of electrons is the primary source of our knowledge of the constituents of the cosmos. The electrons may either be bound in atoms, ions or molecules, or be free particles, moving in the diffuse magnetised plasmas in stars, or in interstellar and intergalactic space. Thus, astrophysics and astrophysical cosmology can be considered to be the application of the physics of the electron on the grandest scales we can study in the Universe. It has not, however, been a one-way process – astronomy has provided a number of the crucial pieces of evidence needed to establish some of the basic features of electron physics.[*]

Joseph von Fraunhofer is credited with the discovery of the multitude of dark absorption lines in the solar spectrum during his experiments on the refractive indices of glasses in 1814 at the Benediktbraun glass works in Prussia. He labelled the ten strongest lines A, a, B, C, D, E, b, F, G and H, and recorded 574 fainter lines between the B- and H-lines. The same notation is still used for the prominent lines in stellar spectra. Fraunhofer noted that the dark D-lines in the solar spectrum coincided with the bright double lines observed in the flame spectrum of sodium. In 1849, Jean-Bernard-Léon Foucault performed a key experiment in which sunlight was passed through a sodium arc so that the two spectra could be compared precisely. To his surprise, he found that the solar spectrum displayed even darker D-lines when passed through the arc than without the arc present. Ten years later, Gustav Kirchhoff repeated the experiment and made the crucial discovery that, to observe an absorption feature, the temperature of the source of light must be greater than that of the absorbing flame. These results were immediately followed up in 1859 with his understanding of the relationships between the emission and absorption properties of

---

[*] For more details of the history of astronomy's contribution to fundamental physics, see Longair (1995).

any substance, and these are now known as Kirchhoff's laws of the emission and absorption of radiation. In the same year, Julius Plücker identified the F-line with the Hβ-line of hydrogen and saw that the C-line was more or less coincident with Hα, indicating the presence of hydrogen in the solar atmosphere. These studies culminated in Robert Bunsen and Kirchhoff's studies of the solar spectrum. In 1861, Kirchhoff published his great papers entitled 'Investigations of the solar spectrum and the spectra of chemical elements', in which the solar spectrum was compared with the spark spectra of 30 common elements. He concluded that the cool outer regions of the solar atmosphere contained iron, calcium, magnesium, sodium, nickel and chromium, and probably cobalt, barium, copper and zinc as well.

William Huggins was inspired to take up astronomical spectroscopy by Kirchhoff's great discovery of the chemical composition of the Sun. In his words:

This news came to me like the coming upon a spring of water in a dry and thirsty land. Here, at last presented itself the very order of work for which in an indefinite way I was looking for – namely, to extend his novel methods of research upon the Sun to the other heavenly bodies.

In collaboration with William Miller, Huggins began a programme of stellar and nebular spectroscopy. In 1864, they published their first results on stellar spectroscopy and found the same strong absorption lines in the stars as in the Sun. Huggins concluded:

One important object of this original spectroscopic investigation of the light of the stars and other celestial bodies, namely to discover whether the same chemical elements as those of our Earth are present throughout the Universe, was most satisfactorily settled in the affirmative; a common chemistry, it was shown, exists throughout the Universe.

In the same year, Huggins turned his attention to the nebulae, the nature of which was unknown at that time. The common view was that they consisted of associations of unresolved stars, in which case their spectra would be expected to display the common stellar absorption features. By 1868, Huggins and Miller had observed about 70 nebulae, and roughly two-thirds of these displayed stellar absorption lines. In the remaining one-third, prominent bright emission lines were observed, quite unlike those found in stellar spectra. The four commonest emission lines were the Hβ- and Hγ-lines of hydrogen and two strong unidentified lines at wavelengths of 500.7 and 495.9 nm. The latter two lines could not be associated with any of the common absorption lines, and they became known as 'nebulium lines'. Huggins and Miller correctly concluded that these emission-line nebulae were not associations of unresolved stars but rather '. . . must be regarded as enormous masses of luminous gas or vapour'.

The solution of the problem of the nebulium lines and the nature of the star-like nebulae had to await the 1920s. In 1927, Ira Bowen showed that the nebulium lines were the forbidden lines of doubly ionised oxygen, which can be emitted by an ionised gas at low densities – the emission-line nebulae are associated with massive ionised gas clouds within our Galaxy. The nature of the nebulae with essentially stellar spectra was conclusively resolved by Edwin Hubble's observations in the early 1920s, in which he showed that the sample included star clusters belonging to our own Galaxy, as well as galaxies, the light of which is the integrated emission of billions of stars.

Huggins took up stellar spectroscopy again in 1876 and was soon converted to the use of dry gelatin plates which had high sensitivity for photographic spectroscopy. By 1880, he had obtained excellent photographic spectra for about a dozen of the brightest stars. Of particular significance were the spectra of the 'white stars', which extended into the ultraviolet region of the spectrum. In these, he found 12 strong absorption lines extending from Hγ into the ultraviolet region of the spectrum. He identified these as lines of hydrogen, the first four of them being identified by H. W. Vogel in his laboratory studies in Berlin. It is intriguing that Johann Balmer, in his two famous papers of 1885, used Huggins' observations to demonstrate the accuracy of his formula for what is now known as the Balmer series of hydrogen up to transitions originating from levels with principal quantum number $n = 16$

$$\nu_{n \to 2} = 3.29 \times 10^{15} \left(\frac{1}{2^2} - \frac{1}{n^2} \right) \text{ Hz.} \tag{1}$$

This was the first formula containing features which can only be accounted for by the introduction of the concept of quantisation.

Another major contribution of these early spectroscopic studies came from the spectra of solar prominences observed during solar eclipses. In August 1868, Georges Rayet discovered a prominent emission line from these regions which he identified with the sodium D-lines. The line was reobserved by Norman Lockyer in October of that year, and he showed that the wavelength of the line did not correspond to either of the D-lines but rather had a wavelength of 587.6 nm. This feature did not correspond to any of the Fraunhofer lines in the Sun or to any absorption feature of any of the known elements. It was assumed that the line was emitted by some element which had not been isolated in the laboratory, and it was named 'helium'. The element was discovered in the mineral cleveite by W. Ramsey in 1895. When the spectrum of the gas was observed in a discharge tube, the line at 587.6 nm was observed as well as five other lines, some of which Lockyer showed were also present in solar chromospheric spectra. The same lines were also present in some of the blue stars observed in the constellation of Orion. On the basis of laboratory experiments of arc and the hotter spark spectra, Lockyer recognised that these stars had to be hotter than stars such as Vega.

In 1896, Edward Pickering discovered a sequence of absorption lines in the star ζ Puppis which resembled the Balmer series and which became known as the Pickering series. He showed that these lines could be accounted for using Balmer's formula, provided half-integral values of the principal quantum number n were used. In 1912, Ralph Fowler showed that these lines could be observed in laboratory spectra of mixtures of hydrogen and helium. In 1913, in his first great paper on the quantum theory of the hydrogen atom, Niels Bohr showed that the lines of the Pickering series were not associated with half-integral quantum numbers, but rather were due to the recombination lines of fully ionised helium, which has twice the nuclear electric charge of hydrogen.

This somewhat lengthy preamble makes the point that astronomical observations played an important role in elucidating the physics of the electron and its interaction with matter. This is a continuing process, but it is perhaps less well known the extent to which the pioneers of electron physics kept looking over their shoulders for clues from astronomical observations.

The subject of the electron in the cosmos is vast, and there is space here only to consider some of the more important aspects of the role of electrons in astrophysics. There are many books which describe the physics and astrophysics of electrons under cosmic conditions, and I have made liberal reference to my own treatment in my three-volume work *High Energy Astrophysics* (1992, 1995, 1997), where more details of most of the physical processes discussed here will be found.

In astrophysics, electrons are generally found in cosmic plasmas. It is convenient to distinguish the plasmas found inside the stars from what might be called diffuse cosmic plasmas. The physics of these cases is quite different, largely because the densities found inside the stars are many orders of magnitude greater than those found in, say, the interstellar or intergalactic medium.

11.2 The internal structures of the stars

This is an enormous subject, and I only intend covering those aspects in which the physics of electrons is of particular interest. Inside normal stars, the material is generally in the form of a plasma. The structures of the stars are determined by the four equations of stellar structure, together with the equation of state for the stellar material as a function of density and temperature. The four equations of stellar structure for a spherically symmetric star are as follows:

(1) The equation of hydrostatic equilibrium,

$$\frac{dp}{dr} = -\frac{GM\rho}{r^2},$$ (2)

where $M = M(\leq r)$ is the mass contained within radius r, and p and ρ are the pressure and density, respectively, of the material at radius r.

(2) The equation of conservation of mass,

$$\frac{dM}{dr} = 4\pi r^2 \rho.$$ (3)

(3) The equation of radiative energy transfer,

$$\frac{dp_r}{dr} = -\kappa\rho \frac{L(r)}{4\pi c r^2},$$ (4)

where $p_r = (1/3)aT^4$ is the radiation pressure and $L(r)$ is the luminosity passing through the radius r. This expression may be understood in terms of the outward diffusion of the radiation of the star at radius r. The outward radial pressure associated with the luminosity of the star is $L(r)/4\pi r^2 c$, and so eq. (4) describes the fraction $\kappa\rho$ of the outward pressure which is scattered per unit length. κ is known as the opacity of the stellar material. The opacity of stellar material is large, and the radiation is scattered many times by electronic processes before it escapes through the surface of the star. In the case of the Sun, it takes radiation about 10^7 years to diffuse from the centre to the surface of the star.

(4) The fourth equation describes the rate of energy generation as a function of position within the star:

$$\frac{dL}{dr} = 4\pi r^2 \rho\epsilon(r),$$ (5)

where $\epsilon(r)$ is the energy generation rate per unit mass as a function of radius. The source of energy is nuclear energy generation derived from the reactions involved in the conversion of hydrogen into helium and subsequently into heavier elements if the star is massive enough.

These equations are primarily concerned with the bulk properties of stars, but the electrons enter at two essential points into their solution. The first is in determining the equation of state of the stellar material, that is the dependence of the pressure p upon the local density ρ and temperature T of the stellar material. Evidently, the electrons are involved in determining the local ionisation state of the gas as well as the degree of degeneracy of the plasma. The second is in determining the opacity κ of the stellar material. Different processes are important at different densities and temperatures, but they all involve electronic processes. A simple representation of the opacity of stellar matter at different temperatures is shown in Fig. 11.1, which is taken from R. J. Tayler's admirable introductory text *The Stars: Their Structure and Evolution* (1995). At high temperatures, the principal source of opacity is electron

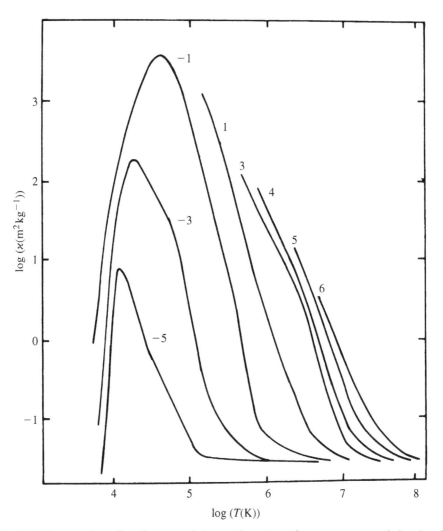

Fig. 11.1 The opacity of stellar material as a function of temperature and density for the transfer of radiation. Each curve represents a different density and is labelled by numbers which are $\log(\rho(\mathrm{kg\,m^{-3}}))$.

scattering by free electrons since the material is fully ionised (see section 11.2.2). At intermediate temperatures, the opacity of stellar material increases by several orders of magnitude because of free–free, bound–free and bound–bound transitions of typical stellar material. At low temperatures, the opacity decreases as the stellar material, in particular the hydrogen, recombines.

Fig. 11.1 is, of course, a gross simplification of the opacity curves used in accurate work. One of the most important pieces of atomic physics of the last decade for these studies has been the redetermination of the opacity of stellar material by two independent programmes, one carried out at the Lawrence Livermore National Laboratory (Rogers and Iglesias 1994) and the other, known as the Opacity Project, carried out by a large consortium of atomic physicists from France, Germany, the UK, the USA and Venezuela (see, for example, Seaton (1993)). New opacity calculations have been made using everything which is known about the atomic physics of the chemical elements. This has been a massive undertaking. In the case of iron, for example, millions of lines have to be taken into account in working out accurate opacities at the temperatures of interest in stellar interiors. The endeavour has been very successful and has resulted in some important changes in key aspects of stellar modelling. For example, studies of Beta Cephei variable stars have shown that their pulsational stability is sensitive to the properties of the ionisation zones within these stars. The effect of the new computations has been to increase the opacity of typical stellar material by a factor of between 2 and 5 at the crucial temperature $T \sim 2 \times 10^5$ K. As a result, a number of groups have shown that the radial fundamental model for pulsation is unstable for these stars and gives excellent quantitative agreement with the properties of these stars.

More generally, the determination of accurate opacities for normal stars has become of the greatest interest because of the accuracy with which the internal structure of the Sun is now known from *solar* or *helio-seismology*. By measuring precisely the resonant modes of the Sun and the fine-structure splittings of these lines, the variation of the speed of sound within the Sun as a function of radius can be measured with quite remarkable accuracy (Fig. 11.2). The speed of sound is known to better than a fraction of 1% throughout the body of the Sun, and opacities of corresponding precision are required to make precise comparisons between theory and observation.

Equations (1) to (4) determine the static properties of all stars. In their classic text *Stellar Structure and Evolution*, Kippenhahn and Weigert (1990) show how the internal properties of stars depend upon their internal densities and temperatures (Fig. 11.3). The majority of the visible stars in the Galaxy are what are known as main-sequence stars, not too different from our Sun. Their masses range from about one-tenth of to about 100 times the mass of the Sun.

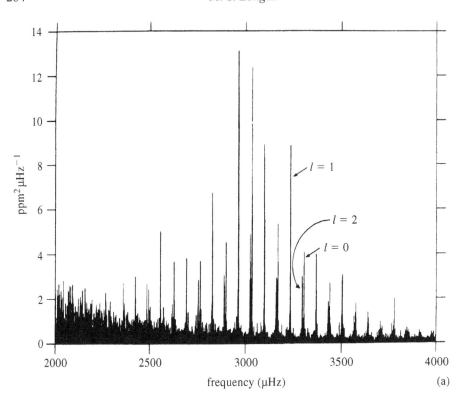

$l = 1$

$l = 2$

$l = 0$

(a)

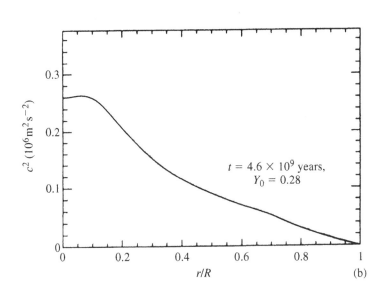

$t = 4.6 \times 10^9$ years,
$Y_0 = 0.28$

(b)

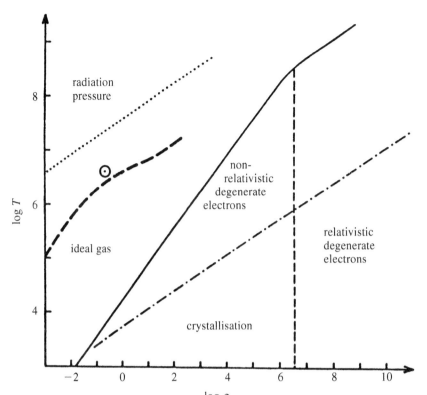

Fig. 11.3 A sketch of the temperature–density plane for stellar material showing the regions in which different forms of the equation of state are applicable. The diagram shows the regions in which the electrons become degenerate and relativistic, as well as the regions in which the radiation pressure of the hot stellar material becomes dominant. Also shown are the regions in which the stellar material is expected to become a solid, the locus indicating the conditions under which the solid melts. The heavy dashed line shows the properties of the material inside the Sun from its core to its envelope and the encircled point is an average value for the Sun as a whole. (After Kippenhahn and Weigert (1990, p. 130).)

Fig. 11.2 (a) An example of the power spectrum of the oscillations of the brightness of the Sun, as measured by the IPHIR experiment on board the Phobos spacecraft, as it travelled between the Earth and Mars (Toutain and Frölich 1992). (b) The variation of the square of the speed of sound inside the Sun as a function of radius r as determined by inverting the helioseismic observations. The measured dependence of the speed of sound upon radius agrees with the predictions of the standard solar models to within the thickness of the line on the diagram, corresponding to an uncertainty of less than 1%. The age of the Sun is taken to be $t = 4.6 \times 10^9$ years, and the helium abundance is 28% by mass. (Courtesy of Prof. D. O. Gough.)

The characteristic of main-sequence stars is that their source of energy is the burning of hydrogen into helium in their cores. For stars with masses less than about 1.5 times the mass of the Sun, the energy generation process is the formation of helium through the process known as the *proton–proton* or *p–p chain*, in which hydrogen is built up out of protons by successive nuclear reactions involving the isotopes of hydrogen and helium. For greater masses, the formation of helium takes place through the sequence of nuclear reactions known as the *carbon–nitrogen–oxygen* or *CNO cycle*, in which these heavy elements act as catalysts for the formation of helium. The internal properties of the Sun on the density–temperature diagram are indicated in Fig. 11.3 by the heavy dashed line, which shows the run of temperature and density from its core to the stellar atmosphere. It can be seen that the Sun lies wholly within the region of the diagram in which the plasma can be considered to be an ideal gas, the pressure being contributed by the electrons and the various ions which are found in thermodynamic equilibrium at that temperature.

There are several regions in Fig. 11.3 in which the electrons play a particularly important role. First of all, the electrons play a key role in determining the maximum luminosity of the stars. If the radiation pressure acting upon the electrons becomes too great, the outer layers of the star are blown off. The physical process involved is radiation pressure acting upon the electrons, which are tied electrostatically to the protons, which contain most of the gravitating mass. The force acting on each electron due to radiation pressure is

$$f = \frac{\sigma_T L}{4\pi r^2 c},$$
(6)

where σ_T is the Thomson cross-section for electron scattering (see section 11.4.1). At the same time, the gravitational pull on each proton is $f = GMm_p/r^2$. The intriguing aspect of these expressions is that they both depend upon r in the same way, that is as r^{-2}. Equating these forces, we find the upper limit to the luminosity of the source, which is known as the *Eddington limiting luminosity* L_E:

$$L \leqslant L_E = \frac{4\pi c GM m_p}{\sigma_T} = 1.3 \times 10^{31} \left(\frac{M}{M_\odot}\right) \text{ W.}$$
(7)

This is a rather firm upper limit to the luminosity of any star, since, by adopting the Thomson cross-section, we have used the minimum cross-section for the escape of radiation from the star. In the most massive stars, and in the X-ray emission of X-ray binary systems, luminosities approach the Eddington limiting luminosity. Equation (7) is a very important result for all classes of astronomical object. Not only stars of all types, but also binary X-ray sources and active galactic nuclei, are subject to a similar limit. The basic physical result is that sources of radiation must not be so intense that they blow away their outer layers or infalling matter, which is their source of fuel.

Moving across Fig. 11.3 from left to right, it can be seen that, for large enough densities at a given temperature, the electrons are expected to become degenerate and ultimately relativistically degenerate. It is inevitable that eventually the central regions of stars move to the right across the diagram. The ultimate fate of stars depends upon their masses. In general, once main-sequence stars have converted about 10% of their hydrogen into helium, they become unstable, the core contracting to a high density, while the outer regions expand to form a giant envelope. In stars with masses roughly the mass of the Sun, the core eventually contracts to form a *white dwarf*, while the envelope is blown off to form the objects known as *planetary nebulae*. In more massive stars the contraction of the core can continue to higher densities and tempera-tures, and subsequent phases of nuclear burning can take place, helium burning resulting in the formation of carbon through the rare triple-α reaction. Ultimately, in the most massive stars, dense iron cores can form before they too collapse in a catastrophic explosion to form neutron stars or black holes.

Electron degeneracy pressure provides the pressure support for the white dwarfs. The condition that degeneracy pressure provides the pressure support for the star can be worked out by a simple order of magnitude calculation. According to Heisenberg's uncertainty principle, the uncertainty in the momen-tum and position of an electron is given by $\Delta p \Delta x \approx \hbar$. The Pauli exclusion principle ensures that two fermions cannot occupy the same quantum mechani-cal state, and so, if the electrons are separated by Δx, they must have momenta $p \approx \Delta p$, according to the uncertainty principle. The density corresponding to this momentum is $\rho \approx m_{\mathrm{p}}/(\Delta x)^3 \approx m_{\mathrm{p}}(p/\hbar)^3$. Therefore, the momentum of an electron associated with the degeneracy pressure is equal to the Maxwellian momentum for a classical gas, $p = mv \approx (3\,mkT)^{1/2}$ at density

$$\rho \approx m_{\mathrm{p}}\left(\frac{p}{\hbar}\right)^3 \approx m_{\mathrm{p}}\left(\frac{3\,mkT}{\hbar^2}\right)^{3/2}. \tag{8}$$

This is the origin of the line in the density–temperature diagram which separates the *ideal gas* region from that labelled *non-relativistic degenerate electrons*. A similar calculation can be made to determine the conditions under which the electron gas becomes relativistically degenerate. In this case, the appropriate form for the momentum is $p \approx mc$, and so the critical density is

$$\rho \approx \frac{m_{\mathrm{p}}}{(\Delta x)^3} \approx m_{\mathrm{p}}\left(\frac{mc}{\hbar}\right)^3 \sim 10^{10}\ \mathrm{kg\,m^{-3}}. \tag{9}$$

The equations of state for degenerate electrons in the non-relativistic and relativistic limits are (where p is now the pressure):

$$\text{non-relativistic} \quad p \propto \rho^{5/3}, \tag{10}$$

$$\text{relativistic} \quad p \propto \rho^{4/3}. \tag{11}$$

The key point is that these equations of state are independent of temperature. This results in an enormous simplification in solving the equations of stellar structure and leads to one of the most important results for stellar evolution.

In the first instance, we can solve the equations using the non-relativistic equation of state (10), and it can be readily shown that the masses of these stars are proportional to R^{-3}, where R is the radius of the star. This means that, as the mass of the degenerate star increases, the central density increases, and so there is a critical mass above which the relativistic equation of state has to be used. When the relativistic equation of state (11) is adopted, the remarkable result is found that there is a maximum mass which a fully degenerate relativistic star can have. This value turns out to be

$$M_{\text{Ch}} = \frac{5.836}{\mu_e^2} M_{\odot}, \tag{12}$$

which μ_e is the mean molecular weight per electron. In white dwarf stars, the chemical abundances have evolved through to helium, carbon and oxygen, and so $\mu_e = 2$. Thus, the upper limit to the mass of stars supported by electron degeneracy pressure is $M_{\text{Ch}} = 1.46 M_{\odot}$, which is known as the *Chandrasekhar mass*.

This is a key result for astrophysics, and it is worthwhile illustrating its origin from the following order of magnitude calculation which describes the essential physics. An approximate expression for the equation of state of the relativistic degenerate gas can be derived as follows. Suppose the interelectron spacing is Δx. Then, the energy of the electron is $E \approx pc$, and, from Heisenberg's uncertainty principle, $p \sim \hbar/\Delta x$. Therefore, the energy density of the relativistic degenerate gas is roughly $\epsilon \approx E/\Delta x^3 \approx \hbar c/\Delta x^4$. Since $\rho \sim m_p/\Delta x^3$, on eliminating Δx it follows that, where p is now the pressure,

$$p = \frac{1}{3}\epsilon \approx \frac{\hbar c}{3}\left(\frac{\rho}{m_p}\right)^{4/3}. \tag{13}$$

This is a simple derivation of the equation of state for a fully relativistic degenerate electron gas.

Now, the total internal thermal energy of the star can be found as

$$U = V\epsilon = 3Vp \approx V\hbar c \left(\frac{\rho}{m_p}\right)^{4/3}. \tag{14}$$

One of the standard results of stellar structure is that, in dynamical equilibrium, the internal energy is half the gravitational potential energy Ω_g, a result known as the *virial theorem*. Therefore, $2U = |\Omega_g|$ and so

$$2V\hbar c \left(\frac{\rho}{m_p}\right)^{4/3} = \frac{1}{2}\frac{GM^2}{R}, \tag{15}$$

where R is some suitably defined radius for the star. Now, $V \approx R^3$ and $\rho V = M$. Therefore, the left-hand side of eq. (15) becomes

$$\frac{2\hbar c}{R}\left(\frac{M}{m_p}\right)^{4/3}. \tag{16}$$

This is the key result. Because we have used a fully relativistic equation of state, the dependence of the gravitational potential energy upon radius is exactly the same as that of the internal energy, both depending upon R^{-1}. If we equate the two terms, the radius R cancels out on either side, and we find

$$M \approx \frac{1}{m_p^2}\left(\frac{\hbar c}{G}\right)^{3/2} \approx 2M_\odot. \tag{17}$$

This is an expression of exactly the same form as the result quoted in eq. (12). This must be an upper limit to the mass of the degenerate star, as can be seen from inspection of eqs. (15) and (16). The gravitational potential energy is proportional to M^2, whereas the internal energy is proportional to $M^{4/3}$. Thus, if the gravitational potential energy ever becomes greater than twice the internal energy, it will always dominate, since both terms depend upon radius in the same way. Hence, there is no equilibrium state for masses greater than that given by eq. (17).

This is a very famous calculation – it indicates that, in the final collapse of a star, there is no stable configuration as a fully degenerate star supported by electron degeneracy pressure if the mass of the star is greater than about $2M_\odot$. Notice the remarkable feature that the mass only depends upon fundamental constants of physics. Exactly the same considerations apply for the case in which stars are held up by neutron degeneracy pressure, in which case the objects formed are neutron stars. A similar limiting mass is found for neutron degeneracy pressure, but, in this case, the effects of general relativity can no longer be neglected. The inference is that, in the final collapse of stars, no physical force can prevent collapse to a black hole if the mass of the remnant is greater than about twice the mass of the Sun.

The electrons play a key role in the formation of neutron stars. With increasing density, the degenerate electron gas becomes relativistic, and so, when the total energy of the electron exceeds the mass difference between the neutron and the proton, $E = \gamma mc^2 \geq (m_n - m_p)c^2 = 1.29$ MeV, the inverse β-decay process can convert protons into neutrons:

$$p + e^- \rightarrow n + \nu_e. \tag{18}$$

In a non-degenerate electron gas, the neutrons would decay into protons and electrons in about 11 minutes, but this is not possible if the electron gas is highly degenerate – there are no available states for the electrons to occupy. The condition that stabilisation takes place is that the Fermi energy of the degenerate electron gas is greater than the kinetic energy of the emitted electron. For the case of a hydrogen plasma, the critical Fermi momentum is

$$p_F = \gamma mv = \left(\frac{E_{tot}^2}{c^2} - m^2 c^2\right)^{1/2}. \tag{19}$$

The number density of the degenerate electron gas is given by the standard formula $n_e = (8\pi/3h^3)p_F^3$, and so the total density is $\rho = n_e m\mu_e$. Taking $\mu_e = 1$, we find that this process takes place at densities $\rho \approx 1.2 \times 10^{10}$ kg m^{-3}. This process is often referred to as *neutronisation* and is an essential part of the process of formation of neutron stars, which eventually collapse to average densities $\rho \sim 10^{17}$ kg m^{-3}.

11.3 Dynamics of electrons in diffuse plasmas

Cosmic plasmas are electrically neutral, the scale over which there can be fluctuations from charge neutrality being the *Debye length* l_D:

$$l_D = \left(\frac{\epsilon_0 k_B T}{Ne^2}\right)^{1/2} = 69 \left(\frac{T}{N}\right)^{1/2} \text{ m}, \tag{20}$$

where N is the number density of the electrons in particles per cubic metre and T is the temperature in kelvin. It is clear that, in cosmic plasmas, charge equality is preserved. In typical cosmic plasmas, the mean free path of the electrons is very long, and so the plasmas can be considered collisionless with infinite conductivity for many purposes. In the case of a plasma, the mean free path is defined in terms of the random deflections which the particle suffers in encounters with the ions and electrons of the plasma. In each encounter, the electron acquires a small random component of momentum perpendicular to its direction of travel, and it is the statistical sum of these encounters which eventually results in the electron acquiring a sufficiently great transverse velocity that it loses all memory of its original direction. If $\langle v_\perp^2 \rangle$ is the mean squared velocity acquired per second by the electron as it travels through the plasma, then a collision is deemed to have occurred after a time t_c such that

$$\langle v_\perp^2 \rangle t_c = v^2, \tag{21}$$

where v is the initial velocity of the particle.

The details of these processes are important for cosmic plasmas because t_c is also roughly the time-scale for the interchange of energy between the particles in the plasma. The details of these processes are given in the monograph *Physics of Fully Ionised Gases* by Spitzer (1962). I have given a simple derivation of the essential forms of these results in Longair (1992–7, vol. 1). The time it takes a group of particles of atomic mass $m_a = Am_p$ to relax to a Maxwellian distribution by random electrostatic interactions is roughly t_c, which is given by the expression

$$t_c = 11.4 \times 10^6 \frac{T^{3/2} A^{1/2}}{NZ^4 \ln \Lambda} \text{ s}, \tag{22}$$

where T is the temperature of the plasma, A is the relative atomic mass number of the particles and $\ln \Lambda$ is a slowly varying Gaunt factor which typically has value $\ln \Lambda \approx 28$. It can be seen that, at the same temperature, the electrons, with $A = 1/1836$, relax in only $1/43$ the time it takes the protons to relax. On the other hand, since the protons are moving at only $1/43$ times the velocities of the electrons, the mean free paths for relaxation are the same. The slowest of the exchange processes is between the protons and electrons because only a small energy exchange takes place if the interacting particles have very different masses. In fact, the relaxation time is about 43 times greater than the relaxation time for the protons. The consequence of these calculations is that one should always check whether or not cosmic plasmas have time to come into equipartition by the relatively slow processes of energy exchange by random electrostatic encounters between particles.

Because cosmic plasmas are collisionless, the electrical conductivity of the plasma can be taken to be infinite, and this greatly simplifies the dynamics of the plasmas. One of the standard results of magnetohydrodynamics is that, under these circumstances, any magnetic field threading the plasma behaves as if it were frozen into the plasma, the phenomenon known as *magnetic flux freezing*. As the plasma moves, the magnetic field lines follow the bulk motion of the plasma. There is a simple interpretation of this phenomenon in microscopic terms. The electrons and ions move in spiral paths along the magnetic field lines with *cyclotron radius* or *gyroradius* r_g:

$$r_g = \frac{\gamma m v}{ze} \frac{\sin \theta}{B}, \tag{23}$$

where γ is the Lorentz factor of the particle $\gamma = (1 - (v^2/c^2))^{-1/2}$, θ is the pitch angle of the particle's orbit with respect to the magnetic field direction, B is the magnetic flux density and z is the charge of the particle. A convenient way of expressing this result is in terms of the *gyrofrequency* or *cyclotron frequency* ν_g:

$$\nu_g = \frac{zeB}{2\pi\gamma m}. \tag{24}$$

For the case of electrons, the gyrofrequency can be conveniently written as $\nu_g = 28/\gamma$ GHz T^{-1}, where the magnetic flux density is measured in tesla. In the simplest case of a uniform magnetic field, the axis of the particle's trajectory is parallel to the magnetic field direction, and this axis is known as the *guiding centre* of the particle's motion. In more complicated geometries, it is convenient to work out the *guiding centre motion* of the particle's trajectory. Provided the magnetic flux density does not change greatly during one gyro-period, the particle moves in such a way that it follows the mean direction of the magnetic field and conserves the magnetic flux within its orbit, that is, as it moves in the magnetic field,

$$\Delta(Br_g^2) = 0. \tag{25}$$

This is known as the *first adiabatic invariant* of the particle's motion, and it can be seen that it is the microscopic equivalent of magnetic flux freezing. An important consequence of this relation is that, although the plasma is collisionless, the electrons are strongly tied to the magnetic field lines, and, even if the magnetic field is very weak indeed, the particles are constrained to gyrate about the magnetic field lines within the source region. Any tangling of the magnetic field confines the electrons effectively within the source.

11.4 Radiation processes for electrons in cosmic plasmas

Many of the properties of the radiation of electrons under cosmic conditions can be derived from the simplest expression for the radiation of an accelerated electron. There are three key properties of such radiation:

(1) The total radiation loss rate for an accelerated electron is

$$-\left(\frac{dE}{dt}\right) = \frac{e^2|\ddot{r}^2|}{6\pi\epsilon_0 c^3}, \tag{26}$$

  where $\ddot{r}$ is the acceleration of the electron.
(2) The *polar diagram* of the radiation is of dipole form, that is if θ is the angle between the acceleration vector $\ddot{r}$ and the direction in which the radiation is emitted, the electric field strength varies as $\sin\theta$ and the power radiated per unit solid angle varies as $\sin^2\theta$.
(3) The radiation is polarised, with the electric field vector of the radiation lying in the direction of the acceleration vector as projected onto a sphere at a large distance.

Many of the key results for the radiation of electrons in cosmic circumstances can be derived from these results.

11.4.1 Thomson scattering

A good example of the use of eq. (26) is the derivation of the expression for the *Thomson scattering* of electromagnetic radiation by free electrons. This calculation was first carried out by J. J. Thomson in 1906 in order to interpret his classic experiments on the scattering of X-rays by the electrons bound in atoms of different atomic masses. The electron is accelerated in the electric fields of the incident electromagnetic waves, and it is straightforward to show that the rate of re-radiation of an incident flux of radiation by a single electron is

$$-\left(\frac{\mathrm{d}E}{\mathrm{d}t}\right) = \sigma_{\mathrm{T}} S, \tag{27}$$

where S is the incident flux density of electromagnetic radiation and σ_{T} is the Thomson cross-section, which is given by the expression

$$\sigma_{\mathrm{T}} = \frac{8\pi}{3} r_{\mathrm{e}}^2 = \frac{e^4}{6\pi\epsilon_0^2 m^2 c^4} = 6.653 \times 10^{-29} \ \mathrm{m}^2. \tag{28}$$

In this expression, $r_{\mathrm{e}} = e^2/4\pi\epsilon_0 mc^2$ is the classical electron radius. This process is responsible for the opacity of stellar material at very high temperatures when the plasma is fully ionised (see Fig. 11.1). The polarisation properties of the scattered radiation can be simply derived from rule (3) above – for example, the scattered radiation is expected to be 100% linearly polarised in directions perpendicular to the direction of the incident radiation. This is an important process for producing polarised radiation in cosmic sources.

Among the many applications of Thomson scattering in astrophysics, one of the most important is the scattering of radiation by the hot intergalactic gas at large redshifts. Observations by the Cosmic Background Explorer (COBE) have shown that the cosmic microwave background radiation is spectacularly isotropic over the whole sky and has a perfect black-body spectrum at a radiation temperature of 2.725 K (Mather 1995). In the standard Big Bang picture of the Universe, this radiation is the cooled remnant of the very hot early phases of the Universe, and its radiation temperature T_{r} changes as the Universe expands as $T_{\mathrm{r}} \propto R^{-1}$, where R is the scale factor which describes the expansion of the Universe. Specifically, R is the relative distance between galaxies which partake in the uniform expansion of the Universe. If R is normalised to $R = 1$ at the present epoch, there is a simple relationship between the scale factor and redshift, $R = 1/(1 + z)$, where $z = (\lambda_{\mathrm{obs}} - \lambda_{\mathrm{em}})/\lambda_{\mathrm{em}}$ is defined to be the redshift of the radiation. λ_{em} is the emitted wavelength of the radiation and λ_{obs} is the wavelength with which it is observed. The expression $T_r \propto R^{-1}$ can be thought of as the adiabatic expansion of a gas of electromagnetic radiation for which the ratio of specific heats $\gamma = 4/3$. Thus, at a redshift $z \approx 1500$, corresponding to a scale factor $R \approx 1/1500$, the radiation temperature of the cosmic microwave background radiation was about 4000 K.

The number density of photons of the cosmic microwave background radiation far exceeds the number density of protons and electrons at the present epoch, and their ratio, $N_\gamma/N_p \sim 10^9$, is conserved as the Universe expands during the epochs considered here. As a result, even at a temperature as low as 4000 K, there are as many ionising ultraviolet photons in the tail of a black-body distribution as there are hydrogen atoms, and so, as we travel backwards in time, the intergalactic hydrogen became fully ionised at this redshift. This is a key result because, as soon as the intergalactic gas becomes ionised, the optical depth for Thomson scattering becomes very large. At larger redshifts, the radiation is very strongly scattered, and so all trace of any structures present at larger redshifts is washed out. There is a strong similarity with observing the surface of the Sun. The radiation we observe originates from a last scattering surface at which the optical depth of the solar material is roughly unity, and all information about the inner layers of the Sun is lost because of scattering. In the same way, the last scattering surface for electromagnetic radiation at a redshift of about 1500 acts as the 'photosphere' for the hot intergalactic gas, only what we observe is redshifted to a temperature of only 2.725 K by the present epoch.

The importance of the last scattering surface for astrophysical cosmology is that it provides a direct probe of the process of structure formation on large scales in the Universe. The key result is that small density perturbations grow only slowly in the expanding Universe. If $\delta\rho$ is the density enhancement over the average background density ρ, then the *density contrast*, $\Delta = \delta\rho/\rho$, grows algebraically with increasing scale-factor as $\Delta \propto R$. Thus, since we know that galaxies and larger-scale structures such as clusters of galaxies exist at the present epoch, it is expected that there should be traces of the formation of these structures in the matter distribution on the last scattering surface because of the large optical depth for Thomson scattering.

This is the beginning of an intriguing story, since, if the Universe consisted only of baryonic matter, the above simple picture would result in intensity fluctuations on the last scattering surface which are much greater than those observed by the COBE satellite and by ground-based observations. The preferred solution to this problem is that most of the matter in the Universe is not in the form of baryonic matter but some dark form, a popular model being the *cold dark matter* picture of structure formation. The nature of the cold dark matter is unknown, but a favoured picture is that it consists of massive neutrino-like particles, or some other unknown type of ultra-weakly-interacting particle – the baryonic matter would correspond to only about 1–10% of the total density of the Universe. In this picture, the dominant fluctuations are in the dark matter, and these grow according to the above law $\Delta \propto R$. Prior to the epoch when the matter recombines, perturbations in the baryon density cannot grow because they are tightly bound to the radiation by Compton scattering (see section 11.4.2). As a result, the perturbations in the baryonic matter are expected to be much smaller than those in the dark matter. However, these are crucial because it is the development of the baryonic fluctuations, which are strongly coupled to the background radiation by Thomson scattering as the intergalactic gas recombines, which result in perturbations in the intensity distribution of the cosmic microwave background radiation over the sky. These have now been observed by the COBE satellite and by ground-based observations at a level of about only one part in 10^5 of the total intensity of the radiation (Fig. 11.4). These observations

53 GHz

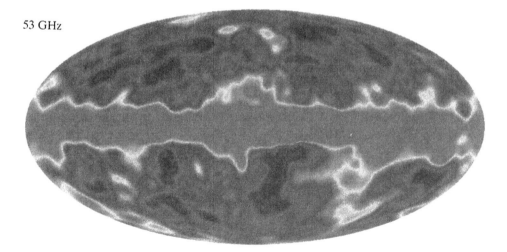

Fig. 11.4 A map of the whole sky as observed in the millimetre waveband at a wavelength of 5.7 mm by the COBE satellite in galactic coordinates, once the dipole component associated with the motion of the Earth through the background radiation has been removed. The centre of the Galaxy is in the centre of the diagram, and the galactic poles are at the top and bottom of the diagram. The residual radiation from the Galaxy can be seen as the bright band across the centre of the picture. The fluctuations seen at high galactic latitudes are predominantly intensity fluctuations on the last scattering surface at a redshift of about 1000. When statistically averaged over the region of sky away from the Galactic plane, the rms amplitude of the intensity fluctuations is $30.0 \pm 1.5 \ \mu K$. (From Bennett *et al.* (1996).)

are of key importance for cosmology since they define the spectrum of the initial fluctuations which must have been present in the earliest phases of the Big Bang.

11.4.2 Compton scattering

In the case of Thomson' scattering, there is no change of frequency of the radiation. In general, so long as the energy of the photon is much less than the rest mass energy of the electron in the centre of momentum frame of reference, $h\nu \ll mc^2$, the scattering can be accurately treated as Thomson scattering. There are, however, cases in which we have to take account of the frequency shift of the radiation. In the classic case of the Compton scattering of photons by a stationary electron, the decrease in frequency of the photon is given by the standard result

$$\frac{\nu'}{\nu} = \frac{1}{1 + (h\nu/mc^2)(1 - \cos\alpha)},\tag{29}$$

where α is the scattering angle of the photon. This represents the transfer of energy from the radiation field to the electron. In general, however, it is a two-way process. If the electron is moving with velocity v, in the low-frequency limit, the change of frequency is

$$\frac{\nu' - \nu}{\nu} = \frac{v}{c}\frac{(\cos\theta - \cos\theta')}{[1 - (v/c)\cos\theta']},\tag{30}$$

where the angles between the velocity vector of the electron and the photon before and after the collision are θ and θ', respectively. To first order, the angles θ and θ' are randomly distributed and there is no net increase in energy of the photons. To second order, however, there is a net increase in energy if the electrons have greater energies than the photons. The average increase in energy of the photons turns out to be

$$\left\langle\frac{\Delta\epsilon}{\epsilon}\right\rangle = \frac{4}{3}\left(\frac{v}{c}\right)^2.\tag{31}$$

The process of the interchange of energy between matter and radiation by Compton scattering is very important in many areas of astronomy. For example, this is the process which keeps the matter at the same temperature as the radiation during the early radiation-dominated phases of the Universe.

The subject of *Comptonisation* is enormous, and it is particularly important in the study of the spectra of X-ray sources in which the emitting plasmas are very hot and so the values of v/c for the electrons are large. The evolution of the spectral energy density of photons $u(\nu)\,d\nu$ acting solely under the influence of Compton scattering is given by the *Kompaneets equation*

$$\frac{\partial n}{\partial y} = \frac{1}{x^2}\frac{\partial}{\partial x}\left[x^4\left(n + n^2 + \frac{\partial n}{\partial x}\right)\right], \tag{32}$$

where

$$y = \int N_e \sigma_T \left(\frac{k_B T}{mc^2}\right) dl \tag{33}$$

and $x = h\nu/k_B T_e$. n is the occupation number of photons, $n = u_\nu c^3/8\pi h\nu^3$. The terms in eq. (32) have the following meanings. The term $\partial n/\partial x$ represents the diffusion of photons along the frequency axis in both directions with respect to their initial energies. The statistical increase in energy of the photons is associated with this term. The first term in round brackets in n represents the cooling of the photons by recoil in Compton collisions, and the term in n^2 describes the effects of induced Compton scattering which are important when the occupation number n is large. It is a useful calculation to show that, when the term in square brackets is zero, the occupation number distribution for the photons is a Bose–Einstein distribution $n = [\exp(x + \mu) - 1]^{-1}$, where μ is the chemical potential. The meaning of this result is that, in equilibrium, when there is a mismatch between the number of photons and the total energy which has to be distributed among them, the equilibrium distribution for photons is a Bose–Einstein distribution with finite chemical potential μ. In the case of the Planck distribution, $\mu = 0$, and the number and energy densities are precisely described by the single parameter, the temperature T.

In some applications of this relation to astrophysical sources, such as X-ray sources, it is assumed that there is a source of low-energy photons and that these are Compton scattered by hot gas in the source region. The results of analytic and Monte Carlo computations by Pozdnyakov *et al.* (1983) are shown in Fig. 11.5, which illustrates how the spectrum of the emerging radiation changes as the Thomson scattering optical depth τ increases. In the cases illustrated, the input photons are of low energy and the temperature of the gas is $k_B T_e = 25$ keV. It can be seen that, as the Thomson optical depth increases, the beginnings of the formation of the Wien peak of the Bose–Einstein distribution can be observed. At smaller optical depths, the emerging spectrum mimics very closely a power-law spectrum up to energies $h\nu \approx k_B T$. This is an intriguing case in which a power-law energy spectrum is produced by thermal processes rather than being ascribed to some 'non-thermal' radiation mechanism involving ultrarelativistic electrons. Processes of this type may be important in forming the spectra of X-ray sources if large amounts of very hot gas

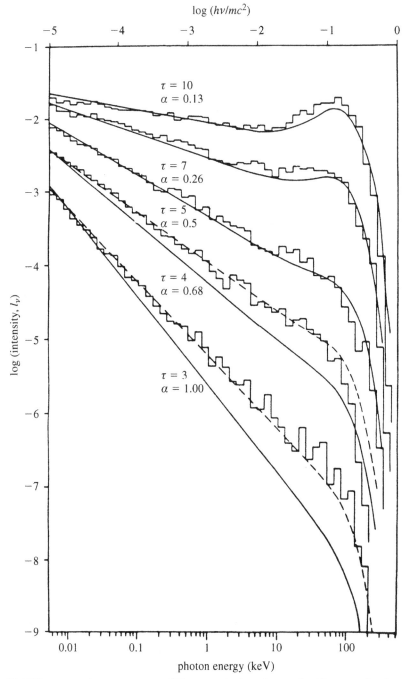

Fig. 11.5 The emerging spectrum of X-ray emission due to the Comptonisation of low-frequency photons in a spherical plasma cloud having temperature $T_e = 25$ keV. The solid lines are analytic solutions of the Kompaneets equation; the results of Monte Carlo simulations are shown by the histograms, and there is good agreement between them. These computations also illustrate the formation of the Wien peak at energies $h\nu \approx k_B T$ for large values of the Thomson optical depth. (From Pozdnyakov *et al.* (1983).)

are present in the source regions. Pozdnyakov *et al.* (1983) describe a number of applications of these results to X-ray sources, for example the source Cygnus X-1, which probably involves the accretion of gas onto a black hole with mass about 16 times the mass of the Sun.

One particularly interesting application of eq. (31) is to the distortion of the spectrum of the cosmic microwave background radiation when it propagates through the hot gas in a rich cluster of galaxies. Then, a certain fraction of the photons of the background radiation is scattered to higher energies by the Compton scattering process. The result is that there should be a small decrement in the intensity of the cosmic microwave background radiation in the direction of the cluster of galaxies in the Rayleigh–Jeans region of the spectrum and a slight excess in the Wien region. This effect, known as the *Sunyaev–Zeldovich effect* (Sunyaev and Zeldovich 1970), has now been observed in the direction of a number of rich clusters of galaxies which are known to be strong X-ray sources. The beautiful example of the decrement in the direction of the cluster Abell 2218 is shown in Fig. 11.6.

The amplitude of the decrement depends upon the Compton scattering optical depth through the cluster. Specifically, the decrement in the Rayleigh–Jeans region of the spectrum amounts to

$$\frac{\Delta I_\nu}{I_\nu} = -2y. \tag{34}$$

The intriguing aspect of this result is that the decrement is proportional to the integral of $N_e T$ along the line of sight through the cluster, that is, the integral of the pressure through the intracluster medium.

Another interesting application of Compton scattering concerns the X-ray spectra of Seyfert galaxies. Although the underlying spectrum is of power-law form, Pounds *et al.* (1990) have found significant distortions in the 5 to 20 keV waveband. Fig. 11.7(a) shows the summed spectra for 12 Seyfert galaxies and the residuals observed when a best-fitting power law is fitted to the data. Fig. 11.7(b) shows that an improved fit to the data is obtained when a reflected X-ray component labelled *b* is added to the spectrum.

The process of X-ray reflection involves illuminating a cool cloud with a power-law radiation spectrum and then working out the reflected X-ray spectrum. There are two competing processes, Compton scattering, which is important at high energies, and photoelectric absorption, which becomes dominant at low energies. In a single Compton scattering of high-energy photons by stationary electrons, the average decrease in energy of each photon is

$$\left\langle \frac{\Delta \epsilon}{\epsilon} \right\rangle = \frac{h\nu}{mc^2}. \tag{35}$$

Thus, the high-energy photons lose more energy per collision than the

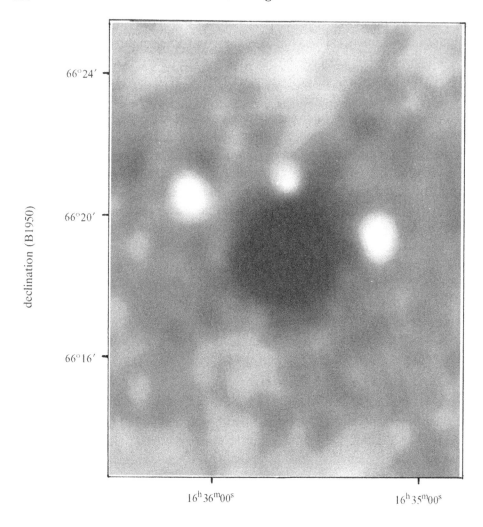

right ascension (B1950)

Fig. 11.6 A radio image of the decrement in the cosmic microwave background radiation in the direction of the rich cluster of galaxies Abell 2218. The decrement is due to the second-order Compton scattering of the photons of the background radiation which pass through the cluster (Jones *et al.* 1993).

lower-energy photons, resulting in a progressive steepening of the scattered spectrum with increasing energy. At low energies, photoelectric absorption is the dominant process and has the characteristic power-law energy dependence of the absorption cross-section $\sigma_{ph} \propto \epsilon^{-3}$ at energies greater than the characteristic absorption edges. The most important of these is the K-edge of almost fully ionised iron which occurs at about 8 keV. Thus, at low X-ray energies,

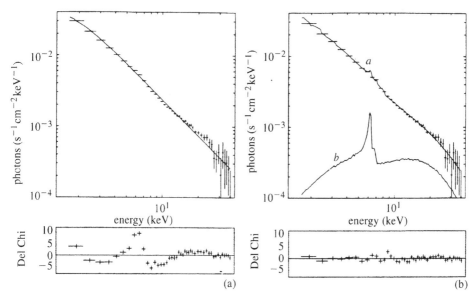

Fig. 11.7 (a) The summed X-ray spectra of 12 Seyfert galaxies compared with a best-fitting power law. The goodness-of-fit to this energy distribution is displayed in the lower panel, in which it can be seen that there are significant deviations from a power law. (b) The addition of a reflection component to the power-law distribution results in a much improved fit to the summed spectra. (From Pounds *et al.* (1990).)

there is strong absorption due to photoelectric absorption and a prominent feature at energies just greater than the K-absorption edge of iron. The net result is that there is a maximum in the reflected spectrum at about 20 keV, at which the combined energy losses due to Compton scattering and photoelectric absorption are at a minimum. To solve the problem properly, it is necessary to carry out a radiative transfer analysis of the competing absorption and scattering processes, and this has been carried out by Lightman and White (1988), whose results are shown in Fig. 11.8. It can be seen that qualitatively the reflected spectrum has the correct signature to account for the features observed in the X-ray spectra of Seyfert galaxies.

There are occasions, particularly in the case of γ-ray astronomy, when it is no longer adequate to use the Thomson cross-section. The key consideration is whether or not the photon has energy $h\nu \geq mc^2$ in the centre of momentum frame of the collision. In this case, the relevant total cross-section is the Klein–Nishina formula

$$\sigma_{\text{K–N}} = \pi r_e^2 \frac{1}{\epsilon} \left\{ \left[1 - \frac{2(\epsilon + 1)}{\epsilon^2} \right] \ln(2\epsilon + 1) + \frac{1}{2} + \frac{4}{\epsilon} - \frac{1}{2(2\epsilon + 1)^2} \right\}, \quad (36)$$

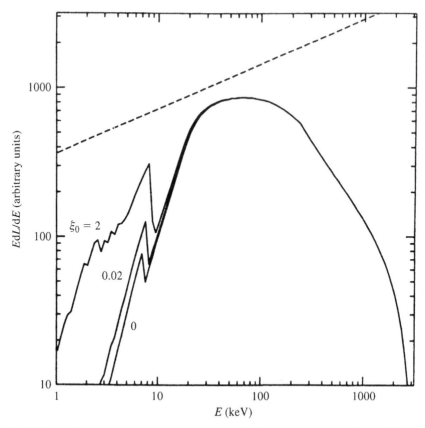

Fig. 11.8 The reflection spectrum from a semi-infinite, plane parallel cold medium assuming the cosmic abundances of the elements. The prominent feature at about 8 keV is associated with the K-absorption edge of iron. The input X-ray spectrum is of power-law form $I(\epsilon) \propto \epsilon^{-0.7}$. The units on the ordinate are $\epsilon I(\epsilon)$. ξ_0 is the differential ionisation parameter which describes how the ionisation state of the cold cloud is modified by the incident X-ray spectrum. In the case $\xi_0 = 0$, the matter remains cold. According to Lightman and White, the value $\xi_0 = 2$ is at the upper limit of acceptable values for active galactic nuclei (Lightman and White 1988).

where $\epsilon = h\nu/mc^2$. For low-energy photons, this cross-section becomes

$$\sigma_{K-N} = \frac{8\pi r_e^2}{3}(1 - 2\epsilon) = \sigma_T(1 - 2\epsilon) \approx \sigma_T. \tag{37}$$

In the ultrarelativistic limit, the cross-section is

$$\sigma_{K-N} = \pi r_e^2 \frac{1}{\epsilon}\left(\ln 2\epsilon + \frac{1}{2}\right); \tag{38}$$

that is, the cross-section for scattering decreases as roughly ϵ^{-1} at high energies.

11.4.3 Bremsstrahlung

Free electrons radiate whenever they are accelerated, and each form of radiation has a specific spectral signature. In the case of a plasma, the electrons are accelerated in electrostatic encounters with the protons or nuclei, and this gives rise to the radiation known as *bremsstrahlung*, the German for braking radiation, or as *free–free emission*, since the electron can be considered to be making a transition in the continuum of unbound states associated with the proton or nucleus. In the classical treatment, the electron suffers a momentum impulse as it passes by the positive nucleus, and the radiation is associated with the accelerations of the electron under the influence of this impulse. The resulting radiation spectrum for a Maxwellian distribution of electron energies at temperature T is flat up to frequency $\nu \approx k_B T / h$, above which the intensity falls off exponentially. Specifically, the emissivity of the plasma is

$$\kappa_\nu = \frac{1}{3\pi^3} \left(\frac{\pi}{6}\right)^{1/2} \frac{Z^2 e^6}{\epsilon_0^3 c^3 m^2} \left(\frac{m}{k_B T}\right)^{1/2} g(\nu, T) N N_e \exp\left(-\frac{h\nu}{k_B T}\right), \qquad (39)$$

where the number densities of electrons, N_e, and nuclei, N, are given in particles per cubic metre. The factor $g(\nu, T)$ is known as the Gaunt factor, and suitable forms for radio and X-ray frequencies are

$$\text{radio: } g(\nu, T) = \frac{3^{1/2}}{2\pi} \left[\ln\left(\frac{128\epsilon_0^2 k_B^3 T^3}{me^4 \nu^2 Z^2}\right) - \gamma^{1/2}\right] \qquad (40)$$

$$\text{X-ray: } g(\nu, T) = \frac{3^{1/2}}{\pi} \ln\left(\frac{k_B T}{h\nu}\right), \qquad (41)$$

where $\gamma = 0.577\ldots$ is Euler's constant.

The general form of the bremsstrahlung spectrum can be understood from a simple argument starting from expression (26) for the radiation of an accelerated electron. We can apply Parseval's theorem to relate the intensity spectrum of the emitted radiation to the Fourier transform of the accleration of the particle. Thus, if we write the Fourier transform of the acceleration in the form

$$\dot{v}(t) = \frac{1}{(2\pi)^{1/2}} \int_{-\infty}^{\infty} \dot{v}(\omega) \exp(-i\omega t) \, d\omega, \qquad (42)$$

Parseval's theorem tells us that

$$\int_{-\infty}^{\infty} |\dot{v}(\omega)|^2 \, d\omega = \int_{-\infty}^{\infty} |\dot{v}(t)|^2 \, dt. \qquad (43)$$

It is a straightforward calculation to show that the spectrum of the emission of the electron is

$$I(\omega) = \frac{e^2}{3\pi\epsilon_0 c^3} |\dot{v}(\omega)|^2. \tag{44}$$

We can now understand why the spectrum of bremsstrahlung is flat. The momentum impulse of the electron is like a delta-function acceleration, and the Fourier transform of a delta-function is a flat frequency spectrum, up to the frequency ν corresponding roughly to the inverse of the duration of the pulse T, $\nu \sim T^{-1}$. In the case of the bremsstrahlung of a thermal plasma at temperature T, the flat spectrum extends up to a frequency $h\nu \approx k_B T$, above which it cuts off exponentially. Two important examples of the bremsstrahlung of cosmic sources are the radio emission of regions of ionised hydrogen at $T \approx 10^4$ K and the X-ray emission of the hot gas in clusters of galaxies at temperatures $T \approx 10^7 - 10^8$ K.

In the case of regions of ionised hydrogen, the radio observations are made at a frequency $h\nu \ll k_B T$ and so the characteristic radio spectrum is a flat $S_\nu \propto \nu^0$ and becomes self-absorbed at low frequencies at which the brightness temperature of the radiation $T_b = (\lambda^2/2k_B)(S_\nu/\Omega)$ becomes 10^4 K, the typical electron temperature within photoionised regions of ionised hydrogen; Ω is the solid angle subtended by the source. Since the brightness temperature of the radiation cannot exceed the temperature of the emitting gas, it follows that the low-frequency spectrum is $S_\nu \propto \nu^2$. The compact regions of ionised hydrogen surrounding hot stars have this characteristic spectrum. In these cases, the gas is photoionised by the intense ultraviolet continuum radiation of the hot O and B stars, or by the central stars of planetary nebulae.

In the case of the hot gas in galaxies and clusters of galaxies, the turnover of the bremsstrahlung spectrum at $h\nu \approx k_B T$ takes place within the observable X-ray waveband, and so the temperature of the gas can be determined from precise measurements of the shape of the X-ray spectrum. Confirmation of the thermal nature of the emission is provided by the observation of lines of highly ionised iron, Fe^{+25}, from the intracluster gas. The bremsstrahlung X-ray emission of clusters of galaxies is of particular importance for astrophysics and cosmology. Once the temperature of the gas has been determined, the total amount of gas in the cluster can be found from expression (39) for the intensity of bremsstrahlung. It turns out that, in clusters of galaxies, the amount of baryonic matter in the intergalactic gas is similar to or greater than that present in the galaxies. Furthermore, the total mass of the cluster must exceed the mass of the intergalactic gas by about a factor of 10. This is a key part of the evidence for dark matter in the Universe. What is of special interest is that the total mass of the cluster can be found from the X-ray data alone, once the spatial and temperature distribution of the gas in the cluster are known.

For simplicity, we assume that the cluster is spherically symmetric and that the total gravitating mass within radius r is $M(\leqslant r)$. The gas is assumed to be in hydrostatic equilibrium within the gravitational potential defined by the total mass of the cluster, that is, the sum of the visible and dark matter, as well as the intracluster gas. If p is the pressure of the gas and ρ is its density, both of which vary with position in the cluster, the requirement of hydrostatic equilibrium is given by expression (2), the pressure being related to the local gas density ρ and temperature T by the perfect gas law

$$p = \frac{\rho k_B T}{\mu m_H}, \tag{45}$$

where m_H is the mass of the hydrogen atom and μ is the mean molecular weight of the gas. Differentiating eq. (45), substituting into eq. (2) and re-organising, we find

$$M(\leqslant r) = -\frac{k_B T r^2}{G \mu m_H}\left[\frac{\mathrm{d}(\log \rho)}{\mathrm{d}r} + \frac{\mathrm{d}(\log T)}{\mathrm{d}r}\right]. \tag{46}$$

Fig. 11.9 shows a beautiful application of this technique to high-sensitivity X-ray observations of the Perseus cluster of galaxies obtained by the ROSAT X-ray Observatory by Böhringer (1995). From the X-ray observations, it is possible to determine both the total gravitating mass $M(\leqslant r)$ and the total mass of gas $M_{gas}(\leqslant r)$ and to compare these with the mass in the visible parts of galaxies. It can be seen that the mass of intracluster gas is typically about five times greater than the mass in galaxies, but that it is insufficient to account for all the gravitating mass which must be present. Some form of dark matter must be present to bind the cluster gravitationally.

One of the more intriguing uses of observations of the hot gas in clusters of galaxies, which is observed both by its bremsstrahlung and by the Sunyaev–Zeldovich effect, is that the size of the hot gas cloud can be determined, independent of knowledge of its distance. The temperature of the cloud can be found from its X-ray spectrum. The intensity of bremsstrahlung determines the quantity $\int N_e^2 T_e^{-1/2}\,\mathrm{d}l$, while the decrement in the cosmic microwave background radiation provides a measure of the quantity $\int N_e T_e\,\mathrm{d}l$. Thus, assuming the cloud is spherically symmetric, it is possible to solve for the size of the cloud, independent of its distance. This technique provides an attractive method of measuring Hubble's constant since the *angular size* corresponding to the characteristic size of the cluster can be measured directly, and so the distance of the cluster measured. For the cluster Abell 2218, Jones *et al.* (1993) have found a value $H_0 = 38^{+18}_{-16}$ km s^{-1} Mpc^{-1}. There is some scatter in the values

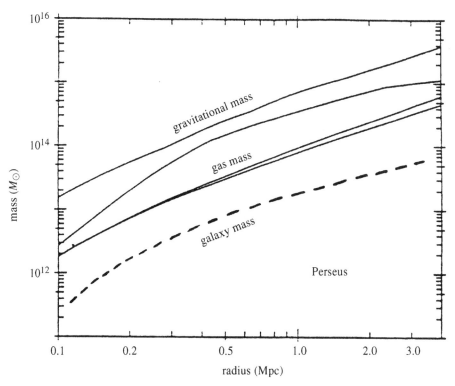

Fig. 11.9 Integrated radial profiles for the mass in the visible parts of galaxies, hot gas and total gravitating mass for the Perseus cluster of galaxies, as determined by observations by ROSAT. The upper band indicates the range of possible total masses, and the central band shows the range of gaseous masses (Böhringer 1995).

found by different authors, but it is encouraging that they are in the range found by the traditional optical techniques, and there is much scope for improving the accuracy of these estimates.

11.5 The radiation of relativistic electrons

11.5.1 Synchrotron radiation

The emission process which dominates a great deal of high-energy astrophysics is the *synchrotron radiation* of ultrarelativistic electrons. It is simplest to begin with the emission of non-relativistic electrons moving in their spiral paths in a uniform magnetic field. The acceleration is directed towards the guiding centre of the particle's motion, and the radiation rate is given by eq. (26), in which the radial acceleration of the electron is $\dot{v}_\perp = (v \sin \theta)^2 / r_g = (ev/m)B \sin \theta$ from eq. (23). Therefore,

$$-\left(\frac{dE}{dt}\right) = \frac{e^2|\ddot{\boldsymbol{r}}^2|}{6\pi\epsilon_0 c^3} = \frac{e^4}{6\pi\epsilon_0 c m^2}\left(\frac{v}{c}\right)^2 B^2 \sin^2\theta. \tag{47}$$

This relation can be rewritten in terms of the Thomson scattering cross-section, σ_T, and the energy density of the magnetic field, $U_{\text{mag}} = B^2/2\mu_0$:

$$-\left(\frac{dE}{dt}\right) = 2\sigma_T c U_{\text{mag}}\left(\frac{v}{c}\right)^2 \sin^2\theta. \tag{48}$$

The radiation is 100% linearly polarised if the electron is observed perpendicular to the magnetic field direction and is 100% circularly polarised if observed along the magnetic field direction, as can be derived from rule (3) for the polarisation of the emission of an accelerated electron. One intriguing application of these results is to the observation of cyclotron absorption features in the hard X-ray spectrum of the binary X-ray source Hercules X-1. The source of energy is the accretion of matter from the primary star onto the poles of a strongly magnetised neutron star. The absorption feature observed in the hard X-ray spectrum is interpreted as cyclotron absorption at 34 keV by hot gas in the vicinity of the poles of the magnetised neutron star (Fig. 11.10). Inserting 34 keV into the expression for the gyrofrequency, the magnetic flux density is found to be about 3×10^8 T, a very strong magnetic field indeed, but within the range of values found from studies of the slow-down rate of pulsars.

As the electrons become significantly relativistic, beaming of the dipole pattern of the emission of the electron has to be taken into account, and this results in the generation of radiation at harmonics of the gyrofrequency. Just as in the case of gyroradiation, the emission in the harmonics is circularly polarised, and so a great deal can be learned about the orientation of the magnetic field to the line of sight in sources in which the cyclotron emission can be observed. Such emission has been observed in the eclipsing magnetic binary stars known as *AM Herculis binaries* or *polars*. In these systems, a red dwarf star orbits a white dwarf star with a strong magnetic field. Accretion of matter from the red star onto the magnetic poles of the white dwarf results in the heating of the matter to temperatures of about 10^7 K so that, in addition to emitting X-rays, these sources are also sources of cyclotron radiation. Fig. 11.11 shows the optical spectrum of the X-ray source EXO 033319-2554.2, in which the thermally broadened cyclotron harmonics can be clearly distinguished. The frequency spacing between the harmonics has enabled an accurate estimate of the magnetic field strength to be made. This turns out to be 5600 T, the largest magnetic flux density measured for an AM Herculis system.

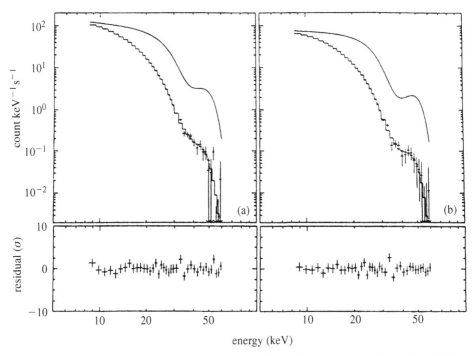

Fig. 11.10 The hard X-ray spectrum of the binary X-ray source Hercules X-1, as observed by the Ginga satellite, showing the 34 keV cyclotron absorption feature. The X-ray source has a pulse period of 1.24 s, and in (a) the spectrum observed at pulse maxima is shown. In (b), the spectrum is derived from the difference between the spectra observed at pulse maximum and minimum, and shows more clearly the absorption feature. The lower panels show the residuals obtained for the fit of the model to the data. (From Mihara *et al.* (1990).)

The next step is to consider the case in which the electrons are moving at speeds very close to the speed of light, that is, we have to take into account the Lorentz factor of the electrons $\gamma = (1 - (v^2/c^2))^{-1/2} \gg 1$. The relativistic generalisation of the radiation loss rate of an accelerated electron is

$$-\left(\frac{dE}{dt}\right) = \frac{\gamma^4 e^2}{6\pi\epsilon_0 c^3}\left[a^2 + \gamma^2\left(\frac{\boldsymbol{v}\cdot\boldsymbol{a}}{c}\right)^2\right], \tag{49}$$

where the three-velocity and three-acceleration vectors are measured in the observer's frame of reference. The energy loss rate of the electron is readily derived from the relativistic expression for the gyroradius of the electron, eq. (23). Recalling that, for an electron gyrating in a magnetic field, $\boldsymbol{v}\cdot\boldsymbol{a} = 0$, the total loss rate is

$$-\left(\frac{dE}{dt}\right) = 2\sigma_\text{T} c U_\text{mag} \gamma^2 \left(\frac{v}{c}\right)^2 \sin^2\theta; \tag{50}$$

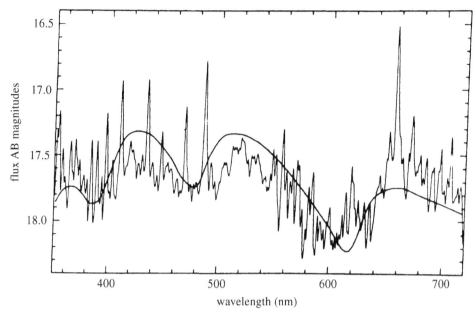

Fig. 11.11 The optical spectrum of the AM Herculis object EXO 033319–2554.2, which is a soft X-ray source. The presence of a strong magnetic field is inferred from the observation of strongly circularly polarised optical emission. The solid line shows a best fit of the cyclotron emission spectrum to the broad cyclotron harmonics at 420, 520 and 655 nm. The inferred strength of the magnetic field is 5600 T. (From Ferrario *et al.* (1989).)

that is, the radiation loss rate is γ^2 times the non-relativistic expression. In addition, the radiation is now very strongly beamed in the direction of motion of the electron. It is straightforward to show that the radiation is beamed within an angle $\theta \sim \gamma^{-1}$ of the instantaneous velocity vector of the electron. The frequency spectrum of the radiation is found from the Fourier transform of the time variation of the intensity of the radiation as measured by the distant observer. This is not a trivial computation since, being ultrarelativistic, the electron almost catches up with the radiation it emits. The details of how the results come about are given in Longair (1992–7, vol. 2, chap. 18). The observed frequency of the radiation is centred roughly at a frequency

$$\nu \approx \gamma^2 \nu_{\mathrm{g}} \sin \theta, \tag{51}$$

where ν_{g} is the non-relativistic gyrofrequency, $\nu_{\mathrm{g}} = (1/2\pi)(em/B)$. The intensity spectrum of the radiation of a single electron is shown in Fig. 11.12 in both linear and logarithmic forms. It can be seen that the continuum radiation is broad-band. This feature is used in powerful terrestrial synchrotron sources of ultraviolet and X-ray emission, such as that at the Daresbury Laboratory in the

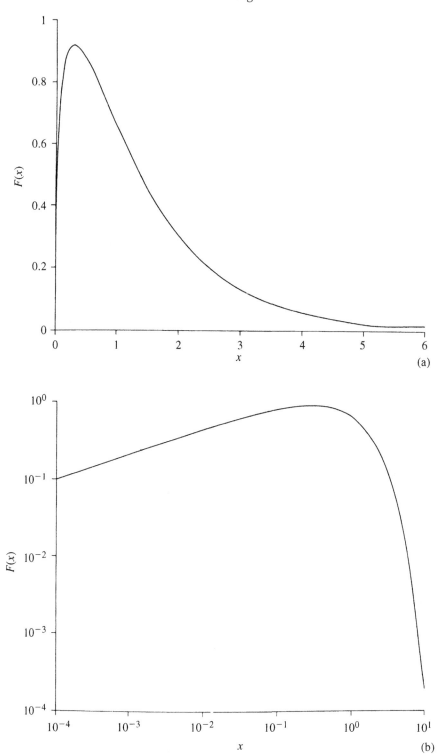

(a)

(b)

UK, to generate a featureless continuum spectrum of radiation of known intensity.

In the astronomical context, the emission is not due to electrons of a single energy but is emitted by a distribution of electrons with a wide range of energies. In the case of our own Galaxy, observations of the interstellar flux of high-energy electrons from satellite measurements show that the spectrum is of power-law form at high electron energies $\gamma \geqslant 2 \times 10^4$:

$$N(E)\,dE = 700E^{-3.3}\,dE \text{ particles } \mathrm{m^{-2}\,s^{-1}\,sr^{-1}}. \tag{52}$$

There are certainly electrons of lower energy, but it is difficult to determine their spectrum directly because of the effects of solar modulation; that is, the primary electron spectrum is distorted by the fact that the electrons have to travel to the Earth from interstellar space through the outflowing solar wind. It is a straightforward calculation to show that, if the energy spectrum of the ultrarelativistic electrons has the form $N(E)\,dE = \kappa E^{-x}\,dE$, then the spectrum of synchrotron radiation emitted by these electrons in a magnetic field of flux density B is

$$I(v) \propto \kappa B^{\alpha+1} v^{-\alpha}, \quad \text{where} \quad \alpha = \frac{x-1}{2}. \tag{53}$$

The spectrum of the radio emission of our Galaxy is of power-law form and can be fitted smoothly onto the high-energy electron spectrum. Indeed, the spectra of cosmic ray protons and nuclei are also of power-law form with $x \sim 2.5\text{--}3$.

The success of the synchrotron hypothesis in accounting for the Galactic radio emission strongly suggests that the emission of many of the strong radio sources which possess the same characteristics – power-law emission spectrum and polarised radio emission – are also sources of synchrotron radiation. For example, the radio spectra of supernova remnants such as Cassiopeia A (Fig. 11.13(a)) are of power-law form, and it is inferred that such sources are major contributors to the interstellar flux of high-energy electrons and cosmic ray protons and nuclei. However, these sources of high-energy electrons are very modest indeed compared with the radio galaxies and quasars. In radio galaxies such as Cygnus A, the radio luminosity of the source is about 10^7 times greater than that of our Galaxy. In addition, the radio emission does not originate from

Fig. 11.12 The intensity spectrum of the synchrotron radiation of a single electron (a) with linear axes and (b) with logarithmic axes. The spectra are plotted in terms of $x = v/v_c$, where v_c is the critical frequency $v_c = (3/4\pi)(c/v)\gamma^2 v_g \sin\theta$, where v_g is the non-relativistic gyrofrequency and θ is the pitch angle of the electron with respect to the magnetic field direction.

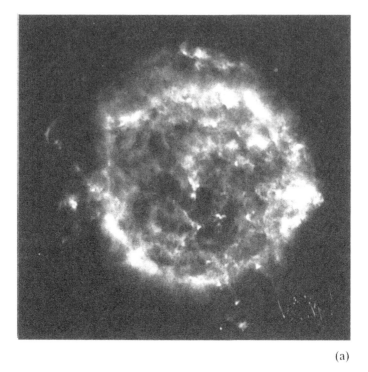

(a)

Fig. 11.13 (a) A radio image of the young supernova remnant Cassiopeia A. The radio emission is the synchrotron radiation of high-energy electrons accelerated within the supernova remnant (Bell 1977).

the galaxy itself but from two enormous radio lobes more or less symmetrically disposed on either side of the galaxy (Fig. 11.13(b)). It can be seen in Fig. 11.13(b) that there are radio jets which link the nucleus of the galaxy to the outer radio lobes; these are responsible for supplying the enormous amounts of energy in relativistic material present in the radio lobes.

It is apparent from expression (53) that the energy present in the electrons cannot be determined without knowledge of the magnetic flux density within the source region. It is, however, possible to work out the minimum amount of energy in the form of magnetic fields and relativistic electrons which must be present within the source region, since a given radio luminosity can be produced by a large flux of electrons in a weak field or vice versa. The minimum energy required corresponds closely to equal amounts of energy in

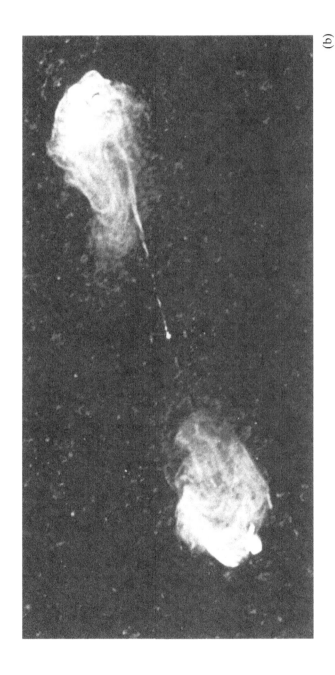

(b)

Fig. 11.13 (b). A radio image of the radio galaxy Cygnus A. The extended radio lobes are powered by jets originating in the nucleus of the galaxy (Perley *et al.* 1984).

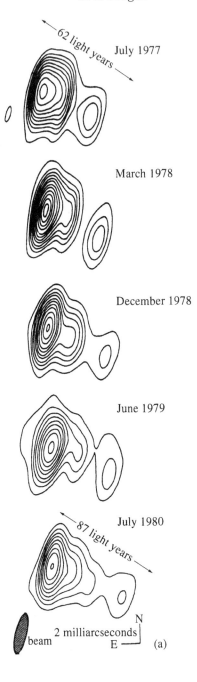

Fig. 11.14 (a) The fine-scale radio structure of the quasar 3C 273 as observed by very long baseline interferometry. It can be observed that the components have separated by 25 light years in only 3 years, corresponding to an observed separation velocity greater than eight times the speed of light (Pearson *et al.* 1982).

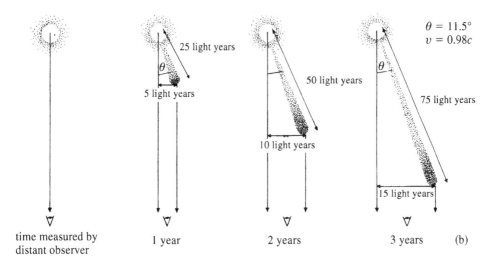

Fig. 11.14 (b). The geometry of a jet of radio-emitting material ejected at a high velocity at an angle θ to the line of sight.

the relativistic electrons and the magnetic field, and in a source such as Cygnus A it corresponds to 2×10^{52} J, the rest mass energy of $2 \times 10^5 M_\odot$ of matter. This is an enormous amount of mass to be converted into high-energy electrons and magnetic fields. Furthermore, in this calculation, it is assumed that no protons and nuclei were accelerated at the same time. If the total energy in relativistic material were η times greater than that present in electrons, the minimum energy requirements would increase by a factor of $\eta^{4/7}$. These arguments were crucial in the 1950s and 1960s, and demonstrated that high-energy particles and magnetic fields are essential ingredients of galaxies.

In the nuclei of active galaxies and quasars, the continuum spectra in the optical and X-ray wavebands are often of power-law form, and it is often assumed that the emission is the synchrotron radiation of high-energy electrons gyrating in the strong magnetic fields which must be present in these regions, if only in order to provide a means of collimating the jets of relativistic electrons observed to be emitted from their nuclei. As if these problems were not bad enough, it has been found that, in certain of the compact radio sources associated with active galaxies, the components are observed to be ejected from the active nuclei at speeds exceeding the speed of light. Fig. 11.14(a) shows the structure of the radio source 3C 273 as observed by very long baseline interferometry with angular resolution of the order of 1 milliarcsec. It can be observed that the components appear to have separated by 25 light years in the matter of only 3 years, corresponding to an apparent motion on the sky of about eight times the speed of light.

The preferred interpretation of these observations is that these superluminal velocities are the result of the ejection of the radio-emitting jet at a velocity close to the speed of light at a small angle to the line of sight (Fig. 11.14(b)). It is a straightforward (non-relativistic) calculation to show that the apparent sideways motion of the source component is

$$v_\perp = v \frac{\sin\theta}{1 - \frac{v}{c}\cos\theta}. \tag{54}$$

$v_\perp$ has a maximum value of γv when the angle of ejection of the component to the line of sight is given by $\cos\theta = v/c$. Thus, there is no problem in principle in accounting for the fact that superluminal velocities are observed, but it does require the radio-emitting plasma to be moving with large Lorentz factors, $\gamma \geqslant 8$ in the case of 3C 273. This is very good news in many ways because it can be shown that the extended radio components decline rapidly in radio luminosity unless they are continuously supplied with energy, and the radio jets observed in the nuclei of active galaxies, and on a larger scale in Cygnus A, show that this indeed takes place.

Synchrotron radiation of high-energy electrons dominates a great deal of thinking in the physics of active galaxies. Not only the continuum emission from the nucleus itself but also the optical emission from jets, such as that observed in the nearby galaxy M87, are identified with this emission process. In all these cases, the key signatures are a power-law radiation spectrum and linearly polarised radiation. In a number of cases, the sources are so compact that the brightness temperature of the radiation $T_b = (\lambda^2/2k_B)(S_v/\Omega)$ exceeds 10^{11} K, and so the electrons responsible for the emission must be relativistic.

11.5.2 Inverse Compton scattering

In the presence of high-energy densities of radiation, the high-energy electrons are also subject to energy losses by the process known as *inverse Compton scattering*. This is the relativistic version of Compton scattering and is 'inverse' in the sense that the electron transfers energy to the photons rather than the other way round, which is the classical form of Compton scattering. The simplest way of understanding the process of inverse Compton scattering is to transform from the laboratory frame of reference to the rest frame of the electron, in which case the photons are observed to approach the stationary electron from the forward direction with typical energies $h\nu' \approx \gamma h\nu$, where γ

is the Lorentz factor of the electron's motion. The photons are scattered with the usual angular distribution in the rest frame of the electron. On transforming to the laboratory frame, the photons are observed with energy $h\nu_{IC} \approx \gamma h\nu' \approx \gamma^2 h\nu$. Thus, the photons are increased in energy by typically a factor γ^2, and this energy is abstracted from the kinetic energy of the high-energy electron. The exact answer for inverse Compton scattering losses by the relativistic electron, where U_{rad} is the energy density of the photons, is

$$-\left(\frac{dE}{dt}\right) = \frac{4}{3}\sigma_T c U_{rad}\left(\frac{v}{c}\right)^2 \gamma^2. \tag{55}$$

It can be seen that the form of the expression for the inverse Compton scattering loss rate is very similar to the expression for synchrotron losses, and this is no coincidence. The key point is that the electron is accelerated by the electric field it experiences. In the case of synchrotron losses, the electron experiences the $E = v \times B$ field associated with motion through the magnetic field, while, in the case of inverse Compton scattering, the electric fields are the fields of the incident electromagnetic waves; in both cases, the rate of radiation depends only on the energy densities of these fields.

There are several important applications of this formula for relativistic electrons in various different cosmic environments. One interesting example concerns inverse Compton scattering of the photons of the cosmic microwave background radiation by relativistic electrons, wherever they are located in the Universe. The cosmic microwave background radiation permeates all space, and so relativistic electrons inevitably lose energy when they scatter these photons. At the present epoch, the energy density of the photons of the cosmic microwave background radiation is $U_{rad} = aT^4$, which for $T = 2.725$ K corresponds to an energy density of 2.6×10^5 eV m^{-3}. Hence, the lifetime of a relativistic electron is at most

$$\tau = \frac{E}{dE/dt} = \frac{E}{(4/3)\sigma_T c\gamma^2 U_{rad}} = \frac{2.3 \times 10^{12}}{\gamma}\text{ years}. \tag{56}$$

Thus, for example, the 100 GeV electrons which are observed at the top of the atmosphere must have lifetimes of less than 1.15×10^7 years.

The relativistic electrons present in the galactic and extragalactic radio sources typically have Lorentz factors of about 10^3 to 10^4. As a result, inverse Compton scattering of photons of the microwave background radiation produces X-rays since

$$\nu = \gamma^2 \nu_0 \approx 10^6 \times 10^{11}\text{ Hz} = 10^{17}\text{ Hz}; \quad \epsilon = 0.4\text{ keV}. \tag{57}$$

The scattering of ambient optical photons produces a flux of γ-rays:

$$\nu = \gamma^2 \nu_0 \approx 10^6 \times 10^{15}\text{ Hz} = 10^{21}\text{ Hz}; \quad \epsilon = 4\text{ MeV}. \tag{58}$$

An important application of the inverse Compton scattering process is to compact high brightness temperature sources of synchrotron radio emission. If the energy density of the radio photons produced by the relativistic electrons becomes too great, the electrons can lose more energy by inverse Compton scattering of the photons they have created than by the synchrotron process itself. Such sources are in danger of undergoing what is known as the *inverse Compton catastrophe* – the sources create large fluxes of X-rays rather than radio waves, which are then scattered to γ-ray energies by inverse Compton scattering and so on. It is a straightforward calculation to show that the upper limit to the brightness temperature of compact radio sources before the inverse Compton catastrophe takes place is $T_b \approx 10^{12}$ K. In fact, many of the compact radio sources for which angular sizes have been measured by very long baseline interferometry satisfy this limit, but a few of them exceed it. In addition, in a number of compact sources, flux densities are observed to vary on such short time-scales τ that, adopting the causality limit $r \leqslant c\tau$, the inferred brightness temperatures would exceed the above limit. The solution of this problem is almost certainly that the source components are moving relativistically, as has been observed directly in the superluminal radio sources. As illustrated in the above example of a source component moving rapidly at an angle close to the line of sight, changes can take place in the intensity and physical size of the source which apparently violate the causality limit $r \leqslant c\tau$ because the source of radiation almost catches up with the radiation it has emitted. In addition, because of its relativistic motion, the observed intensity of the source component is greatly enhanced because of the combined effects of aberration and the Doppler shift of the radiation.

Returning to the inverse Compton scattering process in sources of synchrotron radiation, it is evident that these must also be X-ray sources as indicated by the calculation (53). There are a number of sources in which it is claimed that the X-ray emission from the same population of electrons responsible for the radio emission has been observed, and the combination of these observations enables the magnetic flux density of the source region to be estimated. In addition, it is possible that the continuum emission of compact extragalactic X-ray sources is some form of *synchro-Compton radiation*, namely the self-Compton scattering of low-energy synchrotron photons by the same population of electrons.

11.6 Electrons and positrons

If photons have energies greater than $2mc^2$, then electron–positron pair production is possible in the field of an atomic nucleus. The process cannot take place spontaneously because of the need to conserve both energy and momentum in the pair-production process. For intermediate photon energies, in the case of no screening, the pair-production cross-section can be written as

$$\sigma_{\text{pair}} = \alpha r_e^2 Z^2 \left[\frac{28}{9} \ln \left(\frac{2h\nu}{mc^2} \right) - \frac{218}{27} \right] \ \text{m}^2 \ \text{atom}^{-1}, \tag{59}$$

where Z is the charge of the nucleus and $\alpha = e^2/2\epsilon_0 hc \approx 1/137$ is the fine-structure constant. In general, the cross-section is $\sim \alpha \sigma_T Z^2$. This process turns out to be the principal energy loss mechanism for the interaction of high-energy γ-rays with matter, specifically at energies $\epsilon \geqslant 5$ MeV. In the ultra-relativistic limit, the path length for pair production is the same as that for the loss of energy by the bremsstrahlung process – from the point of view of quantum electrodynamics, these processes have very similar Feynman diagrams. It is the combination of these processes which is responsible for the electromagnetic showers stimulated by the arrival of an ultrarelativistic electron or an ultrahigh-energy γ-ray at the top of the atmosphere or within the body of a particle detector. Indeed, this process, taking place within the body of a cloud chamber, resulted in Carl Anderson's discovery of the positron, which was announced in 1931. These showers can be simply modelled by considering that, in the first path length through the atmosphere, the γ-ray creates an electron–positron pair, each of which, in the next path length, creates a high-energy bremsstrahlung photon, each of which in turn create high-energy electron–positron pairs, and so on. The result is an exponential increase in the numbers of electron–positron pairs and photons as the shower propagates through the atmosphere. The formation of so many electron–positron pairs high in the atmosphere makes the determination of the primary cosmic ray electron spectrum difficult from high-flying balloon observations, even at the top of the atmosphere.

In the cases of active galactic nuclei, which are intense sources of X-rays and γ-rays, electron–positron pair production can take place more efficiently by collisions between the X-ray and γ-ray photons. It is a straightforward calculation to show that the threshold for this process is

$$\epsilon_2 = \frac{2m^2 c^4}{\epsilon_1 (1 - \cos \theta)}, \tag{60}$$

where ϵ_1 and ϵ_2 are the energies of the photons and θ is the angle between their directions of incidence. For head-on collisions, $\theta = \pi$, and so the threshold becomes

$$\epsilon_1 \epsilon_2 \geqslant 0.26 \times 10^{12} \ \text{eV}^2, \tag{61}$$

where the energies of the photons are measured in electron-volts.

The opposite process, *electron–positron annihilation*, can take place in two ways. In the first case, the electron and positron annihilate at rest or in flight by the simple interaction

$$e^+ + e^- \rightarrow 2\gamma. \tag{62}$$

When emitted at rest, each of the photons has energy equal to the rest mass energy of the electron, $\epsilon = 0.511$ MeV. When the particles annihilate in flight, meaning that they suffer a fast collision, the annihilation line is Doppler broadened.

If the velocity of the positron is small, *positronium atoms*, consisting of a bound state of the electron and positron, are formed by radiative recombination; 25% of the positronium atoms form in the singlet $^1S_0$ state and 75% in the triplet $^3S_1$ state. The modes of decay of these states are different. The singlet state has a lifetime of only 1.25×10^{-10} s, and the atom then decays into two γ-rays, each with energy 0.511 MeV. The majority triplet states have lifetimes of 1.5×10^{-7} s, and three γ-rays are emitted, the maximum energy being 0.511 MeV in the centre of momentum frame of reference. Because it is a three-body process, there is a low-energy continuum tail to the 0.511 MeV emission line. These spectral differences are useful in determining the physical conditions under which the annihilations take place. If the annihilations take place in a neutral medium with particle density less than 10^{21} m^{-3}, positronium atoms are formed. If the electrons and positrons collide in a gas at temperature greater than about 10^6 K, annihilation takes place directly without the formation of positronium.

The cross-section for electron–positron annihilation in the extreme relativistic limit is

$$\sigma = \frac{\pi r_e^2}{\gamma} [\ln 2\gamma - 1]. \tag{63}$$

For thermal electrons and positrons, the cross-section is

$$\sigma \approx \frac{\pi r_e^2}{v/c}. \tag{64}$$

One of the most exciting results of γ-ray astronomy has been the detection of the 0.511 MeV electron–positron annihilation line from the direction of the Galactic Centre (Fig. 11.15), indicating that there must be a powerful source of electrons and positrons in the general direction of the Galactic Centre.

If the source of γ-rays is compact enough, the photon–photon pair-production process can result in the destruction of the flux of high-energy γ-rays. Typical γ-ray sources have intensity spectra which decrease with increasing energy, and so the most abundant photons are those close to the threshold

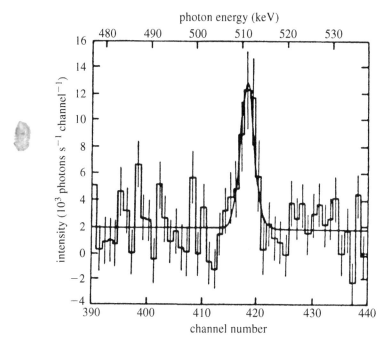

Fig. 11.15 HEAO-3 observations of the 0.511 MeV electron–positron annihilation line from the general direction of the Galactic Centre (Riegler *et al.* 1981).

energy for pair productions, $\epsilon \approx 1$ MeV. At this energy, the cross-section for photon–photon pair production is about $0.2\sigma_T$. If the dimension of the γ-ray source is R, then the number density of photons is $N_\gamma \approx L/4\pi R^2 ch\nu$, and, since $h\nu \approx mc^2$, this is roughly $N_\gamma \approx L/4\pi R^2 mc^3$. Therefore, the probability that the 1 MeV photons undergo a photon–photon pair-creating interaction is

$$\tau \approx 0.2\sigma_T N_\gamma R \approx \frac{L_\gamma \sigma_T}{mc^3 R}. \tag{65}$$

The quantity $C = L_\gamma \sigma_T/mc^3 R$ is known as the *compactness parameter*, where L_γ means the luminosity of the γ-ray source at 1 MeV. If the compactness parameter is greater than unity, then the γ-rays cannot escape from the source and the result is the production of large fluxes of electrons and positrons.

This result is of the greatest interest for the extragalactic γ-ray sources discovered by the Compton Gamma-Ray Observatory (CGRO), which has carried out the first complete sky survey in the γ-ray waveband. One of the most important discoveries has been that the most luminous extragalactic high-energy γ-ray sources are associated with the compact radio quasars, many of which are known to be superluminal sources. The luminosities of these sources are enormous, luminosities of up to 10^{42} W being found, and the variability of these luminosities occurs over a time-scale of the order of days. As a result, the compactness factor $C \sim 10^5$, and so all the γ-rays should have been degraded into electron–positron pairs. The solution to the problem comes from the observation that these intense γ-ray emitters are associated with superluminal radio sources. If the sources are relativistically beamed, as in the case of the superluminal sources, the compactness factor is increased by a factor of $[\gamma(1 - (v/c)\cos\theta)]^5$ over its rest-frame value. For the values found in the superluminal sources, $\gamma \approx 5$–10, it is evident that the rest-frame value of the compactness parameter can be reduced to values less than unity.

These considerations have suggested a model in which the γ-rays are generated within a beam of high-energy electrons which escapes from the nucleus with a velocity close to the speed of light. The mechanism for generating the hard radiation within the beam is not known, but it may be a variant of the inverse Compton scattering process of ambient low-energy photons. The γ-rays cannot originate too close to the active nucleus or else the photon–photon process will degrade them into electron–positron pairs, and so there is expected to be a γ-ray photosphere (Fig. 11.16). Further out, the beam becomes optically thin for synchrotron radiation in the centimetre radio waveband and results in the phenomenon of superluminal motions as illustrated in Fig. 11.14(a).

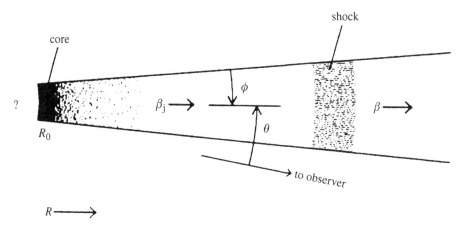

Fig. 11.16 A model for the γ-ray and radio luminosity in superluminal radio sources. The beam is oriented at a small angle to the line of sight. The beamed γ-rays originate close to the γ-ray photosphere. The superluminal motion of the radio components is observed further out from the nucleus where the source components become optically thin to synchrotron radiation. β and β_j are speeds of the material of the beam, as a fraction of the speed of light, in front of and behind the shock front, respectively (Marscher 1993).

11.7 The production of free electrons in cosmic conditions

Normally, cosmic conditions are rather hostile for atoms and molecules. It used to be thought that the interstellar medium was rather quiescent, but this is far from the case – it is a rather violent place. It is continually buffeted by the explosions of supernovae. Close to young hot stars, there are copious supplies of far-ultraviolet radiation, which ionises the surrounding gas and causes ionisation fronts to propagate through the interstellar medium. In addition, the interstellar flux of high-energy particles can ionise the interstellar gas and penetrate even into dense cool giant molecular clouds.

It is therefore not at all surprising that free electrons are found in copious supply, even in as quiescent a galaxy as our own. Evidence for free electrons in our Galaxy comes from a variety of different observations which involve the different ways in which electrons can affect the propagation of electromagnetic waves. For example, the pulsars emit very sharp pulses of radio emission at low radio frequencies. Since there are electrons present in the interstellar medium, the group velocity of the pulse of radiation is

$$v_{gr} = c\left[1 - \frac{1}{2}\left(\frac{v_p}{v}\right)^2\right], \tag{66}$$

where v_p is the *plasma frequency*,

$$v_p = \left(\frac{e^2 N_e}{4\pi^2 \epsilon_0 m}\right)^{1/2} = 8.98 N_e^{1/2} \text{ Hz}. \tag{67}$$

Therefore, the lower frequencies propagate more slowly through the interstellar medium than the high frequencies. Specifically, the delay time of the signal as a function of frequency is

$$T_a = \frac{e^2}{8\pi^2 \epsilon_0 mc} \frac{1}{v^2} \int_0^l N_e \, dl = 4.15 \times 10^9 \frac{1}{v^2} \int_0^l N_e \, dl, \tag{68}$$

where the electron density is measured in electrons per cubic metre, the distance l is measured in parsecs (1 pc $= 3.086 \times 10^{16}$ m) and the frequency v is measured in hertz. The quantity $\int N_e \, dl$ is known as the *dispersion measure* of the pulsar. Because the radio pulses from pulsars are so sharp, it is easy to measure their dispersion measures, and these have been measured in over 500 separate directions through the interstellar gas. If it is assumed that the electron density is uniform in the plane of the Galaxy, the dispersion measure provides an estimate of the distance of the pulsar. Improved distances can be found by adopting a more detailed picture for the distribution of ionised gas in the Galaxy (Taylor and Cordes 1993).

The interstellar medium is permeated by a large-scale magnetic field, and hence it constitutes a magnetised or *magnetoactive medium*. Under typical interstellar conditions, both the plasma frequency ν_p and the gyrofrequency $\nu_g = 2.8 \times 10^{10} B$ Hz, where B is measured in tesla, are much less than typical radio frequencies, $10^7 \leqslant \nu \leqslant 10^{11}$ Hz. Under these circumstances, the position angle of the electric vector of linearly polarised radio emission is rotated on propagating through a region in which the magnetic field is uniform. This phenomenon is known as *Faraday rotation*. The phenomenon arises because the modes of propagation of radio waves in a magnetoactive plasma are elliptically polarised in opposite senses and the phase velocities of the two modes are different. The physical reason for this is that, under the influence of the electric field of the electromagnetic wave, the electrons are constrained to move in spiral paths about the magnetic field direction. Under the above conditions, the refractive indices of the two modes are

$$n^2 = 1 - \frac{(\nu_p/\nu)^2}{1 \pm (\nu_g/\nu)\cos\phi}, \tag{69}$$

where ϕ is the angle between the direction of propagation of the wave and the magnetic field direction. It is a straightforward calculation to show that the angle through which the electric field vector is rotated on passing through a magnetoactive medium is

$$\theta = \frac{\pi}{c\nu^2}\int_0^l \nu_p^2\nu_g \cos\phi \, \mathrm{d}l, \tag{70}$$

or, rewriting the formula in terms of more convenient astronomical units,

$$\theta = 8.12 \times 10^3 \lambda^2 \int_0^l N_e B_{\parallel} \, \mathrm{d}l, \tag{71}$$

where θ is measured in radians, λ in metres, N_e in particles per cubic metre, $B_{\parallel}$ (the component of the magnetic field strength along the line of sight) in tesla, and l in parsecs. The quantity θ/λ^2 is known as the *rotation measure* (in radians per square metre) and thus provides information about the integral of $N_e B_{\parallel}$ along the line of sight. In addition, the sign of the rotation measure provides information about the weighted mean direction of the magnetic field along the line of sight: if θ/λ^2 is negative, the weighted magnetic field is directed away from the observer; if θ/λ^2 is positive, it is directed towards the observer. Both Galactic and extragalactic radio sources emit polarised radio emission, and so rotation measures can be found for many different lines of sight through the Galaxy. Evidently, by combining measurements of the rotation measure and the dispersion measure for individual sources, a weighted mean value of the magnetic field strength along the line of sight can be found:

$$\langle B_{\parallel} \rangle = \frac{\text{rotation measure}}{\text{dispersion measure}} \propto \frac{\int N_e B_{\parallel} \, dl}{\int N_e \, dl}. \tag{72}$$

Fig. 11.17 shows the distribution of the rotation measures of 976 extragalactic radio sources plotted in galactic coordinates. The magnitude of the rotation measure is indicated by the size of the dot: positive rotation measures are indicated by filled circles, and negative values are indicated by open circles. It can be seen that there is some large-scale order in the distribution of the rotation measures. They decrease towards high galactic latitudes more or less as expected if the magnetic field ran predominantly parallel to the plane of the Galaxy. Evidence that there is large-scale order in the field is provided by the clustering of points of the same sense and magnitude in different regions of the diagram, but there must be considerable irregularities as well to account for the spread in rotation measures in neighbouring regions of sky.

In addition to rotation of the plane of polarisation, the radio emission is expected to be *depolarised* with increasing wavelength. If the emission originates from a region of dimensions l in which the magnetic field and the electron density are uniform, the synchrotron radio emission will be highly polarised at high enough frequencies since the internal Faraday rotation within the region varies as λ^2 and so tends to zero as $\lambda \to 0$. At a certain wavelength, however, the Faraday rotation through the source region itself will amount to π radians, and then the radiation is expected to be significantly depolarised. Thus, provided the magnetic field and electron density are uniform, the wavelength at which the radiation is significantly depolarised provides information about the quantity $N_e B_{\parallel} l$ in the source region. This process, which provides information about physical conditions within the source region, is known as *Faraday depolarisation*.

The sum of these observations suggests that the typical magnetic field strength in the interstellar medium is about $(2-3) \times 10^{-10}$ T and that the free electron density is about $(0.1-1) \times 10^6$ m^{-3}. How are these free electrons created?

The principal mechanisms for the production of free electrons are *photoionisation* and *collisional ionisation*. Hydrogen is by far the commonest element in the interstellar medium and it can be photoionised by ultraviolet photons with energies greater than the ionisation potential of hydrogen, 13.6 eV. Under interstellar conditions, the principal sources of ionising photons are hot stars, the hot O and B stars typically having surface temperatures of about 40 000 and 20 000 K, respectively, and also the hot central stars of

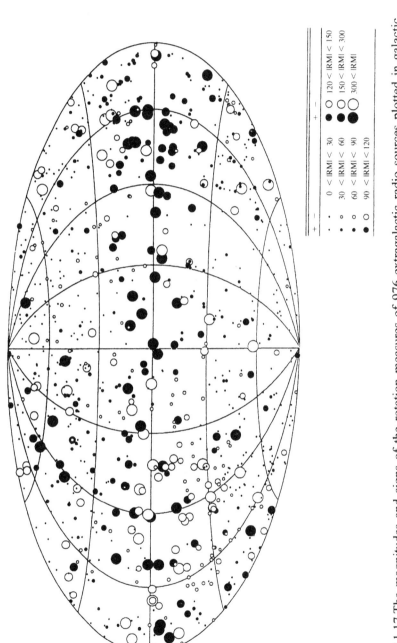

Fig. 11.17 The magnitudes and signs of the rotation measures of 976 extragalactic radio sources plotted in galactic coordinates. The data were assembled by P. P. Kronberg and published by Wielebinski (1993).

planetary nebulae. The main sequence O and B stars are massive and have relatively short lifetimes. Many of them are found within the regions of ionised hydrogen in the giant molecular clouds in which they were formed. In a classic analysis, Bengt Strömgren showed that the presence of hot O and B stars in an interstellar cloud can ionise the surrounding gas forming the characteristic regions of ionised hydrogen, often called *Strömgren spheres*, found in the vicinity of regions of star formation. At the interface between the region of ionised gas and the neutral interstellar gas, an ionisation front is formed which propagates through the interstellar gas, and these are common features of regions of ionised hydrogen (Fig. 11.18). The typical temperature of these regions is about 10^4 K, the energy input from the star being balanced by the line emission of the gas.

As mentioned in section 11.1, one of the principal energy loss mechanisms for regions of ionised hydrogen is the emission of *forbidden lines* of elements such as oxygen and nitrogen. For example, the forbidden transitions of oxygen originate from low-lying levels of singly and doubly ionised oxygen, [OII] and [OIII], respectively, and the transition to the ground state is forbidden according to the selection rules for electric dipole transitions. Under laboratory conditions, the oxygen ions are de-excited by electron collisions, but, if the density is low enough, these collisions may not be frequent enough to depopulate the low-lying levels, which can become very highly populated as the oxygen ions recombine. Under these conditions, the forbidden transition can take place by magnetic dipole or other rare transitions. As a result, the observation of forbidden lines, and, in particular, the ratios of forbidden lines originating from different low-lying states, enables the electron densities in these regions to be determined. The details of these processes are elegantly described in Osterbrock's book *The Astrophysics of Gaseous Nebulae and Active Galactic Nuclei* (1989). As implied by the title of the book, the same techniques are applicable to the emission-line regions observed in active galactic nuclei in which the ionising flux of ultraviolet radiation is of roughly power-law form and extends far into the ultraviolet region of the spectrum, resulting in the production of ions in a very wide range of ionisation states. The dynamics of these gas clouds are the primary diagnostic tool for estimating the masses of supermassive black holes in active galactic nuclei.

The second important heating process for electrons is collisional processes, and these are normally associated with interstellar shock waves. The most spectacular of these are associated with the explosions of supernovae, in which the central regions of a dying star collapse to form a neutron star or black hole, and the binding energy of the neutron star, which amounts to about 10% of its rest mass energy, is released into the stellar envelope. The resulting highly

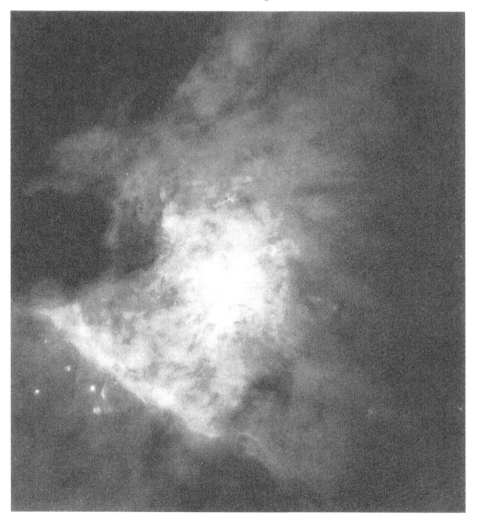

Fig. 11.18 The central region of the Orion Nebula as observed by the Hubble Space Telescope. The Nebula is ionised and excited by four luminous B stars (the Trapezium stars) in the centre of the image. The prominent bar of emission to the bottom left of the picture is an ionisation front associated with the ultraviolet continuum radiation of the Trapezium stars. (Courtesy of Dr C. R. O'Dell, NASA and the Space Telescope Science Institute.)

supersonic shock wave passes out through the stellar envelope and ionises and heats the surrounding gas to a high temperature. Direct evidence for this heating is provided by X-ray images of supernova remnants such as that of Cassiopeia A, the temperature of the gas being about 10^7 K. Typically, the expanding shells of supernovae such as Cassiopeia A have velocities of about 5000 to 10 000 km s^{-1}, and so can communicate a large amount of kinetic energy to the surrounding interstellar gas which is ultimately dissipated as heat. It is estimated that the explosions of supernovae can heat up to about 10% of the interstellar medium to a temperature of about 3×10^6 K, and this is the origin of the soft X-ray emission observed from the interstellar medium.

Supernova explosions and their remnants are the most probable source of cosmic ray protons, electrons and nuclei, and these can also heat the interstellar gas by the process known as *ionisation losses*. The most important aspect of this process is the interaction between high-energy protons and the neutral component of the interstellar gas. As the proton passes by the atom, the electrostatic attractive force between the proton and electron results in a momentum impulse which, in close encounters, is sufficient to remove the electron from the atom. The energy loss rate of the cosmic ray of charge z and velocity v per unit length is

$$-\left(\frac{\mathrm{d}E}{\mathrm{d}x}\right) = \frac{z^2 e^2 NZ}{4\pi\epsilon_0^2 mv^2}\left[\ln\left(\frac{2\gamma^2 mv^2}{\bar{I}}\right) - \frac{v^2}{c^2}\right], \tag{73}$$

where N is the number density of atoms of atomic number Z, γ is the Lorentz factor of the cosmic ray and $\bar{I}$ is the mean ionisation potential of the material through which the cosmic ray passes. This is a very important mechanism for the detection of cosmic ray particles since this is the principal energy loss process for the interaction of cosmic rays with matter, and, by measuring the total energy transferred to the electrons, the total energy of the incoming particle can be estimated.

The interstellar gas is permeated by a flux of cosmic rays, as is beautifully illustrated by the recent γ-ray map of the whole sky as observed by the Compton Gamma-Ray Observatory (CGRO) (Fig. 11.19). The γ-rays are predominantly the result of collisions between high-energy cosmic ray protons and nuclei with the nuclei of atoms of the interstellar gas, in which pions of all sorts are created, the neutral pions decaying to produce a pair of high-energy γ-rays. The cosmic rays are not the most important heating mechanism for the general interstellar medium, but they are thought to be important in the heating and partial ionisation of giant molecular clouds. These clouds contain large amounts of dust, and so the interstellar ionising radiation cannot penetrate into their interiors. Yet there has to be some heating mechanism inside the clouds or else they would cool to a very low temperature. There is evidence for turbulent motions inside the clouds which may contribute to the heating, but the flux of high-energy particles can also penetrate into the clouds and heat them through the process of ionisation losses. In dense molecular clouds, the temperature is very low, typically only about 50 K, and there is a wealth of interstellar molecules present. In addition, various types of molecular ions are found, and

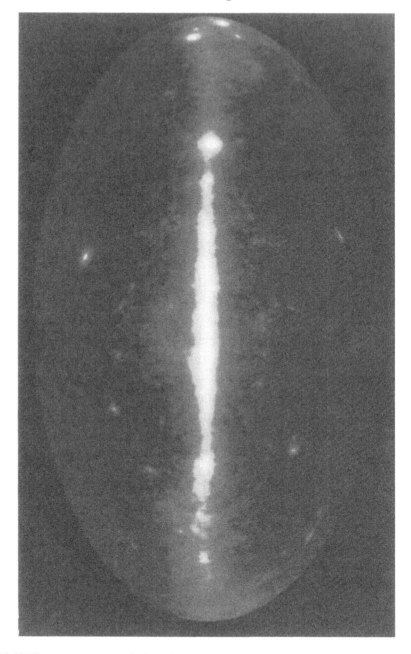

Fig. 11.19 The γ-ray map of the sky as observed by the Compton Gamma-Ray Observatory (CGRO). The γ-rays have energies greater than about 100 MeV and are predominantly associated with the decay of neutral pions created in the collisions of cosmic rays with the nuclei of atoms and molecules of the interstellar gas. This γ-ray image traces the location of cosmic rays and interstellar gas in the disc of the Galaxy. (Courtesy of Dr D. Fichtel, NASA and the CGRO Science Team.)

this provides evidence that ionisation processes must be going on despite the very low temperature of the gas. The abundance of ions is very low, only about one part in 10^8, but they are crucial in enabling gas phase reactions to take place at low temperatures inside the clouds. The scheme of ion–molecular reactions begins with the creation of free electrons and molecular hydrogen ions by the ionisation losses of cosmic rays. The molecular hydrogen ions then interact with other hydrogen molecules to form H_3^+

$$H_2^+ + H_2 \rightarrow H_3^+ + H. \tag{74}$$

H_3^+ readily donates a proton to other species through reactions such as

$$H_3^+ + CO \rightarrow HCO^+ + H_2, \tag{75}$$

$$H_3^+ + N_2 \rightarrow HN_2^+ + H_2. \tag{76}$$

The detection of HCO^+ and HN_2^+ in roughly their expected abundances is confirmation that this type of reaction, initiated by the production of free electrons, does take place in cool dense molecular clouds.

11.8 The acceleration of high-energy electrons

There are thus very good reasons why there should be copious supplies of electrons in the interstellar medium, but we have to explain the origin of the fluxes of relativistic electrons which have Lorentz factors of 10^5 and greater. The case is even worse for the cosmic ray protons and nuclei – the highest energy cosmic rays have energies up to 10^{20} eV or greater. Not only have the very high energies to be explained, but we also need to account for the fact that the spectrum of the particles is of power-law form, the typical spectrum having the form

$$N(E)\,dE = \kappa E^{-x}, \tag{77}$$

where $x \approx 2.5-3$. It is interesting to contrast the enormous complexity of the particle accelerators needed to accelerate protons and nuclei to high energies with the rarified conditions of interstellar space. Plainly, nature has found a much simpler way of accelerating high-energy particles.

Of the many suggestions, current interest has centred on two processes. The first of these is related to one of the earliest proposals due to Fermi, that cosmic rays can be accelerated in collisions with interstellar clouds. Fermi's idea was that, in random collisions with interstellar clouds, there is a greater probability of head-on as opposed to following collisions and so the particle gradually gains energy. The problem with this proposal was that it is second order in the velocity of the interstellar clouds; that is, the rate of gain of energy is proportional to $(v/c)^2$, and so the acceleration takes place over a very long time-scale. Although Fermi found a power-law energy spectrum, there was nothing in the theory which explained why the value of x was always about 2.5–3.

The modern version of the theory involves what is called *first-order Fermi acceleration* in strong shock waves. This proposal was developed independently by a number of authors in 1977–1978 (Axford *et al.* 1977; Krymsky 1977; Bell 1978; Blandford and Ostriker 1978). The idea is to study the dynamical behaviour of high-energy particles in the vicinity of strong shock waves. The shock is assumed to have a high Mach number, and there are scattering media in front of and behind the shock front. The key point about the acceleration process is that the scattering processes maintain an isotropic distribution of particles on either side of the shock front. The physical situation is illustrated in Fig. 11.20. The shock travels at velocity U, but the gas velocities in front of and behind the shock, in the limit of strong shocks, are as shown in Fig. 11.20(b). Thus, both on the upstream and the downstream sides of the shocks, the isotropic distribution of particles is about to be swept up by gas moving at velocity $0.75U$. As the particles pass through the shock, they

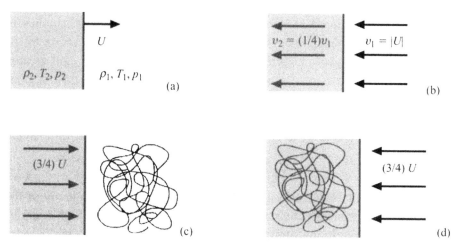

Fig. 11.20 The dynamics of high-energy particles in the vicinity of a strong shock wave. (a) A strong shock wave propagating at a supersonic velocity U through the stationary interstellar gas with density ρ_1, pressure p_1 and temperature T_1. The corresponding quantities behind the shock are ρ_2, p_2 and T_2. (b) The flow of interstellar gas in the vicinity of the shock front in the reference frame in which the shock front is at rest. For the case of a fully ionised gas, for which the ratio of specific heats is $\gamma = 4/3$, the ratio of these velocities is $v_1/v_2 = 4$, as shown. (c) The flow of gas as observed in the frame of reference in which the upstream gas is stationary and the velocity distribution of the high-energy particles is isotropic. (d) The flow of gas as observed in the frame of reference in which the downstream gas is stationary and the velocity distribution of the high-energy particles is isotropic.

suffer first-order Fermi acceleration, on average receiving an increment of energy ΔE such that

$$\left\langle \frac{\Delta E}{E} \right\rangle = \frac{4}{3}\frac{U}{c}. \tag{78}$$

The clever point about this mechanism is that it is first order because the particle receives the same energy increment when it passes through the shock in either direction. Particles are also lost from the vicinity of the shock by being swept downstream. The fraction of particles lost per cycle by this mechanism is small, being only U/c. Combining these results, it is found that the predicted spectrum is of power-law form, as required, but with spectral index $x = 2$. It may be objected that we have obtained a spectral index of 2 rather than 2.5, and this has been the subject of many studies. The great attraction of this picture is that there are excellent physical reasons why a power-law spectrum with a unique spectral index should be found in very diverse astrophysical environments – the only requirement is the presence of strong shocks and that the velocity vectors of the particles are randomised on either side of the shock wave. It is certain that there are strong shocks in the regions in which large fluxes of ultrarelativistic electrons are known to be present – supernova remnants, active galactic nuclei and the diffuse components of extended radio sources. The other great advantage of this mechanism is that the electrons are accelerated in exactly the regions where they are needed. One of the long-standing problems of the simplest picture of the evolution of an expanding sphere of relativistic electrons is that the electrons lose most of their energy by adiabatic cooling as the remnant expands, and this is avoided if the particles are accelerated at the strong shock fronts. The electrons derive their energy from the bulk kinetic energy of the outflowing gas behind the shock front.

There has been an enormous amount of interest in this mechanism, and the results of a large number of studies are summarised in Longair (1992–7, vol. 2, chap. 21). There are mechanisms for steepening the power-law energy spectrum, the simplest being synchrotron or inverse Compton scattering losses, which steepen the energy spectral index by one if there is continued acceleration of the electrons. One of the problems with this picture is that the particles can only be accelerated to energies of about 10^5 GeV per nucleon in supernova remnants because of the finite size of the accelerating region. It is just possible that the highest energy cosmic rays could be accelerated in the extended components of double radio sources, but it may be that another acceleration mechanism is needed for the most powerful sources.

One possibility is acceleration in the vicinity of neutron stars and black holes. Pulsars are known to possess very strong surface magnetic fields, and it is quite conceivable that rotating black holes are threaded by magnetic fields. In these cases, very strong induced electric fields are created and particle acceleration is possible in them. It is certainly possible in principle that very-high-energy particles could be created in such circumstances, but it is not clear how they could escape from these regions without suffering large energy losses.

11.9 Conclusion

This chapter has covered only a few of the many ways in which electrons are central to the study of the Universe at large. We have mentioned, however, most of the important physical mechanisms by which they interact with matter and magnetic fields and so result in observable phenomena. It is the imaginative application of these tools in very diverse astrophysical environments that has led to many of the most important discoveries of modern astronomy and cosmology. In turn, the understanding of the physics of electrons in circumstances very different from those found in the laboratory has been greatly consolidated. One can be optimistic that this symbiosis between laboratory and astronomical studies will be as fruitful in future centuries as it has been during the first century of the electron.

References

Figures in parentheses at the end of each reference show the pages on which that reference is cited.

Abdullah, K. *et al.* 1990. *Phys. Rev. Lett.* **65**, 2347. (35)

Aharonov, Y. and Bohm, D. 1959. *Phys. Rev.* **115**, 485. (221)

Alexandrov, A. S. and Golubov, A. A. 1992. *Phys. Rev.* **B45**, 4769. (76)

Amato, A., Jaccard, D., Flouquet, J., Lapierre, F., Tholence, J. L., Fisher, R. A., Lacy, S. E., Olsen, J. A. and Phillips, N. E. 1987. *J. Low Temp. Phys.* **68**, 371. (184)

Anderson, C. D. 1933. *Phys. Rev.* **43**, 491. (57)

Anderson, P. W. 1961. *Phys. Rev.* **124**, 41. (68, 189)

Anderson, P. W. 1984. *Basic Notions of Condensed Matter Physics* (Benjamin Cummings, Menlo Park, CA). (103)

Anderson, P. W. 1991. *Phys. Rev. Lett.* **67**, 2092. (106)

Anderson, P. W. 1995. *Phys. World* **8** (12), 37. (106)

Anderson, P. W. 1997a. To be published. (135)

Anderson, P. W. 1997b. Forthcoming book. (141)

Anderson, P. W. and Schrieffer, J. R. 1991. *Phys. Today* **44** (6), 54. (105)

Ando, T. 1974. *J. Phys. Soc. Japan* **37**, 1233. (251)

Andrei, E. Y., Deville, G., Glattli, D. C. and Williams, F. I. B. 1988. *Phys. Rev. Lett.* **60**, 2765. (254)

Andres, K., Graebner, J. E. and Ott, H. R. 1975. *Phys. Rev. Lett.* **35**, 1779. (182)

Annett, J. F., Goldenfeld, N. and Renn, S. R. 1990. In *Physical Properties of High Temperature Superconductors*, vol. II, ed. D. M. Ginsberg (World Scientific, Singapore). (180)

Annett, J. F., Goldenfeld, N. and Leggett, A. J. 1996. In *Physical Properties of High Temperature Superconductors*, vol. V, ed. D. M. Ginsberg (World Scientific, Singapore). (178, 180)

Aoki, H., Uji, S., Albessard, A. K. and Onuki, Y. 1993. *Phys. Rev. Lett.* **71**, 2110. (139)

Argyres, P. N. 1958. *J. Phys. Chem. Solids* **4**, 19. (251)

Axford, W. I., Leer, E. and Skadron, G. 1977. *Proc. 15th International Cosmic Ray Conference*, vol. II (Bulgarian Academy of Sciences, Plovdiv), p. 132. (312)

Bardeen, J., Cooper, L. N. and Schrieffer, J. R. 1957. *Phys. Rev.* **108**, 1175. (103, 154, 236)

Barkhausen, H. 1919. *Phys. Zeits.* **20**, 401. (21)

Barone, A. and Paterno, G. 1982. *Physics and Applications of the Josephson Effect* (World Scientific, Singapore). (169, 170)

Barrett, S. E., Tycko, R., Pfeiffer, L. N. and West, K. W. 1994. *Phys. Rev. Lett.* **72**, 1368. (255)

Baym, G. and Pethick, C. 1991. *Landau-Fermi Liquid Theory* (Wiley, New York), chap. 3. (114, 115, 123, 138)

Béal-Monod, M. T., Ma, S.-K. and Fredkin, D. R. 1968. *Phys. Rev. Lett.* **20**, 929. (138)

Becker, R. and Döring, W. 1935. *Ann. d. Phys.* **24**, 719. (12)

Bednorz, J. G. and Müller, K. A. 1986. *Z. Phys.* **B64**, 189. (104, 117)

Bell, A. R. 1977. *Mon. Not. R. Astron. Soc.* **179**, 574. (292)

Bell, A. R. 1978. *Mon. Not. R. Astron. Soc.* **182**, 147. (312)

Bennett, C. L., Banday, A., Gorski, K. M., Hinshaw, G., Jackson, P., Keegstra, P., Kogut, A., Smoot, G. F., Wilkinson, D. T. and Wright, E. L. 1996. *Astrophys. J. Lett.* **464** (1), L1. (275)

Bergmann, G. 1984. *Phys. Rep.* **107**, 1. (209, 218)

Berk, N. F. and Schrieffer, J. R. 1966. *Phys. Rev. Lett.* **17**, 433. (138)

Bernhoeft, N. R. and Lonzarich, G. G. 1995. *J. Phys. Condensed Matter* **7**, 7325. (124)

Bernhoeft, N. R., Hayden, S. M., Lonzarich, G. G., Paul, D. McK. and Lindley, E. J. 1989. *Phys. Rev. Lett.* **62**, 657. (124)

Bernreuther, W. and Suzuki, M. 1991. *Rev. Mod. Phys.* **63**, 313. (37)

Bethe, H. 1931. *Z. Phys.* **71**, 205. (191)

Blackett, P. M. S. 1960. *Biographical Memoirs of Fellows of the Royal Society* (*see entry for* C. T. R. Wilson), vol. 6 (Royal Society, London), p. 269. (11)

Blandford, R. D. and Ostriker, J. P. 1978. *Astrophys. J.* **221**, L29. (312)

Bloch, F. 1929. *Z. Phys.* **57**, 545. (75)

Böhringer, H. 1995. *Ann. N. Y. Acad. Sci.* **759**, 67. (285)

Bohr, N. 1972. *Collected Works*, ed. J. R. Nielson (North Holland, Amsterdam). (16)

Bonevich, J. E., Harada, K., Kasai, H., Matsuda, T., Yoshida, T., Pozzi, G. and Tonomura, A. 1995. In *Electron Holography, Proceedings of the International Workshop on Electron Holography*, eds. A. Tonomura, L. G. Allard, G. Pozzi, D. C. Joy and Y. A. Ono (Elsevier Science BV, Amsterdam), pp. 135–44. (229)

Brinkman, W. F. and Engelsberg, S. 1968. *Phys. Rev.* **169**, 417. (138)

Brommer, K. D., Needels, M., Larson, B. E. and Joannopoulos, J. D. 1992. *Phys. Rev. Lett.* **68**, 1355. (83)

Buchwald, J. Z. 1985a. In *Wranglers and Physicists*, ed. P. M. Harman (Manchester University Press), chap. 9. (14)

Buchwald, J. Z. 1985b. *From Maxwell to Microphysics* (Chicago University Press), chap. 4. (14)

Buhman, H., Joss, W., von Klitzing, K., Kukushkin, I. V., Plaut, A. S., Martinez, G., Ploog, K. and Timofeev, V. B. 1991. *Phys. Rev. Lett.* **66**, 926. (254)

Bullen, K. E. 1976. *Dictionary of Scientific Biography* (*see entry for* Wiechert), vol. 14 (Scribner, New York), p. 327. (9)

Byers, N. and Yang, C. N. 1961. *Phys. Rev. Lett.* **7**, 46. (166)

Ceperley, D. M. 1978. *Phys. Rev.* **B18**, 3126. (74, 76)

Ceperley, D. M. and Alder, B. J. 1980. *Phys. Rev. Lett.* **45**, 566. (74)

Chaikin, P. M. and Lubensky, T. C. 1995. *Principles of Condensed Matter Physics* (Cambridge University Press). (62)

Chakravarty, S., Gelfand, M. P. and Kivelson, S. 1991. *Science* **254**, 970. (172)

Chakravarty, S., Sudbo, A., Anderson, P. W. and Strong, S. P. 1993. *Science* **261**, 337. (179)

Chambers, R. G. 1960. *Phys. Rev. Lett.* **5**, 3. (226)

Clifford, W. K. 1882. *Mathematical Papers* (Macmillan, London), Paper XLIII. (46)

Cohen, E. R. and Taylor, B. N. 1986. *Rev. Mod. Phys.* **59**, 1121. (26, 34)

Coleman, P. 1984. *Phys. Rev.* **B29**, 3035. (194)

Coleman, P. 1995a. *Physica* **B206&207**, 872. (116)

Coleman, P. 1995b. *Phys. World* **8** (12), 29. (106, 116)

Coleman, P., Schofield, A. J. and Tsvelik, A. M. 1996a. *Phys. Rev. Lett.* **76**, 1324. (107)

Coleman, P., Schofield, A. J. and Tsvelik, A. M. 1996b. *J. Phys. C* Special Issue, 'Non Fermi Physics in Metals', eds. P. Coleman, M. B. Maple and A. Millis. (107)

Cornu, A. 1898. *Comptes Rendus* **126**, 181. (9)

Cowan, R. D. 1981. *The Theory of Atomic Structure and Spectra* (University of California Press, Berkeley). (69)

Cox, D. L. and Maple, M. B. 1995. *Phys. Today* **48** (Feb.), p. 32. (104, 174)

Davies, M. 1970. *Biographical Memoirs of Fellows of the Royal Society* (*see entry for* Debye), vol. 16 (Royal Society, London), p. 175. (21)

Davisson, C. and Germer, L. H. 1927a. *Nature* **119**, 558. (209)

Davisson, C. and Germer, L. H. 1927b. *Phys. Rev.* **30**, 705. (209)

de Haas, W. J. and van Alphen, P. M. 1930a. *Proc. Netherlands Roy. Acad. Sci.* **33**, 680. (198)

de Haas, W. J. and van Alphen, P. M. 1930b. *Proc. Netherlands Roy. Acad. Sci.* **33**, 1106. (198)

de Haas, W. J. and van Alphen, P. M. 1932. *Proc. Netherlands Roy. Acad. Sci.* **35**, 454. (198)

de Haas, W. J., de Boer, J. H. and van den Berg, G. J. 1934. *Physica* **1**, 1115. (188)

Deaver, B. S. Jr and Fairbank, W. M. 1961. *Phys. Rev. Lett.* **7**, 43. (236)

Dederichs, P. H., Akai, H., Blügel, S., Stefanou, N. and Zeller, R. 1989. In *Alloy Phase Stability*, eds. G. M. Stocks and A. Gonis (Kluwer Academic Publishers, Dordrecht). (75, 84)

Dirac, P. A. M. 1928. *Proc. Roy. Soc. London A.* **117**, 610. (40, 43)

Dirac, P. A. M. 1931. *Proc. Roy. Soc. London A.* **133**, 60. (221)

Dirac, P. A. M. 1977. In *History of Twentieth Century Physics; Proceedings of the International School of Physics 'Enrico Fermi'*, ed. C. Weiner (Academic Press, New York), p. 109. (43)

Doniach, S. 1977. *Physica* **91B**, 231. (192)

Doniach, S. and Engelsberg, S. 1966. *Phys. Rev. Lett.* **17**, 750. (138)

Dreizler, R. M. and Gross, E. K. U. 1990. *Density Functional Theory* (Springer-Verlag, Berlin). (72, 76, 78)

Drude, P. 1900. *Ann. d. Phys.* **1**, 566. (15)

Du, R. R., Stormer, H. L., Tsui, D. C., Yeh, A. S., Pfeiffer, L. N. and West, K. W. 1994. *Phys. Rev. Lett.* **73**, 3274. (253)

Ducastelle, F. 1991. *Order and Phase Stability in Alloys* (North Holland, Amsterdam). (63)

Dungate, D. G. 1990. A theoretical study of the magnetic and superconducting properties of strongly interacting d and f electron systems. Ph.D. Thesis, University of Cambridge. (146)

Dushman, S. 1923. *Phys. Rev.* **21**, 623. (18, 19)

Ebert, H. (ed.) 1996. *Spin Polarised Spectroscopies* (Springer-Verlag, Berlin). (78)

Edwards, D. M. and Lonzarich, G. G. 1992. *Phil. Mag. B* **65**, 1185. (128)

Edwards, D. M. and Wohlfarth, E. P. 1968. *Proc. Roy. Soc. London A.* **303**, 127. (113)

Ehrenberg, W. and Siday, R. E. 1949. *Proc. Phys. Soc. London B* **62**, 8. (221)

Engel, E., Müller, H., Speicher, C. and Dreizler, R. M. 1995. In *Density Functional Theory*, eds. E. K. U. Gross and R. M. Dreizler, ASI Series Physics vol. 337 (Plenum, New York), p. 65. (78)

Esfarjani, K. and Chui, S. T. 1990. *Phys. Rev.* **B42**, 10 758. (254)

Falconer, I. 1987. *Brit. J. Hist. Sci.* **20**, 241. (4)

Falicov, L. M. 1960 In *The Fermi Surface*, eds. W. A. Harrison and M. B. Webb (John Wiley & Sons, New York), p. 39. (200)

Farnham, D. L., van Dyck, R. S. and Schwinberg, P. B. 1995. *Phys. Rev. Lett.* **75**, 3598. (31)

Fermi, E. 1927. *Rend. Accad. Naz. Licencei.* **6**, 602. (70)

Ferrario, L., Wickramsinghe, D. T., Bailey, J., Tuohy, I. R. and Hough, J. H. 1989. *Astrophys. J.* **337**, 832. (289)

Fertig, H. A., Brey, L., Cote, R. and MacDonald, A. H. 1994. *Phys. Rev.* **B50**, 11 018. (254)

Fetter, A. L. and Walecka, J. D. 1971. *Quantum Theory of Many-Particle Systems* (McGraw-Hill, New York). (69)

Fey, D. and Appel, J. 1980. *Phys. Rev.* **B22**, 3173. (145)

Feynman, R. P. 1949. *Phys. Rev.* **76**, 769. (32)

Feynman, R. P. 1985. *QED: The Strange Theory of Light and Matter* (Princeton University Press). (111)

Feynman, R. P. and Hibbs, A. R. 1965. *Quantum Mechanics and Path Integrals* (McGraw-Hill, New York). (111)

Fitzgerald, G. F. 1881. *Proc. Roy. Dublin Soc.* **3**, 250. (3)

Fitzgerald, G. F. 1896. *Nature* **55**, 6. (5)

Fitzgerald, G. F. 1897. *Electrician* **39**, 103. (8)

Fleming, J. A. 1910. 'Telegraphy', in *Encyclopaedia Britannica,* 11th edn. (20, 21)

Fleming, J. A. 1922. 'Wireless Telegraphy', in *Encyclopaedia Britannica,* 12th edn. (Encyclopaedia Britannica International Ltd., London). (20, 21)

Foulkes, I. F. and Gyorffy, B. L. 1977. *Phys. Rev. B* **15**, 1395. (146)

Fowler, A. B., Fang, F. F., Howard, W. E. and Stiles, P. J. 1966. *Phys. Rev. Lett.* **16**, 901. (238)

Franklin, A. 1986. *The Neglect of Experiment* (Cambridge University Press), p. 170. (14)

Frenkel, J. 1946. *Kinetic Theory of Liquids* (Oxford University Press), p. 390; reprinted by Dover Publications, New York, 1955. (12)

Friedel, J. 1971. In *Physics of Metals: Electrons,* ed. J. M. Ziman (Cambridge University Press). (86)

Fulde, P. 1991. *Electron Correlations in Molecules and Solids* (Springer Verlag, Berlin). (205)

Fulde, P., Keller, J. and Zwicknagl, G. 1988. *Sol. State Phys.* **41**, 1. (139)

Gabor, D. 1949. *Proc. Roy. Soc. London A* **197**, 454. (214)

Gabor, D. 1951. *Proc. Phys. Soc. London B* **64**, 449. (214)

Gee, D. 1996. *D. Phil. Thesis*, Oxford University. (240)

Gell-Mann, M. and Pais, A. 1955. *Phys. Rev.* **97**, 1387. (107)

Ginzburg, V. L. and Landau, L. D. 1950. *J. Exp. Theor. Fiz.* **20**, 1064; translated in *Men of Science*, ed. D. ter Haar (Pergamon, Oxford, 1965). (164, 235)

Goldberg, B. B., Heiman, D., Pinczuk, A., Pfeiffer, L. and West, K. 1992. *Surf. Sci.* **263**, 9. (254)

Goldman, V. J., Santos, M., Shayegan, M. and Cunningham, J. E. 1990. *Phys. Rev. Lett.* **65**, 2189. (254)

Goldman, V. J., Su, B. and Jain, J. K. 1994. *Phys. Rev. Lett.* **72**, 2065. (250)

Gonis, A. 1992. *Green's Functions for Ordered and Disordered Systems* (North Holland, Amsterdam). (84)

Gordon, W. 1926. *Z. Phys.* **40**, 117. (43)

Grewe, N. and Steglich, F. 1991. In *Handbook on the Physics and Chemistry of Rare Earths*, vol. 14, eds. K. A. Gschneider Jr and L. Eyring (Elsevier Science BV, Amsterdam), chap. 97. (193)

Grimes, C. C. and Adams, G. 1979. *Phys. Rev. Lett.* **42**, 795. (253)

Grosche, F. M., Pfleiderer, C., McMullan, G. J., Lonzarich, G. G. and Bernhoeft, N. R. 1995. *Physica* **B206&207**, 20. (140, 142)

Grosche, F. M., Julian, S. R., Mathur, N. D. and Lonzarich, G. G. 1996. *Physica B* **223&224**, 50. (147)

Gross, E. K. U., Kurth, S., Capelle, K. and Lüders, M. 1995. In *Density Functional Theory*, eds. E. K. U. Gross and R. M. Dreizler (Plenum Press, New York), p. 431. (77)

Gunnarson, O., Harris, J. and Jones, R. O. 1977. *J. Chem. Phys.* **67**, 3977. (78, 79)

Gyorffy, B. L. 1993. *Physica Scripta* **149**, 373. (64)

Haldane, F. D. M. 1983. *Phys. Rev. Lett.* **51**, 605. (243)

Halperin, B. I. 1984. *Phys. Rev. Lett.* **52**, 1583. (243)

Halperin, B. I., Lee, P. A. and Read, N. 1993. *Phys. Rev.* **B47**, 7321. (244, 253)

Heaviside, O. 1889. *Phil. Mag.* **27**, 445. (3)

Heilbron, J. L. 1985. In *Niels Bohr*, eds. A. P. French and P. J. Kennedy (Harvard University Press, Cambridge, MA). (16)

Heine, V., Robertson, I. J. and Payne, M. C. 1991. *Phil. Trans. Roy. Soc. London A* **334**, 391. (64)

Heitler, W. and London, F. 1927. *Z. Phys.* **44**, 455. (63, 66)

Heitmann, D. and Kotthaus, J. P. 1993. *Phys. Today* **46** (6), 56. (73)

Herring, C. 1966. In *Exchange Interaction Among Itinerant Electrons: Magnetism*, vol. IV, eds. G. T. Rado and N. Suhl (Academic Press, New York). (125)

Hertz, J. 1976. *Phys. Rev.* **B14**, 1164. (120, 135)

Hess, V. F. 1911. *Akad. Wiss. Wein, Ber.* **120**, 1575. (12)

Hewson, A. C. 1993. *The Kondo Problem to Heavy Fermions* (Cambridge University Press). (188, 189)

Hlubina, R. and Rice, T. M. 1995. *Phys. Rev.* **B51**, 9253. (135, 140)

Hohenberg, P. and Kohn, W. 1964. *Phys. Rev.* **136B**, 864. (70, 77)

Holstein, T., Norton, R. E. and Pincus, P. 1973. *Phys. Rev.* **B8**, 2647. (140)

Hoogeveen, F. 1990. *Nucl. Phys. B* **341**, 322. (36)

Hubbard, J. 1963. *Proc. Roy. Soc.* **A276**, 238. (68)

Hume-Rothery, W. 1936. *The Structure of Metals and Alloys* (Institute of Metals, London). (65)

Imry, Y. 1996. *Introduction to Mesoscopic Physics* (Oxford University Press). (209, 218, 230)

Ishikawa, Y., Noda, Y., Wemura, Y. J., Majhrzah, C. F. and Shirane, G. 1985. *Phys. Rev.* **B31**, 5884. (124)

Itzykson, C. and Zuber, J. B. 1980. *Quantum Field Theory* (McGraw-Hill, New York). (34, 52)

Izuyama, T., Kim, D. J. and Kubo, R. 1963. *J. Phys. Soc. Japan* **18**, 1025. (138)

Jain, J. K. 1989. *Phys. Rev. Lett.* **63**, 199. (244, 245)

Jarlskog, C. 1985. *Phys. Rev. Lett.* **55**, 1039. (37)

Jiang, H. W., Willet, R. L., Stormer, H. L., Tsui, D. C., Pfeiffer, L. N. and West, K. W. 1990. *Phys. Rev. Lett.* **65**, 633. (254)

Joachim, H. 1919. *Ann. d. Phys.* **60**, 570. (21)

Jones, M., Saunders, R., Alexander, P., Birkinshaw, M., Dillon, N., Grainge, K., Hancock, S., Lasenby, A. N., Lefebvre, D. C., Pooley, G. C., Scott, P. F., Titterington, D. J. and Wilson, D. M. A. 1993. *Nature* **365**, 320. (280, 285)

Jones, R. O. 1991. *Phys. Rev. Lett.* **67**, 224. (80)

Jones, R. O. and Gunnarson, O. 1989. *Rev. Mod. Phys.* **61**, 689. (79)

Josephson, B. D. 1962. *Phys. Lett.* **1**, 251. (168)

Julian, S. R., Mathur, N. D., Grosche, F. M. and Lonzarich, G. G. 1997a. To be published. (139)

Julian, S. R. *et al.* 1997b. To be published. (147)

Kagan, Yu., Kikoin, K. A. and Prokof'ev, N. V. 1992a. *Physica B* **182**, 201. (203)

Kagan, Yu., Kikoin, K. A. and Prokof'ev, N. V. 1992b. *JETP Lett.* **56**, 219. (203)

Kang, W., Stormer, H. L., Pfeiffer, L. N., Baldwin, K. W. and West, K. W. 1993. *Phys. Rev. Lett.* **71**, 3850. (250)

Karanikas, A. I., Ktorides, C. N. and Stefanis, N. G. 1992. *Phys. Lett.* **B289**, 176. (144)

Kasuya, T. 1956. *Prog. Theor. Phys.* **16**, 45. (192)

Kaufmann, W. 1987. *Ann. d. Phys.* **61**, 544. (9)

Kinoshita, T. and Lindquist, W. B. 1990. *Phys. Rev.* **42**, 636. (32)

Kippenhahn, R. and Weigert, A. 1990. *Stellar Structure and Evolution* (Springer-Verlag, Berlin). (263)

Kittel, C. 1986. *Introduction to Solid State Physics*, 6th edn (Wiley, New York); see also previous editions dated 1967 and 1976. (72)

Kivelson, S., Lee, D. H. and Zhang, S. C. 1992. *Phys. Rev.* **B46**, 2223. (254)

Klein, O. 1926. *Z. Phys.* **37**, 895. (43)

Kohn, W. and Luttinger, J. M. 1964. *Phys. Rev. Lett.* **15**, 524. (76)

Kohn, W. and Sham, L. J. 1965. *Phys. Rev.* **140**, A1133. (71, 77, 205)

Kondo, J. 1964. *Prog. Theor. Phys.* **32**, 37. (186, 188)

Krymsky, G. F. 1977. *Dok. Acad. Nauk. USSR* **234**, 1306. (312)

La Rue, G. S., Phillips, J. D. and Fairbanks, W. M. 1981. *Phys. Rev. Lett.* **46**, 967. (13)

Lam, P. K. and Girvin, S. M. 1984. *Phys. Rev.* **B30**, 473. (254)

Landau, L. 1930. *Z. Phys.* **64**, 629. (199)

Landau, L. 1956. *JETP (USSR)* **30**, 10; translation in *Sov. Phys. JETP* **3**, 920. (197)

Landau, L. D. and Lifshitz, E. M. 1980. *Statistical Physics* (Pergamon, Oxford). (122, 127)

Landauer, R. 1970. *Phil. Mag.* **21**, 863. (209, 230)

Lang, N. D. 1973. In *Solid State Physics*, vol. 28, eds. F. Seitz and D. Turnbull (Academic Press, New York), p. 225. (75)

Laughlin, R. B. 1983. *Phys. Rev. Lett.* **50**, 1395. (241)

Leadley, D. R., Nicholas, R. J., Foxon, C. T. and Harris, J. J. 1994. *Phys. Rev. Lett.* **72**, 1906. (252)

Leadley, D. R., van der Burgt, M., Nicholas, R. J., Foxon, C. T. and Harris, J. J. 1996. *Phys. Rev.* **B53**, 2057. (251)

Lee, P. A. and Nagaosa, N. 1992. *Phys. Rev.* **B46**, 5621. (140)

Leggett, A. J. 1975. *Rev. Mod. Phys.* **47**, 331. (117, 172)

Leggett, A. J. 1980. *Prog. Theor. Phys. Suppl.* **69**, 80. (168)

Leggett, A. J. 1995. In *Bose-Einstein Condensation,* eds. A. Griffin, D. W. Snoke and S. Stringari (Cambridge University Press). (163)

Leinaas, J. M. and Myrheim, J. 1977. *Nuovo Cimento* **37**, B1. (177)

Levin, K. and Valls, O. T. 1983. *Phys. Rep.* **98**, 1. (145)

Levy, L. P., Dolan, G., Dunsmuir, J. and Bouchiat, H. 1990. *Phys. Rev. Lett.* **64**, 2074. (209, 231)

Lieb, E. H. 1976. *Rev. Mod. Phys.* **48**, 553. (60, 62)

Lifshitz, E. M. and Pitaevskii, L. P. 1980. *Statistical Physics* (Pergamon, Oxford). (125)

Lifshitz, I. M. and Kosevich, A. M. 1955. *Zh. Eksp. Theor. Fiz.* **29**, 730; English translation in *Sov. Phys. JETP* **2**, 636. (199)

Lifshitz, I. M. and Kosevich, A. M. 1958. *J. Phys. Chem. Solids* **4**, 19. (251)

Lightman, A. and White, T. R. 1988. *Astrophys.* **335**, 57. (281)

Lindemann, F. A. 1915. *Phil. Mag.* **29**, 127. (17)

Liu, A. Y. and Cohen, M. L. 1989. *Science* **245**, 841. (82)

London, F. 1950. *Superfluids*, vol. I (Wiley, New York; reprinted by Dover Publications, New York, 1961). (103, 219, 221)

Longair, M. S. 1992–7. *High Energy Astrophysics*, vols. 1, 2 and 3 (Cambridge University Press); vol. 3 in press. (260, 271, 289, 313)

Longair, M. S. 1995. In *Twentieth Century Physics*, vol. 3, eds. L. M. Brown, A. Pais and A. B. Pippard (IOP Publications, Bristol), pp. 1691–821. (257)

Lonzarich, G. G. 1984. *J. Mag. Magn. Mater.* **45**, 43. (142)

Lonzarich, G. G. 1986. *J. Mag. Magn. Mater.* **54–7**, 612. (128, 139, 142)

Lonzarich, G. G. and Taillefer, L. 1985. *J. Phys.* **C18**, 4339. (142)

Lonzarich, G. G., Bernhoeft, N. R. and Paul, D. McK. 1989. *Physica* **B156&157**, 699. (124)

Lorentz, H. A. 1909. *The Theory of Electrons* (B. G. Teubner, Leipzig; reprinted by Dover, New York, 1952). (60)

Lorentz, H. A. 1927. *4th Solvay Conference* (1924) (Gauthier-Villars, Paris). (16)

Loucks, T. L. 1967. *Augmented Plane Wave Method* (W. A. Benjamin, Inc., New York). (84)

Lozovik, Y. E. and Yudson, V. I. 1975. *JETP Lett.* **22**, 11. (253)

Luttinger, J. M. 1960. *Phys. Rev.* **119**, 1153. (185, 197)

Luttinger, J. M. 1961. *Phys. Rev.* **121**, 1251. (200)

MacLaren, J. M., Pendry, J. B., Rous, J. B., Saldin, P. J., Somorjai, G. A., Van Hove, M. A. and Vvedeusky, D. D. 1987. *Surface Crystallographic Information Service* (Reidel, Dordrecht). (82)

McMullan, G. J. 1989. Magnetism in metals close to the ferromagnetic instability, Ph.D. Thesis, University of Cambridge. (132, 136, 146)

Manoharan, H. C., Shayegan, M. and Klepper, S. J. 1994. *Phys. Rev. Lett.* **73**, 3270. (253)

March, R. 1992. *Physics for Poets* (McGraw-Hill, New York). (88)

Marscher, A. P. 1993. In *Astrophysical Jets*, eds. D. Bugarella, M. Livio and C. O'Dea (Cambridge University Press), p. 73. (302)

Mather, J. 1995. In *Current Topics in Astrofundamental Physics: The Early Universe*, eds. N. Sanchez and A. Zichichi (Kluwer Academic Publishers, Dordrecht), p. 357. (273)

Mathon, J. 1968. *Proc. Roy. Soc. London A* **306**, 355. (140)

Matsuda, T., Hasegawa, S., Igarashi, M., Kobayashi, T., Naito, M., Kajiyama, H., Endo, J., Osakabe, N. and Tonomura, A. 1989. *Phys. Rev. Lett.* **62**, 2519. (217)

Maxwell, J. C. 1873. *A Treatise on Electricity and Magnetism*, 2 vols. (Clarendon Press, Oxford). (2)

Mermin, N. D. 1965. *Phys. Rev.* **137**, A1441. (77)

Mermin, N. D. 1978. *J. Phys. Colloq.* **39**C-6, 1283. (167)

Michelson, A. A. 1898. *Phil. Mag.* **45**, 348. (9)

Migdal, A. B. 1957. *JETP (USSR)* **32**, 399; translation in *Sov. Phys. JETP* **5**, 333. (197)

Migdal, A. B. 1958. *Sov. Phys. JETP* **7**, 996. (61)

Mihara, T., Makashima, K., Ohashi, T., Sakao, T., Tashiro, M., Nagasee, F., Tanaka, Y., Kitamoto, S., Miyamoto, S., Deeter, J. E. and Boynton, P. E. 1990. *Nature* **346**, 250. (288)

Millikan, R. A. 1916. *Phys. Rev.* **7**, 355. (25)

Millikan, R. A. 1917. *The Electron* (University of Chicago Press). (25)

Millikan, R. A. 1935. *Electrons* (Cambridge University Press), p. 27. (4, 11)

Millikan, R. A. 1951. *Autobiography* (Macdonald, London), p. 91. (13)

Milliken, F. P., Penney, T., Holtzberg, F. and Fisk, Z. 1988. *J. Mag. Magn. Mater.* **76&77**, 201. (146, 187)

Millis, A. J. 1993. *Phys. Rev.* **B48**, 7183. (120, 135)

Millis, A. J., Sachdev, S. and Varma, C. M. 1988. *Phys. Rev.* **B37**, 4975. (145)

Mollenstedt, G. and Bayh, W. 1962a. *Phys. Blaetter* **18**, 299. (226)

Mollenstedt, G. and Bayh, W. 1962b. *Naturwiss* **4**, 81. (226)

Moriya, T. 1985. *Spin Fluctuations in Itinerant Electron Magnetism* (Springer, Berlin), and references cited therein. (120, 125, 133, 135, 140)

Moriya, T. and Kawabata, A. 1973a. *J. Phys. Soc. Japan* **34**, 639. (142)

Moriya, T. and Kawabata, A. 1973b. *J. Phys. Soc. Japan* **35**, 669. (142)

Moruzzi, V. L., Janak, J. F. and Williams, R. A. 1978. *Calculated Electronic Properties of Metals* (Pergamon Press, Oxford). (84)

Mott, N. F. and Jones, H. 1936. *The Theory of the Properties of Metals and Alloys* (Oxford University Press). (66)

Murata, K. K. and Doniach, S. 1972. *Phys. Rev. Lett.* **29**, 285. (142)

Newns, D. M. and Read, N. 1987. *Adv. Phys.* **36**, 799. (194, 197)

Norman, M. R. and Koelling, D. D. 1993. In *Handbook on the Physics and Chemistry of Rare Earths*, vol. 17, eds. K. A Gschneider Jr, L. Eyring, G. H. Lander and G. R. Choppin. (Elsevier Science BV, Amsterdam), chap. 110. (202, 204)

Nozières, P. 1974. *J. Low. Temp. Phys.* **17**, 31. (191)

Nozières, P. 1975. In *Proceedings of the XIV Conference on Low Temperature Physics LT14*, eds. M. Krusius and M. Vuorio, vol. 5 (North Holland/Elsevier, Amsterdam). (191)

Olariu, S. and Popescu, I. I. 1985. *Rev. Mod. Phys.* **57**, 339. (226)

Onnes, H. K. 1911. *Commun. Phys. Lab. Univ. Leiden* no. 119b. (16)

Onsager, L. 1952. *Phil. Mag.* **43**, 1006. (199)

Onsager, L. 1961. *Phys. Rev. Lett.* **7**, 50. (166)

Onuki, Y. and Komatsubara, T. J. 1987. *J. Mag. Magn. Mater.* **63&64**, 281. (189)

Onuki, Y., Goto, T. and Kasuya, T. 1991. *Materials Science and Technology*, vol. 3A (VCH, Weinheim). (203, 206)

Osterbrock, D. E. 1989. *The Astrophysics of Gaseous Nebulae and Active Galactic Nuclei* (University Science Books, Mill Valley, CA). (307)

Ott, H. R., Rudigier, H., Fisk, Z. and Smith, J. L. 1983. *Phys. Rev. Lett.* **50**, 1595. (117, 183)

Ott, H. R., Rudigier, H., Fisk, Z., Willis, J. O. and Stewart, G. R. 1985. *Solid State Commun.* **53**, 235. (186)

Pais, A. 1986. *Inward Bound: Of Matter and Forces in the Physical World* (Oxford University Press), chaps. 1 & 12. (88)

Parr, R. G. and Yang, W. 1989. *Density-Functional Theory of Atoms and Molecules* (Oxford University Press). (80)

Particle Data Tables. 1994. *Phys. Rev. D* **50**, 1172. (24, 34)

Pauli, W. 1926a. *Z. Phys.* **36**, 336. (39)

Pauli, W. 1926b. *Z. Phys.* **37**, 263. (44)

Pauli, W. 1927. *Z. Phys.* **43**, 601. (44)

Pauli, W. 1936. *Ann. Inst. Henri Poincaré* **6**, 109. (45, 47)

Pauling, L. and Wilson, B. L. 1935. *Introduction to Quantum Mechanics* (McGraw-Hill, New York), p. 340. (63)

Pearson, T. J., Unwin, S. C., Cohen, M. H., Linfield, R. P., Readhead, A. C. S., Seielstad, G. A., Simon, R. S. and Walker, R. C. 1982. In *Extragalactic Radio Sources*, eds. D. S. Heeschen and C. M. Wade (D. Reidel and Co., Dordrecht), p. 356. (294)

Peierls, R. E. 1929. *Phys. Zeits.* **30**, 273. (15)

Pendry, J. B. 1974. *Low Energy Electron Diffraction, the Theory and its Application to the Determination of Surface Structures* (Academic Press, London). (209)

Penning, F. M. 1937. *Physica* **4**, 71. (27)

Perdew, J. P. 1995. In *Density Functional Theory*, eds. E. K. U. Gross and R. M. Dreizler, ASI Series B, vol. 337 (Plenum Press, New York), p. 51. (74)

Perley, R. A., Dreher, J. W. and Cowan, J. J. 1984. *Astrophys. J.* **285**, L45. (293)

Perrin, J. 1895. *Comptes Rendus* **121**, 1130. (5)

Peshkin, M. and Tonomura, A. 1989. *The Aharonov-Bohm Effect* (Springer-Verlag, Berlin). (221, 226)

Peshkin, M., Talmi, I. and Tassie, L. J. 1960. *Ann. Phys.* **16**, 426. (225)

Peterman, A. 1957. *Helv. Phys. Acta.* **30**, 407. (32)

Pettifor, D. G. 1988. *Mat. Sci. & Technol.* **4**, 675. (65)

Pippard, A. B. 1987. *Eur. J. Phys.* **8**, 55. (128)

Pfleiderer, C., McMullan, G. J. and Lonzarich, G. G. 1994. *Physica* **B199&200**, 634. (140)

Pfleiderer, C., McMullan, G. J. and Lonzarich, G. G. 1995. *Physica* **B206&207**, 847. (140, 142)

Pines, D. and Nozières, P. 1966. *The Theory of Quantum Liquids*, vol I (Benjamin, New York). (73, 100)

Pounds, K. A., Nandra, K., Stewart, G. C., George, I. M. and Fabian, A. C. 1990. *Nature* **344**, 132. (279)

Pozdnyakov, L. A., Sobol, I. M. and Sunyaev, R. A. 1983. *Astrophys. Space Sc. Rev.* **2**, 263. (277, 279)

Preston, T. 1898. *Trans. Roy. Dublin Soc.* **6**, 385. (9)

Rajagopal, A. K. 1978. *J. Phys. C* **11**, L943. (77)

Ramakrishnan, T. V. 1974. *Phys. Rev.* **B10**, 4014. (132)

Rasul, J. W. 1989. *Phys. Rev.* **B39**, 663. (200)

Reinders, P. H. P., Springford, M., Coleridge, P. T., Boulet, R. and Ravot, D. 1986. *Phys. Rev. Lett.* **57**, 1631. (139, 201)

Reinders, P. H. P., Springford, M., Coleridge, P. T., Boulet, R. and Ravot, D. 1987. *J. Mag. Magn. Mater.* **63&64**, 297. (201)

Reizer, M. 1989. *Phys. Rev.* **B39**, 1602. (140)

Richardson, O. W. 1901. *Proc. Camb. Phil. Soc.* **11**, 286. (17)

Richardson, O. W. 1903. *Phil. Trans. Roy. Soc.* **A201**, 497. (17)

Richardson, O. W. 1916. *The Emission of Electricity from Hot Bodies* (Longman, Green, London). (17, 18)

Riecke, E. 1898. *Ann. d. Phys.* **66**, 353. (15)

Riegler, G. R., Ling, J. C., Mahoney, W. A., Wheaton, W. A., Willet, J. B., Jacobson, A. S. and Prince, T. A. 1981. *Astrophys. J. Lett.* **248**, 113. (301)

Ringwood, D. E. 1977. *Geochem. J.* **11**, 111. (82)

Rogers, F. J. and Iglesias, C. A. 1994. *Science* **263**, 50. (263)

Rokhsar, D. S. 1993. *Phys. Rev. Lett.* **70**, 493. (177)

Rowland, H. A. 1878. *Am. J. Sci.* **15**, 30. (2)

Rowland, H. A. and Hutchinson, C. T. 1889. *Phil. Mag.* **27**, 445. (2)

Ruderman, M. A. and Kittel, C. 1954. *Phys. Rev.* **96**, 99. (192)

Ryder, J. D. and Fink, D. G. 1984. *Engineers and Electrons* (IEEE Press, New York). (20, 21)

Sachdev, S. 1996. Preprint. (135)

Sandars, P. G. H. 1965. *Phys. Lett.* **14**, 194. (35)

Sauls, J. D. 1994. *Adv. Phys.* **43**, 113. (174)

Scalapino, D. J. 1995. *Phys. Rep.* **250**, 331. (180)

Schiff, L. I. 1963. *Phys. Rev.* **117**, 2194. (35)

Schonhammer, K. and Gunnarson, O. 1988. *Phys. Rev.* **B37**, 3128. (205)

Schrieffer, J. R. 1964. *Theory of Superconductivity* (Addison-Wesley, Reading, MA). (120)

Schrödinger, E. 1926a. *Ann. Phys.* **79**, 361. (39)

Schrödinger, E. 1926b. *Ann. Phys.* **79**, 489. (39)

Schrödinger, E. 1926c. *Ann. Phys.* **80**, 437. (39)

Schrödinger, E. 1926d. *Ann. Phys.* **81**, 109. (39)

Schuster, A. 1897. *Phil. Mag.* **43**, 1. (3)

Schweber, S. 1994. *QED and the Men Who Made It: Dyson, Feynman, Schwinger and Tomonaga* (Princeton University Press). (60, 61)

Schwinger, J. 1948. *Phys. Rev.* **73**, 416. (32)

Seaton, M. J. 1993. In *Inside the Stars*, IAU Colloquium no 137, eds. W. W. Weiss and A. Baglin (Astronomical Society of the Pacific Conference Series, San Francisco), p. 222. (263)

Sham, L. J. 1991. *Phil. Trans. Roy. Soc. Lond. A* **334**, 481. (84)

Shankar, R. 1994. *Rev. Mod. Phys.* **86**, 129. (112)

Sigrist, M. and Ueda, K. 1991. *Rev. Mod. Phys.* **63**, 239. (117, 146)

Silvestrelli, P.-L., Baroni, S. and Car, R. 1993. *Phys. Rev. Lett.* **71**, 1148. (74)

Skriver, H. L. 1983. *Muffin Tin Orbitals and Electronic Structure* (Springer-Verlag, Berlin). (84)

Slater, J. C. 1975. *Solid-State and Molecular Theory: A Scientific Biography* (John Wiley & Son, New York). (72, 79)

Sokolov, A. A. and Pavlenko, Y. G. 1967. *Opt. Spectrosc.* **22**, 1. (29)

Solontsov, A. Z. and Wagner, D. 1995. *Phys. Rev. B* **51**, 12 410. (133)

Sommerfield, C. M. 1957. *Phys. Rev.* **107**, 328. (32)

Sondhi, S. L., Karlhede, A., Kivelson, S. A. and Rezayi, E. H. 1993. *Phys. Rev.* **B47**, 16 419. (254)

Spitzer, L. 1962. *Physics of Fully Ionised Gases* (Interscience Publishers, New York). (271)

Srivastava, G. P. and Weaire, D. L. 1987. *Adv. Phys.* **36**, 463. (81, 82)

Staunton, J., Gyorffy, B. L., Pindor, A. J., Stocks, G. M. and Winter, H. 1985. *J. Phys. F: Metals Phys.* **15**, 1387. (77)

Steglich, F., Aarts, J. Bredl, C. D., Lieke, W., Meschede, D., Frank, W. and Schafer, H. 1979. *Phys. Rev. Lett.* **43**, 1892. (117, 183)

Stern, A., Aharonov, Y. and Imry, Y. 1990. *Phys. Rev.* **A40**, 3436. (218)

Stern, E. A. 1960. In *The Fermi Surface,* eds. W. A. Harrison and M. B. Webb (John Wiley & Sons, New York), p. 50. (200)

Stewart, G. R., Fisk, Z., Willis, J. O. and Smith, J. L. 1984. *Phys. Rev. Lett.* **52**, 679. (117)

Stich, I., Payne, M. C., King-Smith, R. D., Lim, J.-S. and Clark, L. J. 1992. *Phys. Rev. Lett.* **68**, 1351. (83)

Stojkovic, B. P. and Pines, D. 1996. *Phys. Rev. Lett.* **76**, 811. (106)

Stoney, G. J. 1881. *Phil. Mag.* **11**, 381. (10)

Stoney, G. J. 1891. *Sci. Trans. Roy. Dublin Soc.* **4**, 563. (4)

Stoney, G. J. 1894. *Phil. Mag.* **38**, 418. (4, 22)

Storey, L. R. O. 1953. *Phil. Trans. Roy. A* **246**, 113. (21)

Sunyaev, R. A. and Zeldovich, Y. B. 1970. *Astrophys. Sp. Sci.* **7**, 20. (279)

Sutton, A. P. 1993. *Electronic Structure of Materials* (Clarendon Press, Oxford). (66)

Taillefer, L. and Lonzarich, G. G. 1988. *Phys. Rev. Lett.* **60**, 1560. (139, 201)

Taillefer, L., Lonzarich, G. G. and Strange, P. 1986. *J. Mag. Magn. Mat.* **54–7**, 957. (139)

Taillefer, L., Newbury, R., Lonzarich, G. G., Fisk, Z. and Smith, J. L. 1987. *J. Mag. Magn. Mater.* **63&64**, 372. (201)

Takahashi, Y. and Moriya, T. 1985. *J. Phys. Soc. Japan* **54**, 1592. (142)

Tayler, R. J. 1995. *The Stars: Their Structure and Evolution* (Cambridge University Press). (262)

Taylor, J. and Cordes, J. M. 1993. *Astrophys. J.* **411**, 674. (303)

Tesanovic, Z. and Valls, O. T. 1986. *Phys. Rev.* **B34**, 1918. (194)

Thessieu, C., Stepanov, A. N., Lapertot, G. and Flouquet, J. 1995. *Solid-State Commun.* **95**, 707. (140)

Thomas, L. H. 1927. *Proc. Camb. Phil. Soc.* **23**, 542. (70)

Thomson, J. J. 1881. *Phil. Mag.* **11**, 229. (3)

Thomson, J. J. 1885. British Association Report, p. 97. (3)

Thomson, J. J. 1893a. *Phil. Mag.* **36**, 313. (12)

Thomson, J. J. 1893b. *Notes on Recent Researches in Electricity and Magnetism* (Clarendon Press, Oxford). (2, 5, 15)

Thomson, J. J. 1897a. *Proc. Roy. Inst.* **15**, 419. (5)

Thomson, J. J. 1897b. *Phil. Mag.* **44**, 293. (6)

Thomson, J. J. 1898. *Phil. Mag.* **46**, 528. (25)

Thomson, J. J. 1899. *Phil. Mag.* **48**, 547. (17)

Thomson, J. J. 1900. *Congrés International de Physique*, vol. 3 (Gauthier-Villars, Paris), p. 138. (15)

Thomson, J. J. 1903. *Conduction of Electricity Through Gases* (Cambridge University Press). (5, 9)

Thomson, J. J. 1907. *The Corpuscular Theory of Matter* (Constable, London). (3, 17, 73)

Thomson, J. J. 1914. *The Atomic Theory*, Romanes Lecture 1914 (Clarendon Press, Oxford). (4)

Thomson, J. J. 1915. *Phil. Mag.* **30**, 192. (17)

Thomson, J. J. 1936. *Recollections and Reflections* (Bell, London). (2, 5, 7)

Tillman, J. R. and Tucker, D. G. 1978. In *History of Technology*, vol. 7, ed. T. I. Williams (Clarendon Press, Oxford), p. 1091. (20, 21)

Tonomura A., ed. 1993. *Electron Holography* (Springer-Verlag, Berlin). (213, 214, 221)

Tonomura, A., Osakabe, N., Matsuda, T., Kawasaki, T. and Endo, J. 1986. *Phys. Rev. Lett.* **56**, 792. (227)

Tonomura, A., Allard, L. F., Pozzi, G., Joy, D. C. and Ono, Y. A., eds. 1994. *Electron Holography*, Proceedings of the International Workshop on Electron Holography, Knoxville, TN, USA (North Holland, Amsterdam). (214)

Toutain, T. and Frölich, C. 1992. *Astron. Astrophys.* **257**, 287. (265)

Tsvelik, A. M. 1995. *Quantum Field Theory in Condensed Matter Physics* (Cambridge University Press). (135)

Tsui, D. C., Stormer, H. L. and Gossard, A. C. 1982. *Phys. Rev. Lett.* **48**, 1559. (241)

Usher, A. J., Nicholas, R. J., Harris, J. J. and Foxon, C. T. 1990. *Phys. Rev.* **B41**, 1129. (254)

Van Dyck, R. S., Ekstrom, P. and Dehmelt, H. 1976a. *Nature* **262**, 776. (30)

Van Dyck, R. S., Wineland, D., Ekstrom, P. and Dehmelt, H. 1976b. *Appl. Phys. Lett.* **28**, 446. (30)

Van Dyck, R. S., Schwinberg, P. B. and Dehmelt, H. 1977. *Phys. Rev. Lett.* **38**, 310. (30)

Van Dyck, R. S., Moore, F. L., Farnham, D. L. and Schwinberg, P. B. 1986. *Bull. Am. Phys. Soc.* **31**, 244. (30)

Van Dyck, R. S., Schwinberg, P. B. and Dehmelt, H. 1987. *Phys. Rev. Lett.* **59**, 26. (31)

Varma, C. M., Littlewood, P. B., Schmitt-Rink, S., Abrahams, E. and Ruckenstein, A. 1989. *Phys. Rev. Lett.* **63**, 1996. (106, 140)

Villars, P., Mathis, K. and Hullinger, F. 1989. *The Structure of Binary Compounds* (North Holland, Amsterdam). (65)

von Klitzing, K. 1986. *Rev. Mod. Phys.* **58**, 519. (241)

von Klitzing, K., Dorda, G. and Pepper, M. 1980. *Phys. Rev. Lett.* **45**, 494. (233, 237, 239)

Wang, C. S., Norman, M. R., Albers, R. C., Boring, A. M., Pickett, W. E., Krakauer, H. and Christensen, N. E. 1987. *Phys. Rev.* **B35**, 7260. (202)

Wasserman, A., Springford, M. and Hewson, A. C. 1989. *J. Phys. Cond. Matter* **1**, 2669. (200)

Webb, R. A. and Washburn, S. 1989. *Phys. Today* **41**, 46. (209, 218, 230)

Webb, R. A., Washburn, S., Umbach, C. P. and Laibowitz, R. B. 1985. *Phys. Rev. Lett.* **54**, 2696. (209, 232)

Wehnelt, A. 1904. *Ann. d. Phys.* **14**, 425. (20)

Weyl, H. 1929. *Z. Phys.* **56**, 330. (49)

White, R. M. 1983. *Quantum Theory of Magnetism* (Springer-Verlag, New York). (112)

Whittaker, E. T. 1951. *A History of the Theories of Aether and Electricity*, vol. 1 (Nelson, London). (2, 3)

Wiechert, E. 1897. *Schr. d. Phys. Ökon. Ges. (Königsberg)* **38**, 3. (9)

Wielebinski, R. 1993. In *The Cosmic Dynamo*, ed. F. Krause (Kluwer Academic Publishers, Dordrecht), p. 271. (306)

Wigner, E. P. 1934. *Phys. Rev.* **46**, 1002. (253)

Wigner, E. P. 1938. *Trans. Faraday Soc.* **34**, 678. (76)

Wigner, E. P. and Seitz, F. 1955. *Solid State Physics*, vol. 1, eds. F. Seitz and D. Turnbull (Academic Press, New York), p. 97. (68, 72)

Wilczek, F., ed. 1990. *Anyon Superconductivity* (World Scientific, Singapore). (177)

Willett, R. L., Stormer, H. L., Tsui, D. C., Pfeiffer, L. N., West, K. W. and Baldwin, K. W. 1988. *Phys. Rev.* **B38**, 7881. (254)

Willett, R. L., Ruel, R. R., West, K. W. and Pfeiffer, L. N. 1993a. *Phys. Rev. Lett.* **71**, 3846. (249)

Willett, R. L., Ruel, R. R., Palaanen, M. A., West, K. W. and Pfeiffer, L. N. 1993b. *Phys. Rev.* **B47**, 7344. (249)

Willett, R. L., West, K. W. and Pfeiffer, L. N. 1995. *Phys. Rev. Lett.* **75**, 2988. (249)

Wilson, C. T. R. 1900. *Phil. Trans. Roy. Soc. A* **193**, 289. (11)

Wilson, H. A. 1903. *Phil. Mag.* **6**, 429. (13)

Wilson, K. G. 1975. *Rev. Mod. Phys.* **47**, 773. (189)

Wilson, M. and Gyorffy, B. L. 1995. *J. Phys. Condensed Matter* **7**, 1565. (78)

Wilson, S., Grant, I. P. and Gyorffy, B. L., eds. 1991. *The Effects of Relativity in Atoms, Molecules, and the Solid State* (Plenum Press, New York). (78)

Wineland, D. and Dehmelt, H. 1975. *J. Appl. Phys.* **46**, 919. (30)

Wineland, D., Ekstrom, P. and Dehmelt, H. 1973. *Phys. Rev. Lett.* **31**, 1279. (30)

Woltz, K. 1909, *Ann. d. Phys.* **30**, 273. (25)

Wu, M. K. *et al.* 1987. *Phys. Rev. Lett.* **58**, 908. (117)

Wu, T. T. and Yang, C. N. 1975. *Phys. Rev.* **D12**, 3845. (226)

Yamada, H. 1993. *Phys. Rev.* **B47**, 11 211. (142)

Yamada, K. 1975a. *Prog. Theor. Phys.* **53**, 970. (191)

Yamada, K. 1975b. *Prog. Theor. Phys.* **54**, 316. (191)

Yang, C. N. 1962. *Rev. Mod. Phys.* **34**, 694. (233)

Yang, C. N. and Mills, R. L. 1954. *Phys. Rev.* **96**, 191. (226)

Yin, M. T. and Cohen, M. L. 1982. *Phys. Rev.* **B26**, 5668. (81)

Yip, S. K. 1984. *Phys. Lett. A* **105**, 66. (161)

Yoshioka, D. and Fukuyama, H. 1979. *J. Phys. Soc. Japan* **47**, 394. (253)

Yosida, K. 1957. *Phys. Rev.* **106**, 893. (192)

Zeeman, P. 1897. *Phil. Mag.* **43**, 226. (8)

Zener, C. 1951. *Phys. Rev.* **81**, 440. (188)

Ziman, J. M. 1960. *Electrons and Phonons* (Oxford University Press). (120)

Zwicknagl, G. 1992. *Adv. Phys.* **41**, 203. (205, 206)

Index

absolute system of units (Stoney's) 22
action principles 52
Aharonov–Bohm effect 221
anyons 177

Barkhausen effect 21
BCS theory 154, 163
Bohr–van Leeuwen theorem 16, 150
bonding
  in metals 84
  in molecules 78
  in semiconductors 80
Born–Oppenheimer approximation 62
Bose condensation 153
bremsstrahlung 283
broken symmetry 102

charge on electron 10, 25
charge/mass ratio of electron 5, 25
Clifford algebras 45
cloud chamber 11
coherence
  definitions 208
  in heavy fermion conductors 187
  in mesoscopic conductors 229
  in normal metals 216
  in superconductors, and ODLRO 219, 233
collisional ionisation 305
composite fermions 241
  Fermi surface 248
  masses 250
Compton scattering 276
  inverse 296
Cooper pairs 103, 155
cosmic rays 309

de Haas–van Alphen effect 198
Debye length 270
density functional theory 69
  relativistic 77
  and superconductivity 76
dielectric loss 21
Dirac equation 41

discovery of electron 1, 4
Drude theory of metals 15

effective pair-potential models 63
electric dipole moment of electron 34
electron interference and coherence length 210
electronic devices, early 20
electrons in plasmas 270
  acceleration of 311
  radiation processes 272
exchange field 112
exclusion principle, origin of 57

Faraday rotation 304
Fermi liquid theory 99, 113, 138
  and beyond 101
Fermi surface 95
Fermi–Dirac statistics
  connection with spin 54
  and second quantisation 52
Feynman diagrams 32
fine-structure constant 31, 34
fluctuation dissipation theorem 120, 126
flux quantisation 166
forbidden lines 307
fractional quantum Hall effect 241
Fraunhofer lines 257

gauge principle 49
Grassmann functions 54

heavy fermions 138, 182
  band structure 204
  de Haas–van Alphen effect 198
  superconductivity 172
  transport properties 186
high-temperature superconductivity 104, 174

interference of coherent electrons 211–36
interstellar shock waves 307
  as acceleration mechanism for cosmic rays 312

Josephson effect   168

Klein–Gordon equation   43
Kondo effect   188
Kondo lattice   192

Landau damping   123
Landau Fermi liquid theory   99, 114, 138
local density approximation   72
   in metals   84
   and molecular bonding   78
   in semiconductors   80
   spin-polarised   75
   and surface rearrangement   82
Luttinger liquid   104
Luttinger's theorem   95, 185, 197

magnetic moment of electron   27
   anomalous   31
magnetic properties, phenomenological model
      of   118
   comparisons with experiment   137
   instabilities   141
   mode–mode coupling   129
   overdamped and dissipative modes   127
   susceptibility and resistivity   132
mass of electron   8, 25, 31
mass renormalisation   102
mesoscopic conductors   229

naming of electron by Stoney   1, 22
non-Fermi liquids   104, 139

pair creation and annihilation in space   298
pair-potential models   63
paramagnons   138
Pauli exclusion principle   54, 57, 92
Pauli spin matrices   44
Penning trap   27
periodic Anderson model   193
periodic table and crystal structures   65
phenomenological theories of solids   62

photoionisation   305
Pippard coherence length   159
plasma frequency   303
positronium in space   300
positrons   57

quantised Hall resistance   34
quantum electrodynamics   31
quantum ferromagnets   254
quantum field theory   55
quantum Hall effect   237
   fractional   241

renormalisation   58, 101

skyrmions   254
spin and statistics   54
spin Fermi liquid   203
spin fluctuations   142
spin–spin interactions   115
SQUIDs   168
stars, structure of   260
stellar electrons   266
superconductivity   148
   anisotropic (p-wave, d-wave)   116, 144, 178
   BCS theory   154, 163
   Bose condensation   153, 162
   flux quantisation   166
   heavy fermion   172
   high-temperature   104, 174
   Josephson effect   168
   London phenomenology   149
superluminal velocities   296
supersymmetry   59
synchrotron radiation in space   286

thermionics   17
Thomson scattering   273
tight-binding method   66

whistlers   21
Wigner solid   253